MARINE HYDROTHERMAL SYSTEMS AND THE ORIGIN OF LIFE

Report of SCOR Working Group 91

Edited by

Dr N.G. HOLM

Dept. of Geology and Geochemistry
Stockholm University, Sweden

Reprinted from
Origins of Life and Evolution of the Biosphere,
Volume 22, Nos. 1–4, 1992

KLUWER ACADEMIC PUBLISHERS
DORDRECHT / BOSTON / LONDON

ISBN 0-7923-2018-2

Published by Kluwer Academic Publishers,
P.O. Box 17, 3300 AA Dordrecht, The Netherlands.

Kluwer Academic Publishers incorporates
the publishing programmes of
D. Reidel, Martinus Nijhoff, Dr W. Junk and MTP Press.

Sold and distributed in the U.S.A. and Canada
by Kluwer Academic Publishers,
101 Philip Drive, Norwell, MA 02061, U.S.A.

In all other countries, sold and distributed
by Kluwer Academic Publishers Group,
P.O. Box 322, 3300 AH Dordrecht, The Netherlands.

Printed on acid-free paper

Printed in the Netherlands

Table of Contents

Preface		1
Chapter 1	Why are Hydrothermal Systems Proposed as Plausible Environments for the Origin of Life? Nils G. Holm	5
Chapter 2	Hydrothermal Systems: Their Varieties, Dynamics, and Suitability for Prebiotic Chemistry Nils G. Holm and Remy J.-C. Hennet	15
Chapter 3	Modern Life at High Temperatures Roy M. Daniel	33
Chapter 4	Aqueous Organic Geochemistry at High Temperature/High Pressure Bernd R. T. Simoneit	43
Chapter 5	Chemical Environments of Submarine Hydrothermal Systems Everett L. Shock	67
Chapter 6	Chemical Markers of Prebiotic Chemistry in Hydrothermal Systems James P. Ferris	109
Chapter 7	Hydrothermal Organic Synthesis Experiments Everett L. Shock	135
Chapter 8	An Experimental Approach to Chemical Evolution in Submarine Hydrothermal Systems Hiroshi Yanagawa and Kensei Kobayashi	147
Chapter 9	Mineral Theories of the Origin of Life and an Iron Sulfide Example A. Graham Cairns-Smith, Alan J. Hall and Michael J. Russell	161
Chapter 10	Future Research Nils G. Holm, A. Graham Cairns-Smith, Roy M. Daniel, James P. Ferris, Remy J.-C. Hennet, Everett L. Shock, Bernd R. T. Simoneit and Hiroshi Yanagawa	181
Literature Cited (Chapters 1-10)		191

PREFACE

The present publication is the result of four years of work carried out within the terms of reference of Working Group 91 of SCOR (the Scientific Committee on Oceanic Research). The idea of life's origin in conjunction with hydrothermal systems first appeared in the SCOR annals in 1987 after the Executive Committee Meeting in Zürich, Switzerland. The Swedish SCOR Committee had proposed the initiation of a Working Group entitled 'Neo-Abiogenesis and Origin of Life in Hydrothermal Systems'. As present representative of the proponents at the meeting in Zürich I had to try to answer a number of questions and comments by the Executive Committee members. Whether, for instance, the topic was appropriate for a SCOR Working Group. Many of the tasks were considered chemical rather than oceanographic. The Executive Committee finally agreed that the proposal ought to be forwarded to IUPAC (the International Union of Pure and Applied Chemistry) for consideration as a possible cosponsor. The proposal should also be revised, taking into account the comments of the Executive Committee, for a discussion at the SCOR General Meeting one year later in Acapulco, Mexico.

A proposal like this one has, of course, a history that predates the official record by quite some time. I happened to arrive as a post-doctoral scholar at Scripps Institution of Oceanography (SIO) at about the time the 'black smokers' of the East Pacific Rise at 21°N were discovered in 1979. The San Diego area (where SIO is situated) has strongholds in both Origins of Life and Ocean Science research. My field of interest was then Ocean Sciences, but during work in the laboratories of Gustaf Arrhenius I came into contact with the other field when I started to study details of the structure of some marine authigenic minerals (FeOOH polymorphs). One day in early 1980 Gustaf was visited by two enthusiastic scientists, Michael Dowler and Dale Ingmanson, of the San Diego State University. They had been on a cruise to the Red Sea by the R/V Atlantis II of the Woods Hole Oceanographic Institution. Now they had proposed that the Red Sea deep hydrothermal brines were likely to be fruitful environments for the study of abiotic synthesis of life precursors. Since I had been working on the structure of akaganéite, a hollow FeOOH polymorph that forms readily in the Red Sea hydrothermal brines, I started some experiments together with Michael Dowler on interactions between amino acids and akaganéite. The experiments were carried out in Leslie Orgel's laboratories at the Salk Institute for Biological Studies, but to this date I am not quite sure that Leslie was aware of what we were up to. Later the same year I happened to attend William Schopf's PPRG-symposium (Precambrian Paleobiology Research Group) at the University of California, Los Angeles. During a splinter meeting John Corliss gave us a glimpse of the ideas that were presented the same year at the Geological Congress in Paris and were published as a paper by himself, John Baross and Sarah Hoffman the year after.

The ideas by Corliss and coworkers were discussed extensively during the years to follow, but not much was published. There was an International Conference on the Origin of Life (ICOL) in Mainz, Germany, in 1983, but no contributions there dealt with the origin of life under hydrothermal conditions (Stetter presented some of his work on extreme thermophilic bacteria, though). In 1984 two papers were published, one by

Origins of Life and Evolution of the Biosphere **22**: 1–3, 1992.

Bernhardt and coworkers and one by Robert White, that seemed to refute the hydrothermal origin of life model. At the next ICOL in Berkeley in 1986, however, there was a slight increase in contributions which touched on hydrothermal conditions - two presentations by Corliss and one each by Shock/Helgeson and Yanagawa/Kobayashi/Ogawa, respectively. Back home in Sweden during the cold winter of 1987 I became acquainted with the activities of SCOR as a new member of its national committee, and SCOR's successful method to initiate and assign short-lived international working groups for the study of specific oceanographic problems. Within the Swedish SCOR Committee we came to the conclusion that the connection of hydrothermal systems to the question of life's origin would be one part of science where actual data had to be compiled, where discussion needed to be structured, and efforts should be made with the aim to open up a new research field. The working group proposal for the Executive Committee Meeting in Zürich was thus written and submitted in June, 1987.

It was obvious after the Zürich meeting that the proposal had to be revised and clarified during the following year. Since I had started to defend the idea of a working group on this particular subject outside of the national sphere, I also began to feel some responsibility for the final success of the proposal. I sought an endless amount of advice from the international science community and received invaluable responses and inputs from Harald Bäcker, Egon Degens, Michael Dowler, James Ferris, William Fyfe, Remy Hennet, Dale Ingmanson, Lev Mukhin, Euan Nisbet, Cyril Ponnamperuma, Michael Russell, and Everett Shock to name a few. So when the proposal was presented at the SCOR General Meeting in Acapulco in 1988, it was much better constrained than the year before. Still there persisted a feeling among several of the nominated members of SCOR that the tasks of the proposed working group had little to do with ocean sciences. The right word at the right time then came from the Past-President of SCOR Konstantin Fedorov, who said: 'If seawater circulates through oceanic crust and precursors of life are likely to be formed in that environment today - then it is part of oceanography'. An ad hoc group consisting of Ross Heath, Jarl-Ove Strömberg, Erwin Suess, and myself was formed in order to revise once more the title of the working group, proposed terms of reference and the membership list. The SCOR General Meeting finally initiated its Working Group No. 91 (WG91) entitled: 'Chemical Evolution and Origin of Life in Marine Hydrothermal Systems'. The accepted terms of reference were:

1) To determine likely constituents necessary for neo-abiogenesis according to the state of art of the origin of life sciences and thermodynamic calculations.

2) To review available data concerning primordial organic monomers and polymers already observed in hydrothermal systems (for example, carboxylic acids, amino acids, cyano- and heterocyclic compounds); compile a list of potential substances that have to be searched for; and differentiate compounds formed abiogenically and biogenically.

3) To evaluate the role of different classes of possible inorganic catalysts which may be required for the synthesis of organic compounds in hydrothermal systems.

4) To sponsor a symposium and published set of papers in 1992 summarizing the state of knowledge and identifying research opportunities in this field.

Members of WG91 proposed by the ad hoc group and since accepted by the SCOR General Meeting have been: Nils Holm (chair), Yurij Bogdanov, Graham Cairns-Smith, Roy Daniel, James Ferris, Remy Hennet, Everett Shock, Bernd Simoneit, and Hiroshi Yanagawa. Alexei Kuznetsov, Vice-President of SCOR, has been the rapporteur in charge of WG91 to the Executive Committee.

During the first year of its existence WG91 did not meet officially. However, with the assistance of Zinovij Masinovsky and Stanley Miller, then chairman of the the International Society for the Study of the Origin of Life (ISSOL), James Ferris and I had a chance to organize a one-day symposium on 'Chemical Evolution and Neo-Abiogenesis in Marine Hydrothermal Systems' during the ICOL in Prague in 1989. The first official meeting of WG91 was held in June, 1990, at the Kristineberg Marine Biological Station in Fiskebäckskil on the Swedish west coast; the second meeting at the Eagle Crest Resort in Redmond, Oregon, in June, 1991. The terms of reference of WG91 state that its work should be completed by the publication of a set of papers in 1992 summarizing the state of knowledge and identifying research opportunities in this field. It is my belief that the Working Group has fulfilled the task and that some promising future research areas have been indicated is this report. WG91 will be disbanded in September 1992; however, it is only now that 'real research', i.e. field studies and experimental work, will begin.

I wish to thank all my fellow working group members for their participation in the hard work that has resulted in this report. The efforts condensed into the publication have not been contributed by Working Group members alone, however. First of all there are fifteen reviewers who have examined the manuscripts thoroughly. Several other persons and organizations must be acknowledged as well. Alan Hall, Kensei Kobayashi and Michael Russell have co-authored two of the chapters; Michael also participated in the Redmond meeting. Funding and subsidies have been contributed by SCOR, NASA, the Royal Swedish Academy of Sciences and the Swedish Natural Science Research Council. The NASA funds for the support of this endeavor were allocated by the NASA-Ames Research Center, Moffett Field, California, under Interchange No. NCA2-600. Elizabeth Gross, Jarl-Ove Strömberg and Alexei Kuznetsov, Executive Director, President and Vice-President of SCOR, respectively, have been of great help in the contacts with the SCOR organization and in arranging the working group meetings. It would not have been possible to reorganize the originally very diverse set of manuscripts into a uniform format in such a short time without the skilled transformation work carried out at the College of Oceanography of Oregon State University. The copy editing of Barbara McVicar and Bernd Simoneit is, therefore, especially acknowledged.

Two of the persons that contributed to the development of the program of SCOR Working Group 91 have passed away since it was first proposed - Egon Degens and Konstantin Fedorov. We miss them both as friends and colleagues.

Stockholm June 11, 1992

Nils G. Holm

CHAPTER 1

WHY ARE HYDROTHERMAL SYSTEMS PROPOSED AS PLAUSIBLE ENVIRONMENTS FOR THE ORIGIN OF LIFE?

NILS G. HOLM

Department of Geology and Geochemistry, Stockholm University, S-106 91 Stockholm, Sweden

1. Introduction

The paradigm change in geology by the general acceptance of plate tectonics around two decades ago has brought about an increased interest in geothermal processes at plate boundaries. Thus the enhanced research activity at spreading centers led to the discovery of large spectacular submarine hydrothermal systems of global significance to ocean chemistry and geochemistry. Among the best known such areas are the Galapagos Ridge (Corliss *et al.*, 1979), the East Pacific Rise at 21°N (Francheteau *et al.*, 1979; Spiess *et al.*, 1980), and the Juan de Fuca Ridge (Chase *et al.*, 1985). Rona and coworkers (1983) have compiled the early landmark studies of hydrothermal processes at seafloor spreading centers. Recently Edmond (1991) also reviewed U.S. research on oceanic hydrothermal chemistry for the period 1987-1990. The spectacular nature of marine hydrothermal systems with features such as 'black smokers', 'white smokers' and peculiar ecosystems that are independent of sunlight as a source of reducing power has focused much interest on hydrothermal processes for the explanation of an array of geochemical processes and phenomena. Hydrothermal systems located at global plate spreading centers soon attracted the attention of geochemists as viable environments for chemical evolution and the origin of life (cf. Ingmanson and Dowler, 1981). However, the discovery of the deep-sea hydrothermal systems is only one of several reasons why in the last ten or fifteen years interest has been focused on hydrothermal systems in general as potential sites for life's origin. Some other aspects are listed below:

1) The most primitive organisms found in modern environments are thermophiles.

2) The view of the composition of the early atmosphere has changed. There is no longer a consensus on the prevalence of reducing conditions on the early Earth. Most scientists now envision a 'redox neutral' primitive atmosphere. There is, however, a possibility that the iron vapor and reduced carbon liberated from impacting objects like meteorites would leave the ocean reducing for a long period.

3) Hydrothermal systems and deep sediments are about the only environments where primitive life would have been protected against postulated meteorite impacts and partial vaporisation of the ocean.

4) The presence of supercritical fluids is known (CO_2, H_2O) or postulated (CH_4) in hydrothermal systems. Supercritical fluids like water and carbon dioxide are excellent

Origins of Life and Evolution of the Biosphere **22**: 5–14, 1992.

solvents of organic compounds and would probably be of great potential for several of the chemical reactions eventually leading to the origin of life.

Hydrothermal cells are dynamic systems, which has been one of the main arguments for their potential as sites for abiotic organic geochemical processes in the lithosphere. One of the first to explore igneous activity in the context of chemical evolution was Oparin (1936). He did not support the idea, mainly because the proposed carbon source on the early Earth would have been volcanogenic CO_2. Oparin rejected CO_2 as the carbon source for primitive life processes by citing Pflüger (1875): 'Carbon dioxide is not the beginning but the end of life'. Later on it was common to ignore igneous environments in conjunction with early life by the argument that the existence of volcanoes is too episodic in order to be of importance for chemical evolution. In a more recent paper Sylvester-Bradley (1976) discussed the hypothesis of life's origin as a result of igneous activity. He emphasized oscillation of chemical systems between different environments far removed from equilibrium as one of the main mechanisms for abiogenesis. Indeed, Mukhin and coworkers (Mukhin, 1974; Mukhin *et al.*, 1978) reported the occurrence of thiocyanates and soluble ferrocyanides in hydrothermal systems of volcanoes of the Kurile Islands and the Kamchatka Peninsula in the Far East of Asian Russia (for geographical names mentioned in text, see Fig. 1). This would indicate the primary formation of hydrogen cyanide, which is often considered a likely precursor of abiogenically formed amino acids and nucleotide bases.

2. Early Proposals

The first article that proposed connections between chemical evolution and plate margins was published by Ingmanson and Dowler (1977). They proposed that the hydrothermal brine pools in the deeps of the Red Sea rift valley, especially the Atlantis II Deep, would be fruitful sites for testing whether or not life could evolve in lithospheric environments. Their arguments were that: 1) the brines were reported sterile, 2) the environment is reducing, 3) there is no free oxygen, 4) there is much methane and ethane, 5) montmorillonite is present, 6) residual gases from the Earth's paleo-atmosphere appear to be present, and 7) the temperature is high (about 60°C in the Atlantis II Deep). Ingmanson's and Dowler's subsequent reports of the occurrence of thiocyanates (Dowler and Ingmanson, 1979) and anomalous concentrations of the amino acid glycine (Ingmanson and Dowler, 1980) in the hot brines of the Red Sea supported their ideas, although their conclusions were later questioned by Simoneit and coworkers (1987). However, the Red Sea does not represent the 'standard' type of oceanic plate spreading center. The Red Sea represents an early stage of opening of an ocean basin with restricted water circulation and thus resembles the Atlantic Ocean about 200 Ma (million years ago). This is probably a process that has occurred at regular intervals in Earth's history, but that type of environment can not be expected to have had the same longevity as the ones of the deep ocean spreading centers.

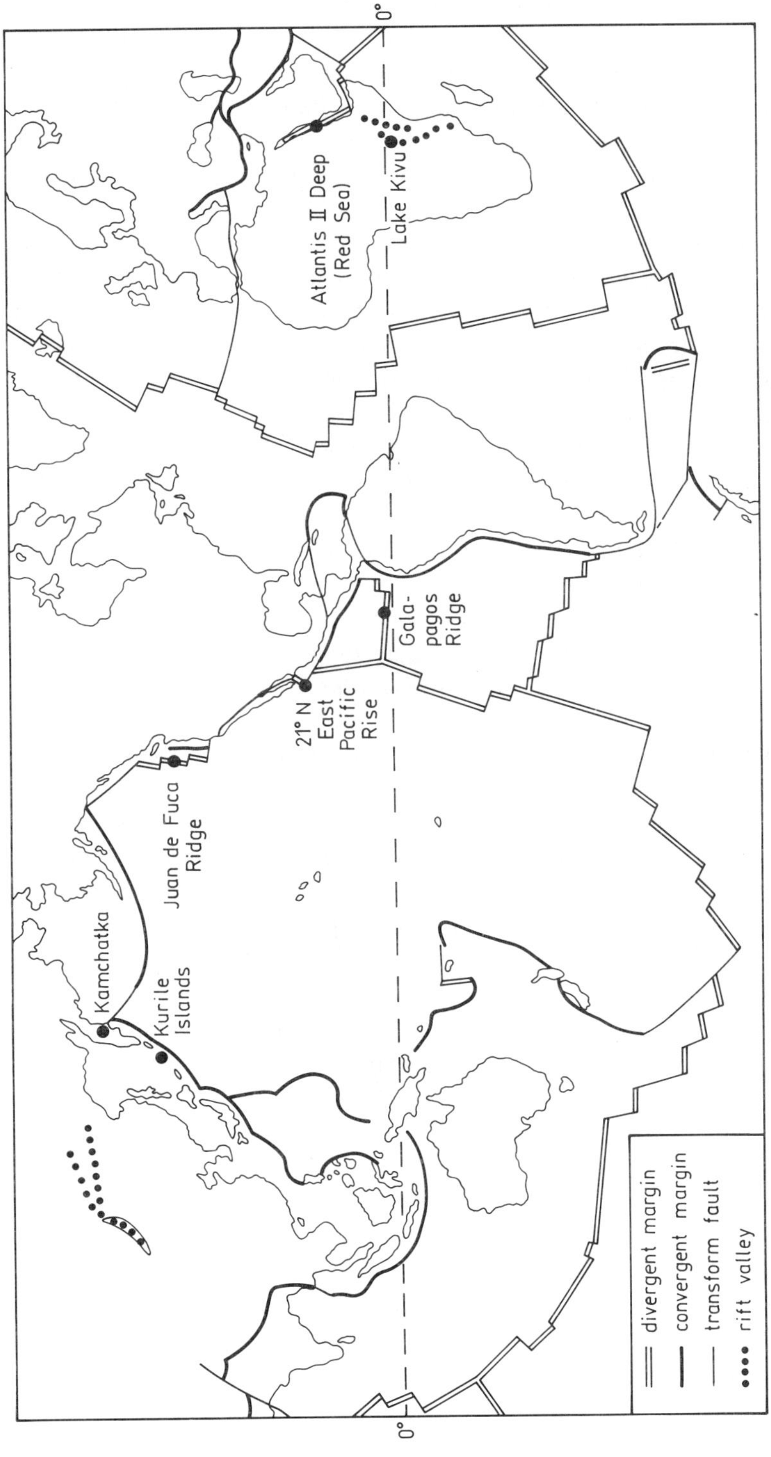

Fig. 1. Outline map showing the major plate tectonic boundaries of the earth's lithosphere and sites mentioned in the text.

The next paper that modelled life's origin in lithospheric hydrothermal systems at plate margins was published by Egon Degens (Degens, 1979). He and Johann Matheja had published articles about ten years earlier that treated the potential of life's origin in lithospheric environments (Degens and Matheja, 1968; Matheja and Degens, 1971). Degens hypothesized that hydrothermal systems, like the ones of Lake Kivu in the East African Great Rift Valley, would be perfect environments for primitive life forms to evolve. He considered three criteria necessary for any model on the origin of life. The system must: 1) be thermodynamically feasible, 2) have a high yield, and 3) allow for chemical evolution towards a primordial cell. Degens stated that biochemical molecules do not arise under equilibrium conditions, mainly because of the inability to separate the newly generated compounds in a quasi stationary phase. He also concluded that the number of non-equilibrium systems established in the lithosphere is great and that there is less occasion in either the hydrosphere or the atmosphere to maintain non-equilibrium conditions for extended periods of time. The East African Great Rift Valley is, however, like the Red Sea, not considered to be a plate spreading center in an ordinary sense. It represents instead the first stage of an 'aulacogen', i.e. a failed arm of a plate triple junction.

3. The Hydrothermal Model of Corliss, Baross and Hoffman

John Corliss led the first expedition that directly observed active submarine hydrothermal vents at an oceanic ridge (the Galapagos Spreading Center). He and his coworkers John Baross and Sarah Hoffman were also the ones who first identified the mid-oceanic rift sutures as likely sites for chemical evolution of organic compounds. The model was presented at the 26th International Geological Congress in Paris in 1980 and was published a year later (Corliss *et al.*, 1981). It has been further developed and presented in later publications (for example, Baross and Hoffman, 1985; Corliss, 1986, 1989, 1990). Corliss and coworkers claimed that the hot springs along the mid-oceanic rifts provided all the conditions necessary for the creation of life on Earth. The scope of the article was wide and had therefore a considerable impact on several disciplines of the natural sciences community. However, because of the wide scope it was also relatively easy for specialized scientists of biochemistry, microbiology, etc. to find weak spots of the model.

In the original hydrothermal origin of life model the dynamic vent systems are thought of as 'life reactors'. This is based on the gradients of temperature, pH and concentrations of various chemical components and the presence of mineral catalysts like montmorillonite clays, iron oxide hydroxides, manganates, sulfides, and zeolites. Water that circulates through the systems transports dissolved substances from the high temperature zone (about 350-400°C) to the intermediate/low temperature zone. Abiogenic organic monomers such as amino acids, sugars, purines, and pyrimidines are proposed to be produced in the high temperature zone from H_2, CH_4, NH_3, H_2S, CO, HCN, etc. This first step of condensation of organic monomers from simple inorganic constituents is followed by transportation to the intermediate temperature zone and polymerization of the monomers to poly-peptides and polynucleotides. Eventually lifelike 'protocells' are formed in a lower temperature window and are released to the deep-sea bottom water. Nisbet (1987, 1991) has elaborated on the hydrothermal origin of life model in his books 'The Young Earth' and 'Living Earth' and elsewhere (Nisbet, 1986). A report summarizing the state of the field was also published after a workshop in Prague, 1989, on 'Chemical Evolution and Neo-Abiogenesis in Marine Hydrothermal Systems' (Holm, 1990).

Why then would the concept of chemical evolution in hydrothermal systems be more attractive than other models? The dynamic characteristics of the hydrothermal 'life reactors' has already been mentioned. This means that non-equilibrium conditions are maintained and that a continuous flow of components connects different chemical environments with each other. Reactions requiring, for instance, quite different temperatures could be going on in the proximity of each other. Oscillations within the system would add to the overall instability and non-equilibrium characteristics. For geochemists one of the main advantages of having the origin of life in hydrothermal water is probably that such systems can be postulated to have existed through the entire length of Earth's history (confer Table I for the ages of the Earth). It is possible that the tectonic style of the Earth changed from 'micro-plate tectonics' to modern plate tectonics (see Chapter 2) during the transition from the Archean to the Proterozoic Eon 2.5 Ga (billion years ago). However, since heat production and heat flow from the planet's interior was much more intense on the young Earth, the hydrothermal activity can only have been more pronounced in the early Archean than now.

4. Modified Models

The generally accepted theory of the redox state of the early Earth not too long ago was that the atmosphere and the oceans were reducing. Most biochemists are of the opinion that life must have originated in reducing environments since there is little evidence for any substantial recovery of reaction products in oxidizing abiotic simulation experiments. This led some geologists to comment that the majority of diagenetic environments are reducing and have always been so (cf. Clemmey and Badham, 1982). However, probably scientists working on problems in conjunction with life's origin are too much focused on the impact on chemistry at the Earth's surface redox conditions. As we shall see later in this chapter, changes in physical parameters of the Earth's interior may cause response by redox pairs that we are not quite used to at surface conditions.

Wächtershäuser (1988a,b; 1990a) and Russell and coworkers (1989) have emphasized the reduced minerals of hydrothermal systems as potential catalysts in chemical processes leading to the origin of life. Wächtershäuser proposed that energy for primitive chemoautotrophic carbon fixation in hydrothermal systems is provided by the redox process of converting ferrous ions and hydrogen sulfide into the mineral pyrite (FeS_2). At the same time pyrite is postulated to serve as a kind of mineral template for the polymer-

TABLE I

The ages of the Earth (modified from Ballard, 1983)

Eon	Era	Period	Epoch	Ma
			Holocene	0.010
		Quaternary	Pleistocene	2
			Pliocene	5
			Miocene	24
			Oligocene	38
			Eocene	55
	Cenozoic	Tertiary	Paleocene	63
		Cretaceous		138
		Jurassic		205
	Mesozoic	Triassic		240
		Permian		290
			Pennsylvanian	330
		Carboniferous	Mississippian	360
		Devonian		410
		Silurian		435
		Ordovician		500
Phanerozoic	Paleozoic	Cambrian		570
Proterozoic				2,500
Archean				3,800
Hadean	Precambrian time			4,600

ization of organic monomers to large structures. The main advantage with mineral surface processes is that slow lateral two-dimensional migration and growth of bonded organic compounds can occur. Small constituents that are weakly bonding will give way to stronger bonding large ones. On the basis of the solubility of salts of organic constituents Wächtershäuser (1988a) ruled out ionic bonding to negatively charged mineral surfaces, for instance clays, as processes of importance for chemical evolution, a statement for which there is little evidence in the literature. He argued that the mineral surfaces must have positive charges and that the organic constituents must be anionic. The most common minerals with positive surface charges at neutral pH in any environment are sulfides and oxides of the transition elements. Both groups of elements (as well as clays) are abundant in hydrothermal systems. However, in most modern marine hydrothermal systems not covered by sediments the pH is several units below neutrality, thus enabling positive surface charge also of the clay minerals.

Wächtershäuser concluded that the surface-bonding function is the oldest function of the phosphate groups. Therefore the insolubility of calcium phosphate and the concomitant low abundance of dissolved phosphate ions poses a major difficulty for all versions of the prebiotic 'broth theory'. In his surface-metabolic scenario, on the other hand, there is a simple solution to this problem if it is assumed that the surface metabolism is attached to a phosphate mineral or to a mineral with adsorbed phosphate ions. Fluids of hydrothermal systems and other geothermally active zones of the Earth's crust are among the few types of environments on our planet where phosphate is a relatively common constituent. Yamagata and coworkers (Yamanaka *et al.*, 1988, Yamagata *et al.*, 1991) have even detected pyrophosphate and tripolyphosphate in condensed water of volcanic gas. Accordingly, Nisbet (1987) has remarked: 'The critical role of phosphates both in the utilisation of energy and in the cell membrane is interesting and can be interpreted as eloquent testimony for a hydrothermal origin. On the Archean planet, perhaps only in such a setting would a non-negligible supply of phosphates be reliably present'.

5. Hydrothermal Systems as Protected Environments

Although Weyl (1968) did not discuss abiotic processes in hydrothermal systems he preferred deep water environments for modelling the origin of life. The reason for doing this was that newly formed organic aggregates would have been shielded from hazardous ultraviolet radiation at depth. Solar ultraviolet light is quenched by only a few meters of water. However, the necessity of a protective screen for early life from radiation and impacting objects from space has become a major argument lately in favour of the hydrothermal model. Maher and Stevenson (1988) proposed a deep marine hydrothermal setting for the origin of life on the basis of the calculated impact flux during the Hadean (pre-3.8 Ga) and Early Archean Eons (3.8-3.4 Ga) (Table I). They concluded that abiogenesis could have happened as early as 4.0 to 4.2 Ga in deep hydrothermal systems without early life forms being totally destroyed by impacts. At the surface of the Earth, on the other hand, it could not have occurred earlier than between 3.7 and 4.0 Ga.

Sleep and coworkers (1989) argued in an independent study that the major way in which the Earth could have been sterilized was by evaporation of ocean water. Impact vaporization of the photic zone as late as 3.8 Ga is probable. Since life was unambiguously present on Earth at about 3.5 Ga, there is therefore a very narrow time window for it to have evolved on the Earth's surface. Ecosystems with chemoautotrophic primary producers in the deep ocean, especially those at hydrothermal vents, would have been much better protected. According to Sleep and coworkers (1989) the last impact that vaporized the ocean completely may have occurred as early as 4.44 Ga.

6. The Hydrothermal Models Challenged

It did not take long after Corliss and coworkers (1981) had published their original paper on the origin of life in hydrothermal systems until articles appeared that seemingly refuted their model. The first two were written by White (1984) and Bernhardt and coworkers (1984). Both articles primarily opposed the report by Baross and Deming (1983) that thermophilic marine bacteria isolated from vents had been grown at 265 bar and 250°C. Their arguments were, however, also applicable to the general ideas of the origin of life in hydrothermal systems. White determined decomposition rates for certain bonds of various biomolecules at 39 bar and 250°C. Bernhardt and coworkers did the same thing at the reported vent conditions, i.e. 265 bar and 250°C. Both articles concluded that organisms can not survive at such high temperatures if they are composed of biomolecules such as proteins and nucleic acids, due to the very rapid rate of decomposition of such molecules. On the other hand Yanagawa and Kojima (1985) and Yanagawa and coworkers (1988) reported the formation of stable silicate derivatives of microspheres from an amino acid mixture at 134 bar and 250-350°C. One of the obvious differences between their experiments and those of Bernhardt and coworkers and White is, first of all, that White worked at relatively low pressure. Another difference is that Yanagawa and coworkers used autoclave tubes made of glass at high pressure whereas Bernhardt and coworkers used nickel tubes. The great difference between the experiments, however, appears to be how and if the problem of presence of oxygen is the experimental systems has been tackled. New results achieved by Yanagawa and Kobayashi (1989) has added the recovery of silicate derivatives of simple amino acids (glycine, alanine, sarcosine) from pressurized CH_4, NH_4^+, CO_2 and silica glass at 260-325°C.

Although the controversy about the existence of the extremely thermophilic bacteria also affected the general view of the Corliss-Baross-Hoffman model, it was not until Stanley Miller and Jeffrey Bada published a couple of articles that their ideas became severely challenged (Miller and Bada, 1988; Miller *et al.*, 1989). Miller's and Bada's main arguments can be divided into four groups: 1) The high temperature of the vents (specified as 350°C) destroys rather than synthesizes organic compounds and polymers (references to Miller and Orgel, 1974; White, 1984; Miller *et al.*, 1976; and Bernhardt *et al.*, 1984); 2) the mere synthesis of polymers is not equivalent to synthesizing a living organism; 3) gradients do not exist near vents except from the mixing of 350°C water with cold sea water; 4) an important role for minerals and clays in prebiotic chemistry still remains to be demonstrated. The last argument would obviously be valid for any model on the origin of life that uses minerals as potential catalysts. Ferris and coworkers (cf. Ferris *et al.*, 1990) have, however, clearly demonstrated the catalytic activity of montmorillonite clays on, for instance, oligomerization reactions of deoxyribonucleotides. Argument no. 1 is not correct in the sense that hydrothermal waters in principle may have any temperature between 2 and 1,200°C; the observed interval in the upper lithosphere thus far is 2-400°C. Therefore vent waters of, for instance, 110°C, which is the maximum registered temperature so far of any living organism (Stetter, 1982), should also be described by the term 'hydrothermal'. Argument no. 3 is not at all true (see Chapter 2). In order to demonstrate further that biomolecules are very unstable at high temperatures and that their stability is not enhanced by high pressures, Miller and Bada carried out several experiments at 250°C, 265 bar and an initial pH that was neutral at room conditions. A mixture of equal molar amounts of aspartic acid, serine, alanine, and leucine was heated for up to 6 hours. All amino acids were decomposed quite rapidly, alanine less so than the others. Glycine, which was not present originally, was produced during the course of the heating experiment. This pattern is almost identical to the one observed for decomposition of L-amino acids by Bernhardt and coworkers (1984). The data presented by Miller and Bada (1988) was, however, used by Shock (1990a,b) in order to show that this type of experiment is normally performed out of equilibrium and that the redox state of the experimental systems is not controlled. Shock showed by simple thermodynamic calculations that Miller's and Bada's experiment probably was carried out somewhere in the predominance field of alanine and that a close approach to metastable equilibrium among leucine, glycine and alanine was attained in their system. Therefore it is not correct, Shock claimed, to use terms like 'instability' and 'decomposition' in this context, when the only thing that is happening is that the aqueous amino acids respond to the changes in external physico-chemical conditions and strive towards metastable equilibrium. The role of temperature is simply to overcome the kinetic barriers that prevent equilibrium or metastable equilibrium from being established.

Shock (1988, 1990a,b) has emphasized that it is the oxidized and metastable carbon and nitrogen compounds that are of special interest in hydrothermal vents rather than the reducing ones. He shares this view with Wächtershäuser (1990). CO_2 and N_2 may be in stable redox equilibrium with organic compounds such as amino and carboxylic acids in the Earth's interior, while CH_4 and perhaps NH_3 are not. This is explained by the fact that the oxidation states of the lithosphere are controlled by mineral buffer assemblages. Shock considered two mineral assemblages to be of primary importance in seawater-basalt systems: the fayalite-magnetite-quartz (FMQ) assemblage and the pyrite-pyrrhotite-magnetite (PPM) assemblage. At great depths within the circulation systems of the oceanic crust the conditions would probably be buffered by the FMQ assemblage. Shallower than about 1.3 km the PPM assemblage is the prevailing redox buffer. This means that for temperatures beyond about 275°C CO_2 and N_2 are the dominant carbon and nitrogen species at stable equilibrium. As a hydrothermal solution cools below 275°C it may still be buffered by the PPM assemblage, although it passes into a stability field where CH_4 and NH_3 predominate. If there exist kinetic hindrances that prevent equilibrium conditions from being established, the organic chemistry of the hydrothermal solution will then be 'locked up' in metastable compounds. Shock has calculated that the activities of several classes of organic metastable compounds, like alkanes, alkylbenzenes, carboxylic acids, amines, and amino acids, are maximized in the temperature interval 150-250°C, i.e. in the intermediate temperature window. The abiotic synthesis of various amino acids at 150°C, under reducing hydrothermal conditions, has been demonstrated experimentally (Hennet *et al.*, 1992). These results are in agreement with the calculations proposed by Shock, as discussed above.

One of the objectives of SCOR Working Group 91 has been to summarize the state of knowledge of hydrothermal systems and the origin of life sciences and to identify research opportunities in the field. This document with its collection of contributions is an attempt to cope with that challenge and has the ambition to open up a scientific area for future research. The following chapters will give you the background to why hydrothermal systems are considered potential environments in the context of life's origin, what the input parameters and starting materials in such systems would be or would have been, what indicator compounds there might be to look for, and what experiments should be conducted in order to verify or falsify models for the origin of life in marine hydrothermal systems.

CHAPTER 2

HYDROTHERMAL SYSTEMS: THEIR VARIETIES, DYNAMICS, AND SUITABILITY FOR PREBIOTIC CHEMISTRY

NILS G. HOLM

Department of Geology and Geochemistry, Stockholm University, S-106 91 Stockholm, Sweden

REMY J.-C. HENNET

S.S. Papadopulos and Associates, Inc., 7944 Wisconsin Avenue, Bethesda, MD 20814, USA

William Fyfe (1978) once remarked that the dominant cooling process of the Earth involves convective circulation of cold seawater. He added that it could hardly be otherwise when magmas are introduced at 1200°C under a thin layer of their own volcanic cover - the oceanic basalts. Convecting water acts as a cooling fluid, carrying thermal energy away from the hot areas of the Earth's crust toward the relatively cold surface. Convective processes dramatically increase the transfer of heat in the Earth's crust towards the surface in comparison with purely diffusive processes. Fluid movement occurs at relatively high velocities within the solid crustal matrix, and the more heat there is to be transferred the faster and more intense circulation has to be. Hydrothermal circulation is likely to have been more intense in the distant past of Earth's history. Probably the best argument for this assumption is that the total heat production of the Earth, based on concentrations of heat-producing radioactive elements, had to be much higher in the Hadean and the Archean than it is at present (see Chapter 1, Table I for the ages of the Earth). For example, assuming the Earth at steady state, the current radioactive heat production for the bulk of the planet corresponds to approximately one half of the observed heat flux (McKenzie and Richter, 1981). In the Hadean and the Archean, heat production by radioactive elements would have been more than 4 times higher (extrapolation based on present concentrations and radioactive decay rates), and therefore the heat flux must have been significantly higher (Davies, 1980; O'Connell and Hager, 1980) and the associated hydrothermal circulation more intense. In addition, other sources may also have contributed significantly to heat production on the early Earth. Among these are accretional energy resulting from planetesimal and meteoritic fragment impacts, and differentiation energy resulting from exothermic chemical reactions, partitioning and fractionation in the interior of the planet (Jeanloz and Morris, 1986). The discovery of Archean komatiites (Viljoen and Viljoen, 1969) is in agreement with a hotter Earth's upper mantle and lower crust, but the existence of high pressure mineral assemblages in Archean metamorphic rocks (Boak and Dymek, 1982) also indicates that horizontal variations of temperature existed in the Archean crust and that temperature anomalies in the mantle

Origins of Life and Evolution of the Biosphere **22**: 15–31, 1992.

might have been higher than today (Jeanloz and Morris, 1986). The global tectonic style of the Archean Earth was probably much different than that observed today. Several models have been proposed, some including a large amount of scattered hotspots (Fyfe, 1978), numerous micro-plates (van Andel, 1985), larger plates subjected to modern plate motion and tectonics (Card, 1990), etc. However, the types of hydrothermal environments in the crust of the Archean Earth must have been quite similar to the ones observed today since the cooling processes were roughly identical (circulation of aqueous fluid in a solid matrix), the big difference being that more hydrothermal activity probably occurred globally, and that the proportions of the different types of environments varied. In this chapter we will discuss briefly the dynamic chemical and physical processes occurring in crustal hydrothermal systems, and the different types of contemporary submarine hydrothermal systems, comparing whenever possible with evidence of fossil ancient systems.

1. Dynamic Processes in Hydrothermal Systems

The primary role of submarine hydrothermal systems can be seen as to allow a smooth transition between masses at different temperatures (molten magma at more than 1000°C and seawater at approximately 0°C). This heat dissipating system works by circulating an aqueous fluid rapidly through fracture and pore pathways within the crustal rock matrix, transporting heat from the hot areas towards the colder ocean water. Recharge by seawater provides the bulk of the cooling mass in motion, and permeable oceanic crust provides the solid matrix supporting and directing the flow of the cooling fluid. The result is complex temperature and flow distributions within the oceanic crust and the existence of sharp physical gradients.

Fluids are released from magmatic intrusions as the crust cools down. Among the components released are H_2O, H_2S, HCl, HF, CO_2, CH_4, SO_2, and H_2 (Burnham, 1979). The release of these volatile components acts as an additional physical force for hydrothermal convection; it might also play a role in the fracturing of the crustal rock above the intrusion producing preferential transport pathways for the cooling fluid (Brimhall and Crerar, 1987). The magmatic components released combine by mixing with the circulating fluids, therefore changing their chemical composition along the flow paths. Water/rock interaction takes place and results in additional chemical variation along the paths. Associated with fluid mixing and water/rock interaction are variations of pH, ionic strength, and redox potentials (see Chapter 5). Therefore, in addition to sharp and complex physical gradients there are no less sharp and complex chemical gradients. Other parameters can influence chemistry independently of temperature or flow patterns (degassing, mixing, mineral precipitation or dissolution, etc.).

Mineral and rock composition gradients also exist along the flow paths. Freshly fractured rock surfaces and recrystallized rock material is found where magmatic melt has been injected. Different mineral assemblages represent different equilibrium conditions along the fluid circulation paths. The pH conditions can be buffered by silicate mineral assemblages (for instance, K-feldspar/muscovite/quartz), and oxygen fugacity (redox conditions) can be buffered by the assemblage pyrite/ pyrrhotite/magnetite in presence of sulfur species, or by magnetite/hematite when no sulfur is present (see Chapter 5). Mineral precipitation or dissolution can occur everywhere in the hydrothermal system in response to chemical disequilibrium. Chemical disequilibrium at a given location in the oceanic crust can be the result of fluid mixing, degassing, change in redox conditions or pH, cooling, high velocity fluid movements, etc.

The pressure and temperature gradients existing in hydrothermal systems have a dramatic effect on the properties of water. This is perhaps best illustrated in Figs. 1 and 2 (for discussion see Brimhall and Crerar, 1987). Fig. 1 illustrates the degree of variation of selected physical properties of water (thermal expansion, compressibility, heat capacity, and viscosity) as a function of temperature. Fig. 2 shows the variation of the dielectric constant of water versus temperature. A decrease of one order of magnitude in the dielectric constant of water between 0 and 374°C (critical temperature of water) can be simply interpreted as meaning that water is a good solvent for charged ions at low temperature and a poor solvent at high temperature. This implies that aqueous chemistry at high temperature is quite different than usually thought; for example, water at 200°C would be expected to have solvation properties more similar to methanol or nitrobenzene than to water at room temperature. This is, however, an over-simplification. It does not take into consideration the changes in chemical speciation of dissolved compounds. It also ignores the effects of higher pressure, which will tend to force the dielectric constant of water towards higher values. Furthermore, high temperature and pressure conditions affect other factors that control the chemical behavior of water and any dissolved molecules (molecular vibration, ligand field stabilization, electron spin-pairing, etc.). For more detailed discussion compare with, for example, Helgeson and coworkers (1981), Brimhall and Crerar (1987) and references listed therein. For our purpose it is important to consider that gradients in the physical and chemical properties exist for water and any dissolved chemical species circulating in hydrothermal systems, and that the changes of these properties under the extreme conditions found in the Earth's crust can be dramatic and are very often ignored in the study of chemical evolution and the origin of life.

Depending on the dimension scale considered for hydrothermal activity, hydrothermal processes appear to be active for periods of millions of years (100 km scale ridge segments), thousands of years (100 m scale hydrothermal vent fields), years (individual hydrothermal chimneys or seeps), days (episodic formation of large hydrothermal plumes), or even minutes (violent hydrothermal discharges). Considering the range of velocities at which the circulating fluids travel through the different parts of an hydrothermal system (m/s where venting, cm/yr or less where recharging), one has to take into account periods of time ranging from seconds to years for studying chemical and physical processes involving contact between mobile and immobile phases. Depending on the time scale considered, a chemical or physical reaction might not have enough time to be carried out to

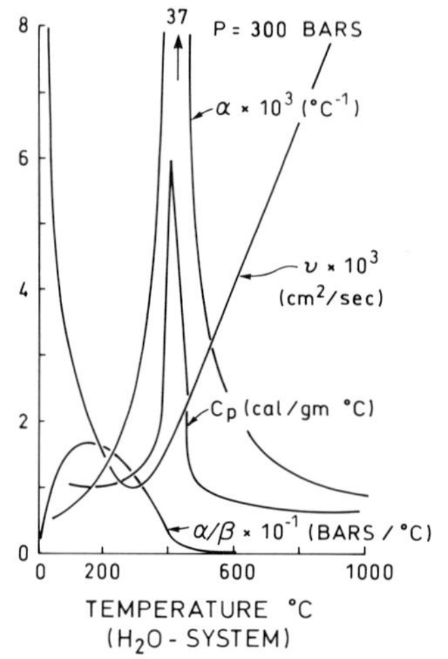

Fig. 1. Physical properties of pure water at 300 bar versus temperature; y-axis gives values in units shown adjacent to each curve in figure. The symbols α, β, Cp, and ν refer to coefficients of thermal expansion and compressibility, heat capacity and viscosity, respectively. From Norton (1984). Reproduced, with permission, from the Annual Review of Earth and Planetary Sciences Vol. 12. Copyright 1984 by Annual Reviews Inc.

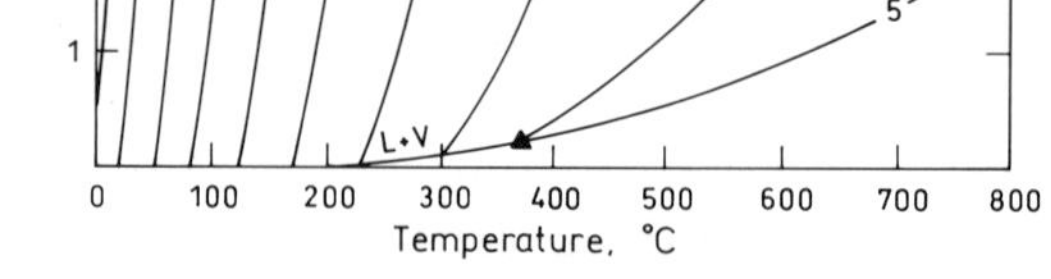

Fig. 2. The dielectric constant of water as a function of temperature and pressure. The triangular spot marks the critical point of water. From Brimhall and Crerar (1987). Copyright 1987 by the Mineralogical Society of America.

completion or equilibrium in hydrothermal systems. This implies that slow chemical and physical processes are expected to take place mostly out of equilibrium, moderately fast chemical and physical processes might only be able to reach equilibrium in some parts of the system, depending on fluid velocity and temperature conditions, and that only very fast chemical and physical reactions are expected to be carried to completion everywhere. The definition of fast and slow reactions varies depending on where a reaction is considered to take place. Thousands of years are available for reactions to approach equilibrium in the recharge zone, whereas minutes or seconds only may be available in a hydrothermal vent. The importance of the kinetics of reaction for chemical and physical processes cannot be overemphasized. In submarine hydrothermal systems a large range of chemical and physical reactions will never have the opportunity to approach equilibrium.

Fluids in submarine hydrothermal systems circulate through pores and fractures in the oceanic crust and its overlying sediments. The entire system can be considered as saturated with water with the exception of localized zones where degassing, boiling, or non-aqueous phases might exist (for example, supercritical CO_2). The porosity (defined as the part of the crustal rock matrix which can be occupied by a fluid, expressed in volume percent) of fresh unaltered oceanic crust is in the order of 10 percent (Davis, 1969; Johnson and Morris, 1962). Chemical and physical hydrothermal alteration, including mineral dissolution and fracture development, can increase porosity to more than 30 or 40 percent, depending on the intensity of the alteration processes. Sediments deposited on the oceanic crust have porosities from as high as 80 percent prior to being compacted through diagenesis, to porosity values of 30 to 50 percent deeper in the sedimentary column. A porosity gradient is therefore in place, and porosity increases towards the surface and within the most hydrothermally altered zones. Therefore, in submarine environments, large amounts of aqueous fluids are integral parts of the oceanic crust and sediments, and exchanges between fluid and solid phases can take place virtually everywhere. The sizes of pore and fracture features can vary by several orders of magnitude, from micrometer-size (interstitial spaces) to meter-size (large open fractures, for example).

For fluids to flow through a block of rock or sediments, pores and fractures have to be opened and connected enough to permit fluid circulation. Tectonic and thermal stresses, which are characteristics of hydrothermal systems, enhance connected porosity through fracturing and weathering of portions of the oceanic crust. Therefore permeability (defined as the ease with which fluids can move through a solid matrix) of the crust can be locally very large, for example, within the most hydrothermally altered and fractured zones connecting deep areas of the hydrothermal system to the seafloor. Away from zones where fluid circulation is facilitated by the presence of open fractures and hydrothermal weathering, anomalies in the temperature distributions are much less pronounced. Slight negative temperature anomalies occur, however, in the areas of the crust where recharge by cooling fluids takes place, and slight positive temperature anomalies occur where returning cooling fluids migrate towards the seafloor (Fig. 3). The main rea-

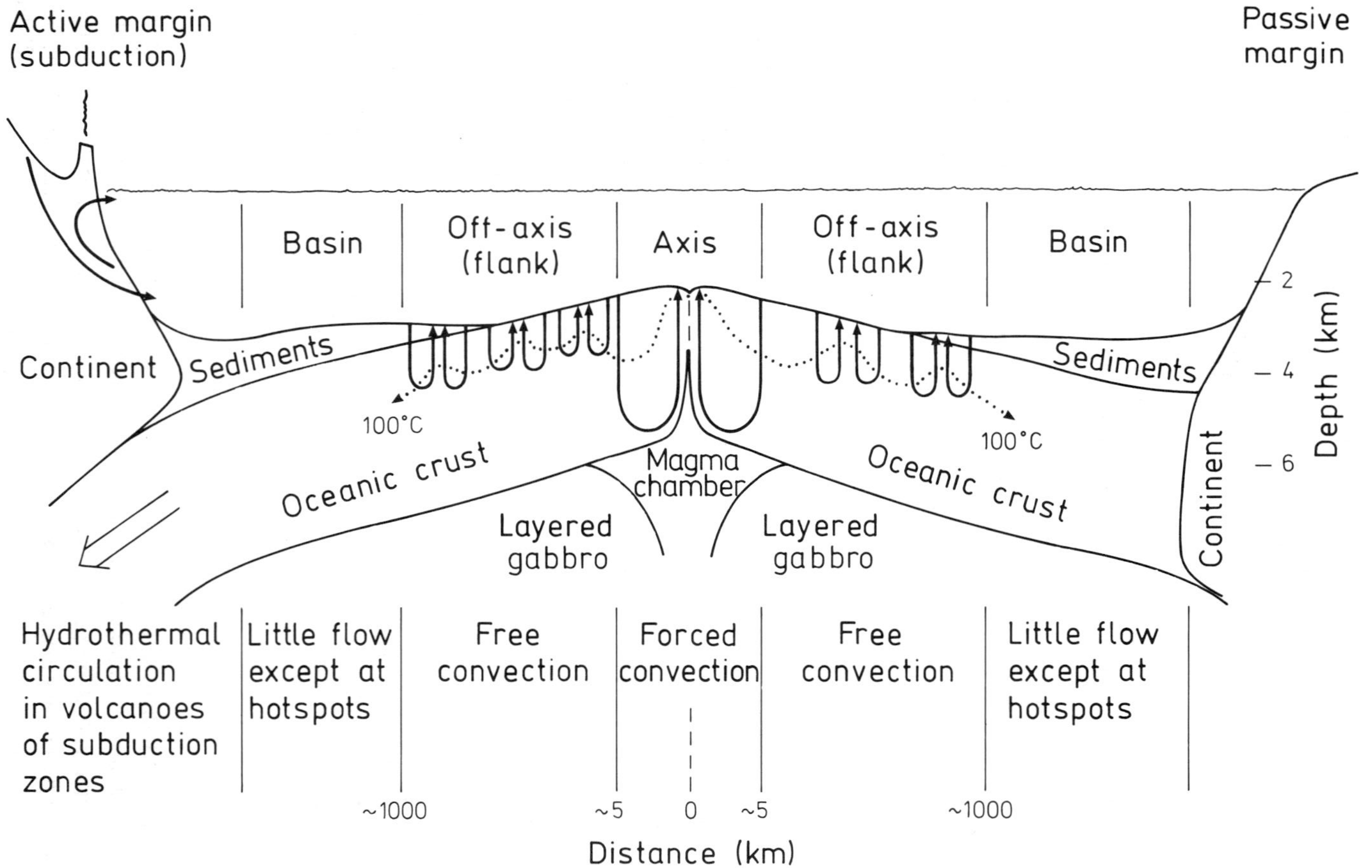

Fig. 3. Cartoon of diverging oceanic plates showing the spreading-center with a sediment-free on-axis hydrothermal system. Some off-axis hydrothermal systems in flank position to the spreading-axis are illustrated, with negative temperature anomalies where the fluids are recharged and positive anomalies where the hydrothermal fluids migrate towards the seafloor. Also included in the picture are both a passive continental margin and an active margin where the oceanic plate is subducted underneath a continent and volcanoes with hydrothermal circulation are formed.

son for the small size of the temperature anomalies away from the center of hydrothermal activity can be explained by a different transport process being dominant. In this case, fluid transport is dominated by the permeability of the porous media, the circulating fluids tend to be dispersed in the solid matrix, rather than channelled in fractures and weathered zones, and the velocity of fluid migration is slower. The interaction between the solid crustal matrix and the fluids is high, and the rate of heat diffusion is significant relative to the rate of circulating fluid migration. Some significant effects associated with fluid transport in porous media (relative to fluid transport in fractured media) are: first, the time available for any chemical reactions to proceed is much longer; second, the amount of interaction between the solid phase and the fluid phase is much more intense; and third, the volume of crust affected by fluid circulation is much larger. It is important to keep in mind that the two fluid flow systems described (fracture flow versus porous flow media) are not separated in the real oceanic crust, but are interconnected. An area dominated by one type of flow regime can be transformed, with time, into a system characterized by the other type of flow regime, with every combination in between.

During transport, the circulating aqueous fluid carries dissolved chemicals and gases within the intricate flow paths of the oceanic crust mineral matrix. Chemical reactions are expected to occur within the circulating fluid (see Chapters 4, 5 and 6), and because of transport, phases with different specific chemistry are juxtaposed and can interact along the flow paths. This results in the possibility for chemical mass transfer between phases to occur. Mass transfer can occur through diffusion, dissolution, precipitation, or volatilization processes. In addition, surface sorption processes can lead to the accumulation of selected chemical species onto the surface of minerals by preferential stripping of chemicals having an affinity for these specific surfaces (cf. Hochella and White, 1990; Holm *et al.*, 1992). This mechanism can lead to relatively high localized concentrations of chemicals which would not be otherwise expected to be found at such levels in the oceanic crust.

The above discussion is only meant to introduce the concept of dynamic processes into the mind of the reader. Dynamic phenomena of potential importance for our understanding of the origin of life on Earth are numerous, among the ones not mentioned in the discussion above are: the formation of fluid inclusions; the formation and transport of colloid-size particulates; the dynamics of chemical exchange and partitioning between the aqueous phase and non-aqueous fluid phases such as supercritical carbon dioxide; and the dynamics of selective chemical stripping in the structure of clay minerals.

2. Classification of Marine Hydrothermal Systems

On the present Earth most of the new oceanic crust is created by basalt production at ocean ridge spreading centers (Fig. 3). Basalt is also produced at oceanic hotspots. Ridge basalts are estimated to make up about 99% of new crust formation (Burke and Wilson, 1976). In both types of geological setting we will, however, expect hydrothermal circulation to exist. Geologists classify hydrothermal systems according to their tectonic setting. Thus, in the marine environment six general classes of hydrothermal systems can be distinguished (cf. Figs. 3 and 4):

1. Sediment-'starved' or sediment-free on-axis systems on plate tectonic spreading centers (mid-ocean ridges).
2. Sediment-covered on-axis systems on plate tectonic spreading centers.
3. Off-axis systems on the flanks of spreading centers and into the ocean basins.
4. Systems associated with backarc basins and backarc spreading centers.
5. Systems at hotspots.
6. Systems associated with subduction zones.

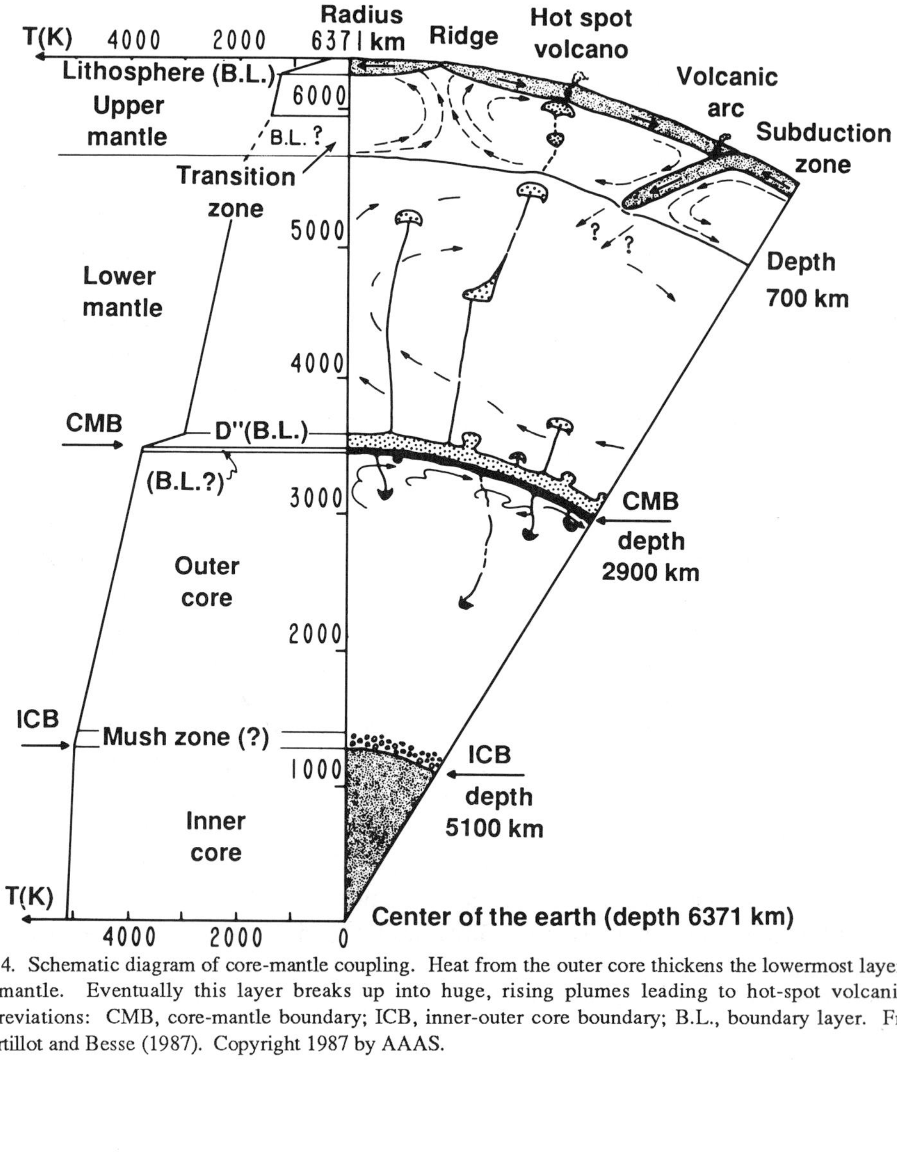

Fig. 4. Schematic diagram of core-mantle coupling. Heat from the outer core thickens the lowermost layer of the mantle. Eventually this layer breaks up into huge, rising plumes leading to hot-spot volcanism. Abbreviations: CMB, core-mantle boundary; ICB, inner-outer core boundary; B.L., boundary layer. From Courtillot and Besse (1987). Copyright 1987 by AAAS.

In addition to the marine systems there also exist continental hydrothermal systems (see below).

During the Second Conference on Scientific Ocean Drilling in Strasbourg in 1987 the different components of fluid circulation in the Earth's crust were thoroughly evaluated (COSOD II Report, 1987). The main contributors to the hydrodynamic flux through the crust appear to be: a) hydrothermal flow at spreading centers and their flanks (points 1, 2 and 3 above), b) upward flow of pore water at active margins due to sediment compaction and other diagenetic processes during subduction, and c) hydrologic flow from continents at passive and active margins (Fig. 3). Hydrothermal activity at subduction zones is probably a minor contributor to the total hydrodynamic flux and should not be confused with the expulsion of cold pore water to the deep sea due to tectonic compaction of subducted sediments.

3. Sediment-Free On-Axis Systems at Spreading Centers

Since the major portion of oceanic basalts is formed at plate tectonic spreading axes, it is not surprising that most hydrothermal systems today are found along the mid-ocean ridges (Fig. 5). Features that most people probably come to think of when the term 'hydrothermal system' is used were first studied on the East Pacific Rise at 21°N (Spiess *et al.*, 1980). This vent field is a typical example of sediment-free on-axis hydrothermal systems at the spreading centers with the black and white 'smokers' and water that is injected into the deep-sea at temperatures of about 350°C. The hydrothermal activity is driven by 'forced convection' due to steep temperature gradients in the crust surrounding the ridge axis (Fig. 3). They were at first believed to exist only in conjunction with fast- and intermediate-spreading ridges (6-12 cm/yr), mainly in the Pacific Ocean. However, such systems have now been found along both the fast- and intermediate- as well as the slow-spreading ridges. They extend all the way from the Pacific through the Indian Ocean into the Atlantic and, most likely, the Arctic Ocean (Campbell *et al.*, 1988). The COSOD II Report (1987) estimated the flow of hydrothermal water at an average temperature of 350°C through on-axis systems to 24 km^3/yr. The time estimated to cool an on-axis hydrothermal system ranges between 10^2 and 10^4 years (Fehn and Cathles, 1986). By definition the spreading centers are very young; the newly formed rock is often referred to as 'zero-age' crust. Therefore, in open settings, normally no sediment cover exists around the vents. The geochemistry and the dynamics of the hydrothermal waters will differ from the fluids of sediment-covered systems in several ways, the two major ones being that pH will stay at a low level (pH 3-4, protons are primarily released to solution through deposition of Mg compounds, cf. Von Damm *et al.*, 1985; Fouquet *et al.*, 1991) and that smooth temperature gradients above the hydrothermal system are less common (often a sudden quenching of 350°C water of their vents to the ambient temperature of about 2°C).

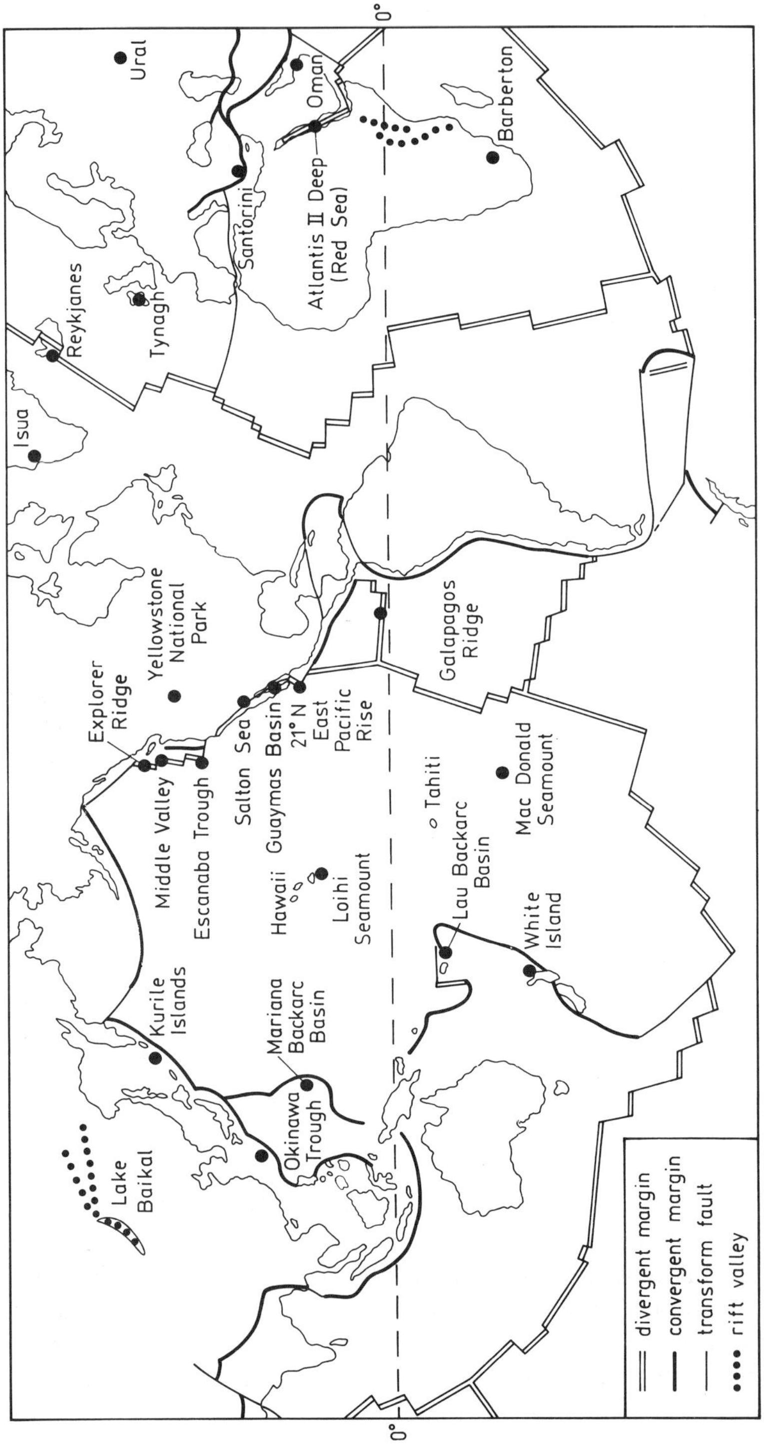

Fig. 5. Outline map of the major plate tectonic boundaries of the Earth and locations mentioned in the text.

4. Sediment-Covered On-Axis Systems at Spreading Centers

Only when spreading zones approach continents may the on-axis type of hydrothermal system be sediment-covered (Fig. 6). Such settings exist, for instance, in the Red Sea (Degens and Ross, 1969), which represents an early stage of opening of an ocean basin, and in the Guaymas Basin of the Gulf of California (Lonsdale *et al.*, 1980), where the East Pacific Rise starts a set of 'transform faults' through western Mexico and southern California. Sediment-covered hydrothermal systems also exist in the NE Pacific Ocean where the extensions of the East Pacific Rise, i.e. the Gorda, Juan de Fuca and Explorer Ridges, run close to the NW parts of the North American continent. Accordingly there exist on the Gorda and Juan de Fuca Ridges the sedimented Escanaba Trough and Middle Valley, respectively (SRDPG, 1990). The Guaymas Basin, as an example, is covered by about 400 m of diatomaceous mud and terrigenous plagioclase-rich sediments (Curray *et al.*, 1979, 1982; Einsele *et al.*, 1980). Conductive heat flow results in hydrothermal circulation and extensive thermal alteration of both organic (2% org. C) and mineral matter (Simoneit, 1983, 1984). Compared with sediment-free on-axis hydrothermal systems the sediment-covered ones may have much more extended thermal gradients, higher pH (about pH 6; Simoneit, 1984) due to buffering by sedimentary $CaCO_3$ (8%) particles and, of course, contain a much more diverse array of potential catalysts for abiotic organic reactions in the form of mineral matter. Since sediment-covered on-axis hydrothermal systems are situated relatively close to continents they are severely 'contaminated' by both terrigenous and autochthonous organic matter (due to rich nutrient supply) of biological origin. Thermal alteration of the biogenic matter leads to formation of petroleum compounds. This has, of course, little to do with abiotic synthesis. However, an understanding of the chemical behaviour of organic compounds in such systems will provide clues for possible abiotic reactions.

5. Off-Axis Systems at Spreading Centers

Off-axis hydrothermal systems in the spreading ridge flanks have maximum temperatures of about 200°C and are driven by 'free convection' due to cooling of the oceanic crust (Fehn, 1986, Fehn and Cathles, 1986, COSOD II Report, 1987). This zone continues between a few kilometers and as much as 1000 km away from the ridge axis (Fig. 3) with a hydrothermal activity that remains over long periods ($>10^5$ years, Fehn, 1986). Free convection cells may under special circumstances migrate laterally through the oceanic crust as separate entities but will mainly move with the plates as if attached. Because the oceanic crust increases in age at a regular rate away from the spreading axis, the sedimentary cover will have accumulated pelagic sediments in direct proportion to the distance from the ridge and in inverse proportion to the spreading-rate. Thus off-axis systems will always be more or less covered by sediments and share the properties of smooth thermal gradients and the presence of an abundance of mineral catalysts with the sedimented on-

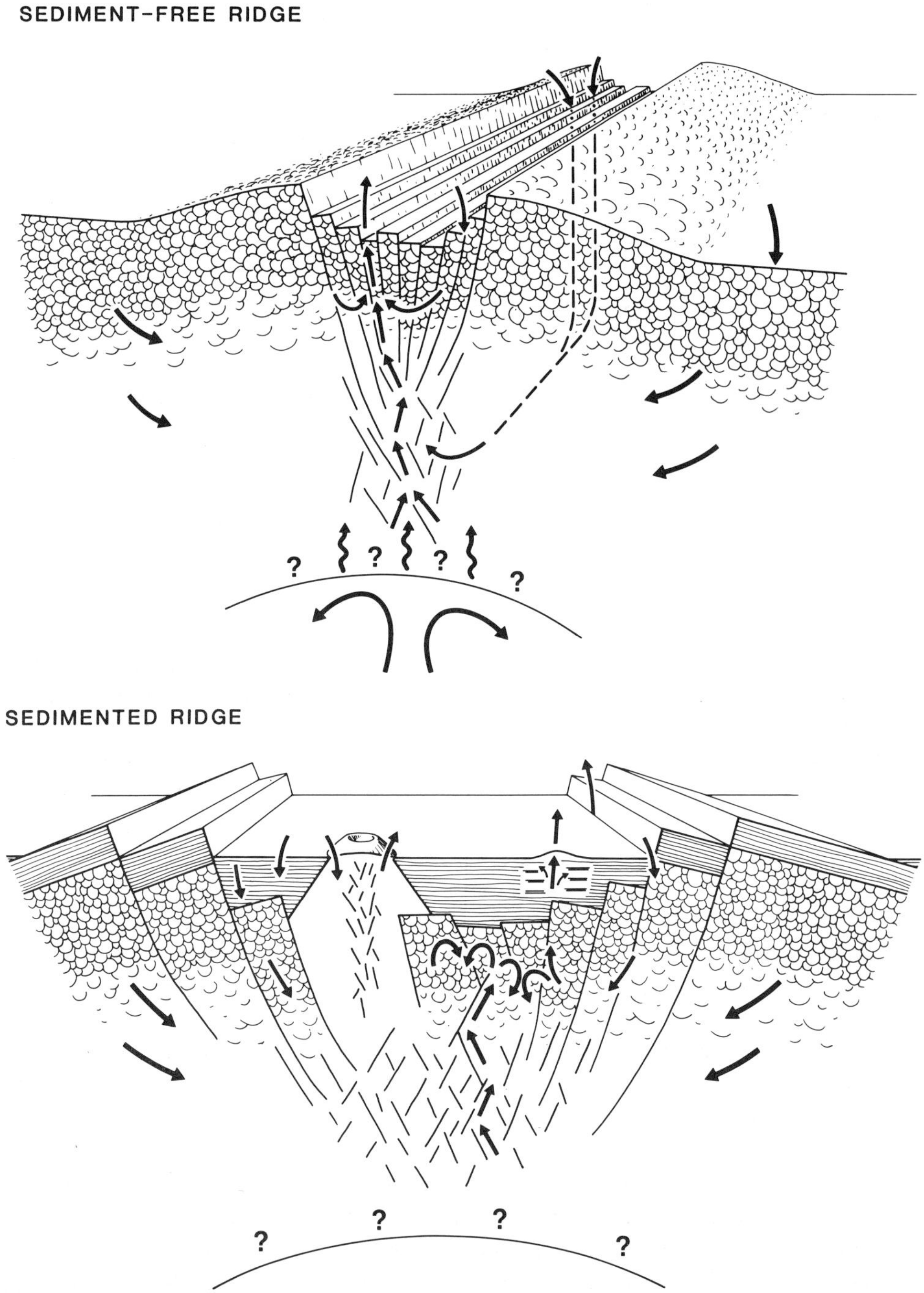

Fig. 6. Schematic illustrations of hydrothermal circulation at sediment-free and sedimented ridges. From COSOD II Report (1987).

axis systems. In general the contamination by biogenic matter is likely to be much lower in the pelagic sediments of deep-sea off-axis hydrothermal systems. Few survey efforts have been directed towards off-axis hydrothermal systems. Fehn (1986), however, reported that the frequency of some hydrothermal low-temperature fields south of the Galapagos spreading center and the distances between them correspond to model calculations of off-axis hydrothermal systems. In its long term plans the Ocean Drilling Program (ODP) recommends a cross-axis flank site array on the East Pacific Rise (EPRDPG, 1991) in order to study off-axis circulation. The volume of water at an average temperature of 150°C that circulates through off-axis systems was estimated by the COSOD II Report (1987) to be more than twenty times greater (560 km^3/yr) than the volume that is circulated through the hot on-axis vents (24 km^3/yr). Sea-floor convection involves large areas of diffuse input of fluids, especially at the off-axis type, and small areas of intense discharge. Fyfe (1978) thus remarked that there is more chance of local chemical equilibrium along paths of input and disequilibrium during discharge. This may have important consequences for the dynamics of both the inorganic and organic geochemistry of hydrothermal systems.

6. Systems Associated with Backarc Basins and Backarc Spreading Centers

Hydrothermal systems associated with backarc basins and backarc spreading centers were not discovered until quite recently. This type of fluid circulation in the Earth's crust was thus never included in the calculations of the COSOD II Report (1987). The type of geological setting is, however, from studies of the Kuroko type of sulfide ores deposits, known to be associated with (reducing) hydrothermal activity. Horibe and coworkers (1986) reported the occurrence of giant plumes of hydrothermal methane-enriched water at 3000 m depth at the Mariana backarc spreading center in the western Pacific Ocean. The great rise height (600-700 m above the sea floor) of the Mariana plume was interpreted to be due to higher effluent temperature, perhaps more than 400°C, than those of mid-ocean spreading centers. Sakai and coworkers (1990) studied hydrothermal chimneys and mounds by submersible at 1300-1500 m depth in the Okinawa backarc basin south of Japan. The highest temperature measured in the vent fields was 320°C. One of their most spectacular findings was the observation of the release of bubbles of liquid CO_2 from the sea floor of active hydrothermal sites and the formation of CO_2 hydrates. The approximate composition of the bubbles was 86% CO_2, 3% H_2S and 11% residual gas, mostly CH_4 and H_2. The isotope ratios of C and S, as well as the He, indicate that the CO_2-rich fluid has a magmatic origin. Fouquet and coworkers (1991) recently reported the occurrence of hydrothermal fields at the spreading center of the Lau backarc basin west of the Tonga Islands in the Pacific Ocean. The main divergences from the hydrothermal fluids of mid-ocean ridges are the acidity (pH 2) and the high temperature (400°C) of the Lau hydrothermal solutions. Increased temperatures in the bottom water are registered several meters away from the vents, something which is not observed with the on-axis hydrothermal systems. Some of the transition elements, such as Mn, Zn and Cd, are also much more abundant in the Lau systems.

7. Hydrothermal Systems at Hotspots

It was mentioned in the beginning of this chapter that ridge basalts make up about 99% of the crust of the modern ocean floor. The remaining percent originates from hotspot basalt production. Hotspots are believed to have their roots deep in the Earth's mantle, maybe in the core-mantle boundary (Fig. 4) (cf. Courtillot, 1990). The magma that forms ridge basalts, on the other hand, is believed to circulate almost entirely in the upper mantle. Some hotspots, like the one heating the Yellowstone National Park, are situated underneath continents. Others, like Iceland and the Axial Volcano of the Juan de Fuca Ridge (Massoth *et al.*, 1989), are integral components of the mid-ocean ridge system. In the Pacific Ocean, as an example, there are three known active hotpots at intraplate positions (Karl *et al.*, 1989). Those are the Loihi Seamount south of the Hawaiian Island group, the Macdonald Seamount of the Austral Island chain and the Mehetia volcano group near Tahiti. Hydrothermal activity has thus far only been observed at low-temperature vents (30°C) of the Loihi Seamount (Karl *et al.*, 1988, 1989), although its existence is postulated at other hotspots as well. Analysis data of the hydrothermal fluids that were sampled from the Loihi vents have been compared with data of other low-temperature vent fields (Karl *et al.*, 1989), such as those of the Galapagos Rift spreading center (Edmond *et al.*, 1979) and the Juan De Fuca Axial Volcano hotspot (Chase *et al.*, 1985). Compared to the the other low-temperature vents the Loihi hotspot systems are especially rich in CO_2 (300 mM) and Fe (1 mM), but the concentrations of NH_4^+ (5.2 μM), PO_4^{3-} (3.5 μM) and CH_4 (7.3 μM) are also elevated. The measured pH was about 5.5.

8. Systems Associated with Subduction Zones

Compared to the marine hydrothermal activity at mid-ocean ridges, backarc spreading centers and hotspots, the circulation and geothermal heating of sea water through the lithosphere at subduction zones is probably of minor importance (Fig. 3). Some well documented systems do exist in the literature, like, for instance, the Santorini caldera in Greece (cf. Holm, 1987) and the systems at the volcanoes of the Kurile Islands off the Pacific Coast of Asian Russia (Mukhin, 1974). For the purpose of studying abiotic organic processes and chemical evolution the subduction zone hydrothermal systems would probably be less suitable than most others because of biotic contamination due to the occurrence of much more organic material. An advantage, on the other hand, would be that they often are easily accessible and require relatively less sophisticated technical equipment.

9. Continental Hydrothermal Systems

Continental hydrothermal systems could be defined by different sets of criteria in addition to their position on continents. Hydrothermal activity in association with continental hotspots was mentioned earlier, but such activity does occur in other geological settings as well. In Chapter 1 the hydrothermal systems of the East African Great Rift Valley was described as a dormant part of the global rift suture system. Tiercelin and coworkers (1989) reported the occurrence of hydrocarbons associated with hydrothermal systems of Lake Tanganyika in the East African Rift. The hydrothermal activity that exists at depth of the Lake Baikal in Siberia, on the other hand, is a result of the opening up of a rift due to the collision between India and the Eurasian continent (Ballard, 1983). The hydrothermal fields of the Salton Sea area in southern California are localized on a transform fault of the spreading center between the Pacific Plate and the North American Plate, but must still be regarded as continental systems. Also the type of activity that occurs intimately associated with subduction in, for instance, southern Europe and New Zealand may be labelled as 'continental' in character. One way of distinguishing the continental systems from the marine ones would be the type of associated petrology. Continental systems are normally characterized by andesitic ('acidic') volcanism, although they may share this criterion with marine systems of the subduction zone type and even some marine hotspots. Several hydrothermal systems on the Icelandic hotspot, for example, occur in relatively andesitic terrain (e.g. Landmannalaugar). Another criterion that could be used is the type of water that circulates through the lithosphere. Continental systems are normally fed by meteoric water, but, again, most hydrothermal areas of Iceland except for those of the Reykjanes Peninsula, are percolated by meteoric water. This criterion would, on the other hand, classify the hydrothermal aquifer of the White Island Volcano off the northern coast of New Zealand as marine even though the volcanism is andesitic and the saline water is isolated from exchange with ocean water. Nisbet (1989) favored continental subaerial hydrothermal systems as likely sites for the origin of life because of their claimed moderate temperatures and fluctuating pH, which is often even mildly alkaline. For chemical evolution purposes it is not clear, however, that there is an advantage in modelling continental hydrothermal systems instead of marine ones. This has yet to be settled. One of the main tasks of this special volume is to inspire the pursuit of such research.

10. Fossil Evidence of Hydrothermal Activity

Fossil evidence of hydrothermal activity through Earth's history is widespread in the sense that hydrothermally altered rocks are common in the geological record. Distinct hydrothermal cells, on the other hand, are perhaps not as easily distinguished due to ageing of unstable mineral phases, at least not in rocks older than the Phanerozoic Eon. Much of the knowledge that we have of the record of ancient hydrothermal activity in the lithosphere is to be found in studies of ore deposits. Hodgson and Lydon (1977) have published a condensed description of the characteristics of different types hydrothermal systems and the geological setting of volcanogenic massive sulfide deposits. When ore geologists talk of 'exhalative sedimentary' ores (Oftedahl, 1958), they are likely to mean metal deposits that have precipitated from hydrothermal solutions. The most obvious evidence for ancient plate tectonics and hydrothermal circulation in oceanic crust is to be found in ophiolites, i.e. slabs of oceanic crust that have not been subducted but have been accreted onto the continents. Oudin and Constantinou (1984) reported the occurrence of hydrothermal vent fragments in ore deposits of Late Cretaceous age of the Troodos ophiolite complex on Cyprus. They also observed fossil worm tubes and other organic remains identical to the modern ones in on-axis deposits of, for instance, the East Pacific Rise and the Juan de Fuca Ridge. Simultaneously with the report by Oudin and Constantinou from Cyprus, Haymon and coworkers (1984, 1989) published their first article on the occurrence of hydrothermal vent worm tubes from 'early Late' Cretaceous (95 million years ago-Ma) sulfide ores of ophiolites in Oman on the Arabian Peninsula. Such worm tubes appeared for a while to be the first preserved hydrothermal eucaryotic fossils of the geologic record. Previously Larter and coworkers (1981) had found Lower Carboniferous (about 350 Ma) hydrothermal pyrite tubes at Silvermines in Ireland but excluded a biological origin, although the 'exhalative sedimentary' deposits were clearly associated with ancient hydrothermal feeder channels. However, Banks (1985) later reported the occurrence of a fossil hydrothermal worm assemblage of about the same age in the Tynagh lead-zinc deposit in Ireland. Kuznetsov and coworkers (1988) found the most ancient and complete fossil hydrothermal benthic fauna reported thus far in the literature. Not only did they find modern type fossil tube worms in Devonian (410-360 Ma) ophiolite formations of the Ural area of Russia, they also recovered fossils of the giant clams that are so intimately associated with present hydrothermal activity. These observations indicate that hydrothermal processes are active also during convergence of continental plates, at least in the final stages.

The oldest (about 3.8 Ga) sedimentary rocks on Earth have been found at Isua on Greenland. The rocks are relatively metamorphosed, but Appel (1979) proposed an 'exhalative sedimentary' origin of the sulfides in the iron-formation of Isua. At the other end of the Atlantic Ocean de Wit and coworkers (1982) described 3.3-3.5 Ga old 'pods' of ironstone (preferred term for 'iron-formation' on the southern hemisphere) in the Archean Barberton greenstone belt of Southern Africa. They interpreted the pods as representing buried, mineralized hydrothermal channels and chimneys. de Wit and coworkers also reported structures that resemble modern subaerial hydrothermal mud pools associated with ferruginous shales, banded iron-formation and stromatolites. Their interpretation of the 'mud pool' structures was, however, later claimed to have been refuted due to the lack of feeder zones (Lowe, pers. comm.). It is probably not so much the specific fossil phenomena but the geological environments as a whole that give us most of the information about the hydrothermal conditions on the young Earth. Thus MacGeehan and MacLean (1980) considered the genesis of all Noranda-type massive sulfide ores of Archean greenstone terrains to be intimately related to sub-seafloor geothermal activity. Similar types of 'exhalative sedimentary' sulfide ores deposited as chemical sediments on the ancient seafloor are also common in the Proterozoic Eon (cf. Rickard *et al.*, 1979). Fripp (1976) pointed to the well known worldwide association of gold deposits with Archean banded iron-formation and the fact that a hydrothermal origin is favored by most geologists for the Archean banded iron-formations (cf. Holm, 1987), although the mere oxidation of some Fe(II) to Fe(III) may have been due to photochemical processes (Braterman *et al.*, 1983). Fripp (1976) proposed that deposition of iron, sulfur, silica, arsenic, carbonate, and trace metals, including gold, took place from hydrothermal solutions on the seafloor as stratiform beds of iron-formation. It must be mentioned in this context that banded iron-formations are probably much more common in the geological record from the Archean than previously believed (Gole and Klein, 1981). Nisbet (1987) noted concerning the information available in Archean terrains: 'The greenstone terrains tell us about the surface of the Earth, the environment in the shallow seas and the beginnings of life'. Many Archean greenstone belts contain mineral deposits in what are probably extinct hydrothermal systems. Characteristically such deposits consist of veins of sulfide minerals, often closely associated with 'graphite' that probably originates from 'cracked' organic residues. There is little proof as yet that the Archean graphite is biogenic, but Nisbet (1987) remarked that the analogy to modern ocean ridges is indeed fascinating.

CHAPTER 3

MODERN LIFE AT HIGH TEMPERATURES

ROY M. DANIEL

Thermophile Research Group, School of Science and Technology, University of Waikato,

Private Bag 3105, Hamilton, New Zealand

1. Introduction

A variety of micro-organisms are now known which grow optimally above 65°C, and are defined as extreme thermophiles. As might be expected they are found in both natural and artificial hot environments. Until comparatively recently the upper optimum temperature for the growth of any living organism was about 85°C. Then in 1982 Stetter described an organism, isolated from the hot sea floor of a submarine solfatara field, which grew optimally at 105°C. Since then several other organisms have been found with optimum growth temperatures at 100°C or above and a few are capable of growth at 110°C (e.g. Huber *et al.*, 1987; Fiala and Stetter, 1986; Zillig *et al.*, 1987; Stetter *et al.*, 1990).

Work on extreme thermophiles is relatively recent. The pioneering work of Brock (1978), who first characterized micro-organisms isolated from hot springs, was carried out during the 1960's. Prior to that the conventional wisdom was that these high temperatures were inimical to life. Indeed one of the most interesting features of work on extreme thermophiles carried out over the last 30 years or so is the way it has challenged established thinking that heat is hostile to life. The use of autoclaves to sterilize solutions and boiled enzyme controls, are entrenched components of conventional scientific thinking.

All extreme thermophiles are eubacteria or archaebacteria, and no eucaryotes are known which can grow at these temperatures. Only archaebacteria are known to have temperature optima for growth above 85°C (Fig. 1).

2. Thermophily and Evolution - The Archaebacteria

Although archaebacteria are a quite recent taxonomic development (Woese and Fox, 1977; Fox *et al.*, 1980), their position as a kingdom separate from both eucaryotes and eubacteria is now well accepted. Recent advances in nucleic acid sequencing have enabled relatively precise genotypic data to be readily gathered for the purpose of determining taxonomic relationships (Wilson *et al.*, 1977; Woese, 1987). The most widely accepted unrooted taxonomic trees based on the sequence of Ribosomal 16S RNA place archaebacteria as a group closer to both the eubacteria and the eucaryotes than either of these are to one another (Figs. 2, 4), although variations on this have also been proposed

Origins of Life and Evolution of the Biosphere **22**: 33–42, 1992.

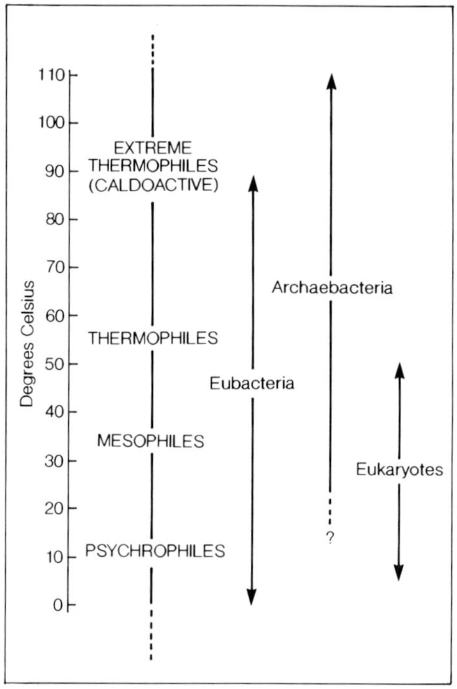

Fig. 1. Growth temperatures of organisms from the three kingdoms.

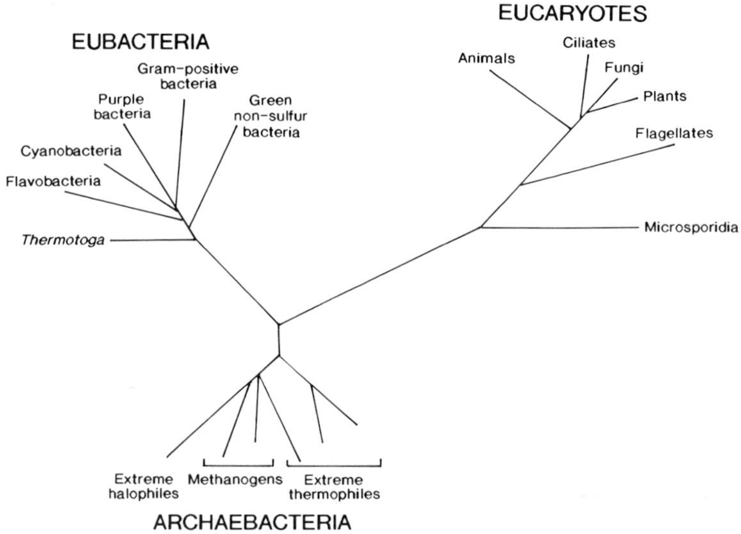

Fig. 2. Universal phylogenetic tree determined from rRNA sequences (Woese, 1987).

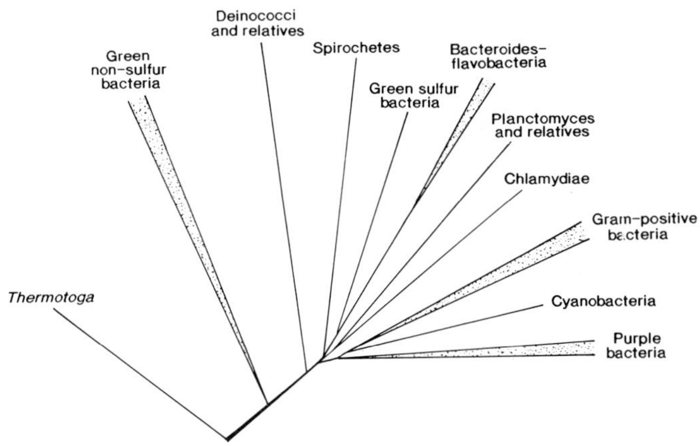

Fig. 3. Eubacteria phylogenetic tree based upon 165 rRNA sequence comparisons (Woese, 1987).

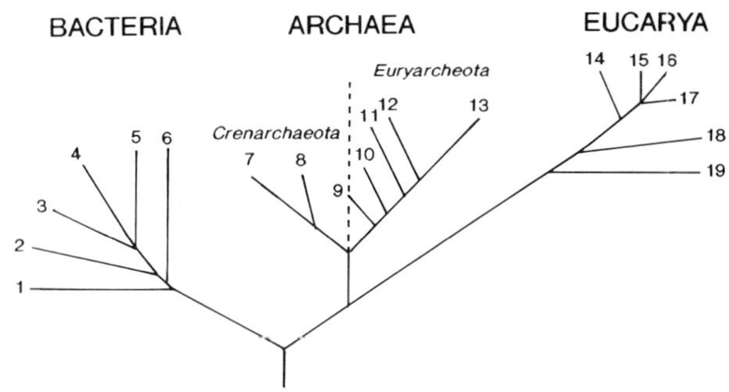

Fig. 4. Universal phylogenetic tree in rooted form, showing the three domains. Branching order and branch lengths are based upon rRNA sequence comparisons. The numbers on the branch tips correspond to the following groups of organisms. Bacteria: 1, the Thermotogales; 2, the flavobacteria and relatives; 3, the cyanobacteria; 4, the purple bacteria; 5, the Gram-positive bacteria; and 6, the green nonsulfur bacteria. Archae: the kingdom Crenarchaeota: 7, the genus *Pyrodictium*; and 8, the genus *Thermoproteus*; and the kingdom Euryarchaeota: 9, the Thermococcales; 10, the Methanococcales; 11, the Methanobacteriales; 12, the Methanomicrobiales; and 13, the extreme halophiles. Eucarya: 14, the animals; 15, the ciliates; 16, the green plants; 17, the fungi; 18, the flagellates; and 19, the microsporidia (Woese *et al*., 1990).

(see Lake, 1991). Furthermore, the common ancestor of the archaebacteria was also more primitive than the eubacterial common ancestor. Finally, the archaebacteria appear to be evolving more slowly than the eubacteria, so that if one accepts that no major change in the rate of evolution has occurred for these groups, the archaebacteria we see today are more primitive than the eubacteria (Woese, 1987).

On this basis the common ancestor of the archaebacteria is likely to be nearest to the earliest form of life. Woese (1987) has argued that the ancestral archaebacterium was an extremely thermophilic anaerobe dependent on sulfur reduction. Extreme thermophily is widespread among the archaebacteria and the most extreme thermophiles of the group appear also to be the slowest evolving. Extreme thermophily is also most strongly represented towards the root of the archaebacteria tree.

Additional evidence is available when we consider the eubacterial evolutionary tree (Fig. 3). The most deeply rooted branch in this tree is that leading to thermotogales (Achenbach-Richter *et al.*, 1987; Woese, 1987). These organisms are the most extremely thermophilic of all the eubacteria (Belkin *et al.*, 1986; Huber *et al.*, 1986; Huser *et al.*, 1986). The next most deeply rooted branch is that leading to the green non-sulfur bacteria containing a number of thermophiles such as *Chloroflexis* and *Thermomicrobium.* The third most deeply rooted group, the deinococci, includes one of the most widely occurring eubacterial extreme thermophiles, *Thermus*. Thermophiles are thus very strongly represented at the root of the eubacterial phylogenetic tree and all these groups are also relatively slowly evolving, strongly suggesting that eubacteria have risen from a thermophilic ancestor (Achenbach-Richter *et al.*, 1987).

Overall then, evidence is gathering that the view of thermophilic bacteria as specialist adaptions to enable colonization of a hostile environment is mistaken. Rather than marvel at life's ingenuity in surviving these harsh conditions, it may be more appropriate to wonder at the sacrifices and adaptions which have to be made by organisms evolving down-temperature into the slow frigidity of sub-80°C temperatures.

If all eubacteria have arisen from extreme thermophiles we might expect to find some evidence of this within a variety of eubacteria. It is tempting to assume that the ability of some thermophiles to grow over an extended temperature range, and the existence of cryptic extreme thermophiles, are manifestations of this (Wiegel, 1990).

Woese *et al.* (1990) have recently proposed that the archaea (archaebacteria) be divided into two groups (Fig. 4), the crenarchaeota, including the genus *Pyrodictium* and *Thermoproteus*, and the euryarchaeota, including the thermococcales, the methanococcales, the methanobacteriales and the methanomicrobiales as well as the extreme halophiles. The general phenotype of the crenarchaeota, it is argued, most closely resembles the ancestral phenotype of the archaea and so is the most primitive of living organisms.

3. Metabolism of the Most Extreme Thermophiles

A number of reviews are available dealing with archaebacterial metabolism (e.g. Danson, 1988), but given the relatively short duration of research on extremely thermophilic archaebacteria there is good reason to suppose that only a small proportion of the species available have been isolated, and that some major groups may yet be undiscovered. In addition, many of thc isolated organisms have not been well enough investigated to be taxonomically placed or validly named with great confidence (and this also applies to extremely thermophilic eubacteria). More research is needed to even establish and confirm the main metabolic pathways in all known extremely thermophilic archaebacteria. We continue to find taxonomic and metabolic "missing links" such as *Archaeglobus* (Stetter, 1988) between known groups of organisms, and we may also expect to find outliers which may extend the taxonomic (and metabolic) framework provided by the achievements of Woese, Stetter, and others. Extrapolation based on current data must therefore be treated with caution.

Within the crenarchaeota, the thermoproteales (comprising *Thermoproteus, Pyrobaculum, Thermophilum, Desulfurococcus,* and *Staphylothermus*) grow in the laboratory at pH's between 2.5 and 8.5 and can be found in both marine and solfataric environments. Some members are heterotrophic and some facultatively autotrophic, but all grow anaerobically and are capable of using sulfur as an electron acceptor. All members of the group grow optimally between 85°C and 100°C. The second group of crenarchaeota, the *Pyrodictiales* has within it two genera, *Pyrodictium* and *Thermodiscus.* All grow optimally between 88°C and 105°C, between pH's of 5 and 7, and have been isolated from marine environments. Most members of the genera are capable of using carbon dioxide as their sole carbon source and of obtaining energy from the oxidation of hydrogen by sulfur, leading to the production of hydrogen sulfide (Fischer *et al.*, 1983). This capability is also found among some members of the thermoproteales. *Pyrodictium abyssum* and *Thermodiscus maritimus*, however, are not dependant on sulfur, but are capable of an unknown mode of fermentation or respiration (Stetter *et al.*, 1990). The relatively closely related *Pyrococcus furiosus*, from within the euryarchaeota, has recently been shown to ferment pyruvate to acetate, CO_2 and H_2 using only three enzymes (Fig. 5) (Schafer and Schonheit, 1991). The key enzyme, acetyl - CoA synthetase (ADP-forming) is also important in *Thermoplasma acidophilum* (Danson, 1988).

Work on archaebacterial metabolic pathways is relatively new, but as Danson (1988) has remarked, all three kingdoms possess variations on a small number of metabolic routes, which must have been established before their separation. The most interesting variation is perhaps the presence in some extremely thermophilic archaebacteria of a modified Entner-Doudoroff pathway with non-phosphorylated intermediates (Fig. 6) (De Rosa *et al.*, 1984; Budgen and Danson 1986; Wood *et al.*, 1987; Danson, 1988), and in the apparent absence of the Emden-Meyerhof pathway.

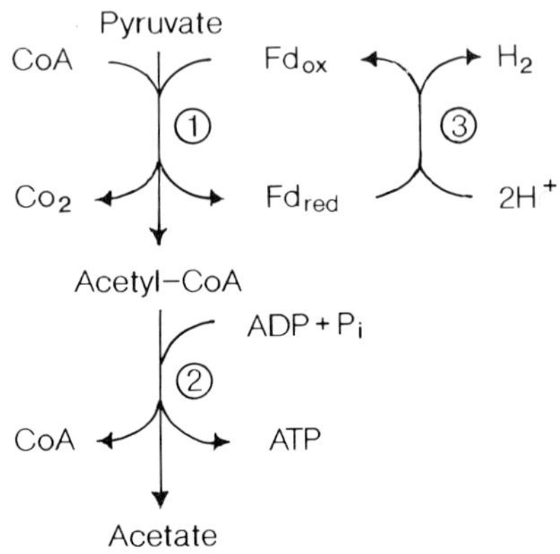

Fig. 5. Proposed scheme of pyruvate fermentation in *Pyrococcus furiosus*. 1: pyruvate: ferredoxin oxidoreductase; 2: acetyl-CoA synthetase (ADP forming), 3: hydrogenase; *CoA*, coenzyme A; *Fd,* ferredoxin (Schafer and Schonheit, 1991).

The study of both extreme thermophiles and of archaebacteria is relatively new and, because all of the most extreme thermophiles are archaebacteria, it is not clear that any of the features found in the most extreme thermophiles are due to their mode of life at high temperatures rather than to their archaebacterial nature. Enzymes whose function is to remove or utilize products of thermal decomposition might be expected, but have not been found.

There is, however, some evidence for the use of iron-sulfur proteins as cofactors in place of NAD or NADP (Mukund and Adams, 1991) in archaebacterial extreme thermophiles. The reduced forms of NAD and NADP have half lives of about 5 min at 80°C *in vitro* (Walsh *et al.*, 1983) and are thus often less stable than the oxido-reductases to which they are linked, with no evidence of stabilization by association with the enzyme (Hudson and Daniel, unpublished). Another feature of archaebacteria is the presence of unusual redox cofactors such as F_{420}, and a decreased specificity for NAD and NADP. While most non-archaebacterial enzymes have an absolute specificity for either NAD or NADP, many archaebacterial enzymes, and especially those from extreme thermophiles, will accept either. These observations can be explained if the use of NAD and NADP, at least as reductants, is a relatively late evolutionary development, possibly associated with evolution "down temperature". The role they now fill could have originally been carried out by non-haem iron proteins and cofactors such as F_{420}. While it could also be argued that the thermal instability of the reduced forms has led to their use having been "evolved out" of the most extreme (archaebacterial) thermophiles over time, if the extremely thermophilic archaebacteria most closely resemble the ancestral phenotype then this is a more complicated explanation. Furthermore, it does not explain the presence of a wider variety of redox cofactors, and decreased specificity between NAD and NADP, throughout the archaebacteria.

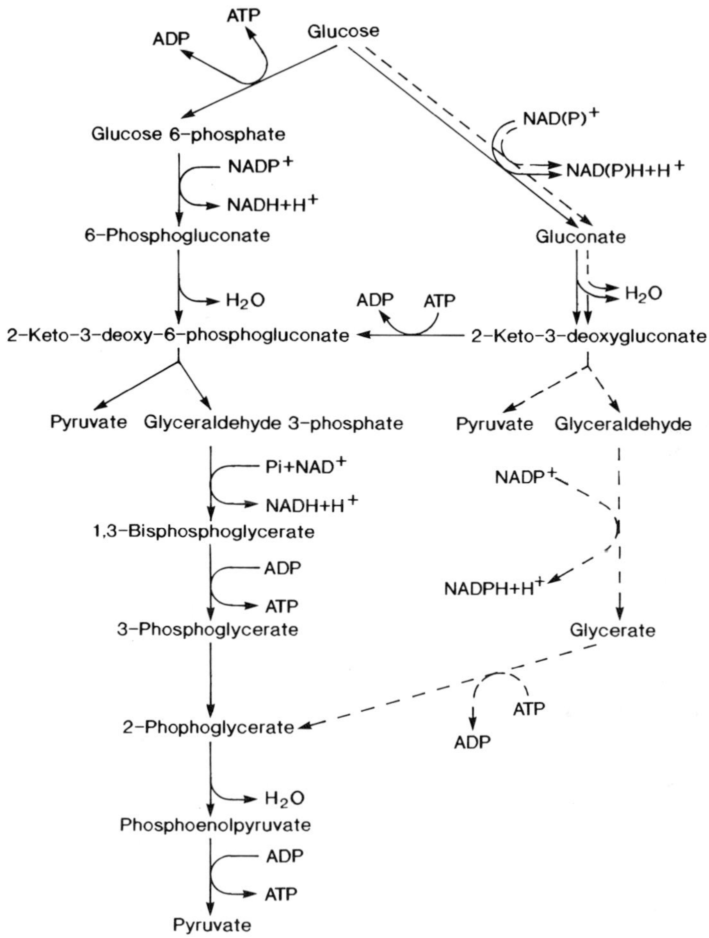

Fig. 6. Pathways of glucose catabolism in halophilic and thermoacidophilic archaebacteria. The modified Entner-Doudoroff pathway of halophiles (⟶) and the non-phosphorylated Entner-Doudoroff pathway of *Sulfolobus solfataricus* and *Thermoplasma acidophilum* (- - ⟶) are shown alongside the classical Entner-Doudoroff pathway of *eubacteria* (⟶). Conversion of glyceraldehyde into pyruvate via glycerate (· · ⟶) has been demonstrated only in *Thermoplasma acidophilum* (Danson, 1988).

4. Thermal Stability in Thermophiles

Among the macromolecules and supramolecular structures found in extreme thermophiles, the stability of lipids and membranes above 100°C is not surprising. There seems no particular reason why the various homeoviscous adaptions which permit survival both of psychrophiles and moderate thermophiles should cease at 65°C. The presence of ester-linked lipids (rather than ether linkages) outside the archaebacteria may be a response to low temperature rather than an advanced characteristic, but the evidence is weak. There is no good evidence that the membrane-spanning tetra-ether lipids found in some archaebacterial extreme thermophiles are required for membrane stability at high temperatures rather than for homeoviscous adaption. (See Russel and Fukunaga (1990) for a general review.)

DNA is apparently stabilized *in vivo* at high temperatures by DNA-binding proteins (Reddy and Suryanarayana, 1988), so that the melting of the double helix (separation of the two strands) takes place at temperatures well above those that might otherwise be expected. There is no evidence for systematic stabilization of DNA by increasing the proportion of the (stronger) G - C interactions compared with A - T interactions, since the percentage of G - C interactions ranges at least from 38 to 62 mole per cent among organisms capable of growth above 100°C (Stetter *et al.*, 1990).

Studies on the nature of the forces stabilizing protein conformation show no obvious upper temperature limit for the maintenance of protein conformation. It has been known for some time (e.g. Brandts, 1967) that the net free energy of stabilization of proteins is of the order of 25 KJ/mol, and is the result of a delicate balance between large stabilizing and large destabilizing forces, both of the order of 1,000 KJ/mol. Individual intramolecular stabilizing interactions can contribute in the region of 10 KJ/mol, so a small number of these can make a considerable difference to the stability of a protein. This is supported by the findings of Langridge (1968), and others (Matthews, 1987; Grutter *et al.*, 1979; Albers, 1989) that single point mutations within a protein can dramatically alter thermostability without significantly affecting the structure. A number of enzymes have now been characterized from extremely thermophilic bacteria which are both stable and functional for tens of minutes above 110°C (Bragger *et al.*, 1989; Simpson *et al.*, 1991; Ruttersmith and Daniel, 1991). It is possible to calculate that four or five additional stabilizing interactions, for example hydrogen bonds or hydrophobic interactions, are all that would be needed to confer conformational stability for useful periods of time to such enzymes at temperatures in the region of 150°C. There is no obvious or clearly delineated upper temperature limit to the conformational stability of proteins.

This leaves unexplained why, if mesophiles have evolved from thermophiles, enzymes from mesophiles are so unstable, with half lives usually of the order of minutes about 25°C above the optimum growth temperature. The answer seems to be that enzymes must be flexible to function effectively (e.g. Daniel, 1986; Vihinen, 1987; Daniel *et al.*, 1990) at the appropriate temperature for the organism. The intramolecular interactions needed to confer adequate stability in thermophiles are likely to render the structure undesirably rigid at lower temperatures in mesophiles. The high thermal stability of enzymes from extreme thermophiles also confers resistance to proteolysis at mesophilic temperatures (Daniel *et al.*, 1982), so high stability may hinder protein turnover. Thus, enzymes from mesophiles evolved to have the necessary degree of instability to maintain molecular flexibility, and protein turnover, at lower temperatures.

However, it is unlikely in fact that the upper limit for the stability of proteins will be governed by conformational stability. There is good evidence to suggest that degradative reactions including hydrolysis and deamidation will become the dominating factor in determining the useful lifetime of proteins above 100°C (Ahern and Klibanov, 1985). It is not clear at what temperatures these degradative reactions will render the lifetime of enzymes too short for them to function *in vivo,* although "modern" proteins will not survive 250°C under the conditions used by White (1984). The work of Klibanov and his colleagues (Zale and Klibanov, 1984; Ahern and Klibanov, 1985, 1986) suggests that the limit may be in the region of 105-110°C, but this work was done on mesophilic proteins at about 100°C and, although relatively little data are available on enzyme stability above 100°C, the stability of some enzymes from archaebacterial extreme thermophiles suggests that the limit may be somewhat higher. The proteinase from *Desulfurococus* has a half life of 8 minutes at 105°C (Cowan *et al.*, 1987) and the immobilized xylanase from *Thermotoga* (Simpson *et al.*, 1991) has a half life of 5 minutes at 115°C in the presence of xylan. This could be raised to at least 10 minutes at 130°C in molten sorbitol, indicating the importance of hydrolytic reactions in the loss of activity. (It has been shown that even mesophilic proteins can display considerable high-temperature stability in anhydrous conditions, e.g. Ahern and Klibanov, 1986.)

There is evidence that some of the reactions, for example deamidation, are strongly dependent on the flexibility of the protein backbone and on the nature of the adjacent amino acid (Aswad, 1990; Stephenson and Clarke, 1989). Deamidation can thus be prevented or slowed, either by substituting different amino acids for either the asparagine (for example) or its neighbor but also by maintaining the protein in a stable (less flexible) conformation. More work of this type needs to be done at 100-150°C on proteins which are stable conformationally at these temperatures before we can make decisions regarding the temperature at which degradation will limit protein stability. It may be, for example, that the combinations of redox and pH buffering, possible surface adsorption, and high pressure, found in hydrothermal systems, can stabilize some types of proteins against degradation. The water activity of the environment will play an important role.

There is no particularly good evidence that macromolecular instability (i.e. the instability of proteins, membranes, DNA, etc.) will set the upper temperature limits for life. Equally good candidates are the stabilities of essential small molecules such as amino acids and enzyme cofactors (Daniel, 1985). Although their stability is likely to be dependent on their particular environment, in aqueous solution some are known to be unstable even around 100°C. The half life of glutamine at 120°C is about 10 minutes (Ratcliffe and Drozd, 1978) and the half life of reduced nicotinamide adenine dinucleotide at 80°C is about 5 minutes (Walsh *et al.*, 1983). There is no evidence to suggest under what conditions, if any, such molecules can be significantly stabilized.

5. Growth Temperature of the Ancestral Organism

On taxonomic grounds there is good support for the ancestral organism being thermophilic (e.g. Woese *et al.*, 1990). On a biochemical basis there would seem to be no objections other than our entrenched psychrocentrism, and some obvious advantages. The question of how thermophilic is more difficult. The temperature range for the growth of the proposed kingdom crenarchaeota (including the genera *Pyrodictium* and *Thermoproteus*), together with the most closely related euryarchaeote genus (*Thermococcus*), is 65°-110°C, with optimum temperatures for growth in the range 85°-105°C, under laboratory conditions. It has to be borne in mind that, at least for the more thoroughly investigated eubacterial thermophiles, there is evidence that organisms growing optimally in the laboratory at 75°C dominate and thrive in some ecosystems at 100°C. Growth experiments under laboratory conditions, with their emphasis on rich media and relatively high growth rates and yields, may underestimate the environmental temperature at which extreme thermophiles thrive. Furthermore, sampling of stable ecosystems containing liquid water above 120°C is difficult and of recent origin, and the techniques for enriching and growing microorganisms above 100°C are poorly developed.

If we could assume that the ancestral organism originated at atmospheric pressure, then 100°C would represent an attractive optimum growth temperature. Given its widespread utilization and the dramatic reduction in surface area which accompanies the melting of sulfur, it has been suggested that its melting point (~113°C) may set an upper temperature limit for the growth of some thermophiles (Stetter, 1982).

If we assume that proteins stable in an aqueous environment are a prerequisite, then, although more research is required, 150°C may be an upper limit, conferred by the susceptibility of proteins to irreversible degradation. It is, however, very much an open question whether a protein with the composition, sequence, and structure to resist degradation at this or higher temperatures and still function as a catalyst, can be found from the enormous variety of possible proteins. Given what we already know of the stability, or potential stability, of most biological macromolecules and metabolites, then 130°C is certainly not an unreasonable upper growth temperature to expect of the ancestral organism. There is currently no evidence to support a higher growth temperature; but 10 years ago there was no convincing evidence for life above 100°C.

CHAPTER 4

AQUEOUS ORGANIC GEOCHEMISTRY AT HIGH TEMPERATURE/HIGH PRESSURE

BERND R. T. SIMONEIT

Petroleum Research Group, College of Oceanography, Oregon State University, Corvallis, OR 97331, U.S.A.

1. Introduction

The discovery in 1977 of active hydrothermal venting and associated mineralization at the seafloor spreading center of the Galapagos (Corliss *et al.*, 1979) initiated extensive surveys of global ridge-crests. Hydrothermal systems associated with ocean spreading centers are now recognized as a relatively common phenomenon and are thus one of the most actively researched topics in geology, biology and geochemistry. The morphology, mineralization, and fluid chemistry of the active systems in spreading centers has been reviewed (e.g. Barrett and Jambor, 1988; Rona, 1988; Rona *et al.*, 1983; Von Damm, 1990) and related to ancient deposits (e.g. Franklin *et al.*, 1981; Scott, 1985). Other non-rift areas with analogous hydrothermal systems are tectonically active subduction zones, fracture zones, and back-arc basins. The active seafloor systems are a major niche for fauna carrying out chemosynthetic biochemistry. This and the characterization of new species is now a main topic in biology (e.g. Childress, 1988; Grassle, 1985; Jones, 1985; Tunnicliffe, 1991). The geochemistry of vent fluids, interstitial fluids in sediments, and organic matter associated with hydrothermal systems are also a major research effort (e.g. Edmond and Von Damm, 1983; Gieskes *et al.*, 1982a,b, 1988; Simoneit, 1990; Simoneit *et al.*, 1984; Von Damm, 1990).

It is the purpose of this chapter to provide a description of the fate and chemical alterations of organic matter under hydrothermal conditions, with a brief overview of the geographic localities where these processes have been studied to date. Two major but overlapping aspects will be considered: (a) Alteration and degradation processes and reactions, both reductive and oxidative, and (b) synthesis processes and reactions that are primarily reductive.

Petroleums generated in high temperature and high fluid flow regimes are defined here as hydrothermal because the agent of thermal alteration and mass transfer, hot circulating water (temperature range – warm to > 350°C), is responsible for organic matter alteration (generally reductive), and product expulsion and migration from the source rocks or unconsolidated sediments (Didyk and Simoneit, 1989, 1990). These oils generally contain disequilibrium reaction products comprised of reduced and oxidized species (e.g. benzene vs. methylcyclopentane, cholestane vs. Diels' hydrocarbon). In contrast, conventional oils are derived products of basin evolution and are generated contemporaneously

Origins of Life and Evolution of the Biosphere **22**: 43–65, 1992.

with sediment compaction and heating (temperature window – warm to ~150°C). Generation of hydrothermal oils and gases is a geologically rapid process, taking place within 100's to 1000's of years (Peter *et al.*, 1991), whereas geothermal oils are generated at a rate that is tied to basin subsidence occurring over millions of years (Tissot and Welte, 1984; Hunt, 1979; Hunt *et al.*, 1991).

2. Alteration/Degradation Processes and Reactions

2.1. GEOLOGICAL LOCALES WITH HYDROTHERMAL ORGANIC PRODUCTS

2.1.1. Marine sedimented systems

The locations with known hydrothermal activity and associated mineralization at seafloor spreading centers (divergent plate boundaries) currently number about 100 and are catalogued in the reviews by Rona (1984, 1988). Those with associated organic matter alteration are indicated on the tectonic sketch map in Fig. 1. Both of these lists are expected to expand as exploration continues. Two continental systems are also shown in Fig. 1.

Guaymas Basin (Fig. 1) is an actively spreading oceanic basin (2000 m water depth in the rifts) where sedimentation is rapid (> 2 m/1000 yr) covering the rift floors to a depth of ~300-500 m (Curray *et al.*, 1982). The organic matter of these recent sediments is derived primarily from diatomaceous and microbial detritus and averages about 2% organic carbon. Numerous hydrothermal mounds rise to 20-30 m above the south rift floor and most are actively discharging vent fluids with water temperatures ranging from warm up to 350°C at ~200 bars (e.g. Fig. 2; Lonsdale, 1985; Lonsdale and Becker, 1985; Sayles *et al.*, unpublished data). The mounds are composed of complex deposits of sulfide, sulfate, silicate and carbonate minerals, and petroleum (Peter, 1986; Koski *et al.*, 1985; Simoneit, 1985a,b). They are covered with colonies of tube worms, bacterial mats and other chemosynthetic organisms (Jones, 1985). Typical oils from this basin are depleted in aliphatic hydrocarbons compared to normal crude oils (Didyk and Simoneit, 1990). The *n*-alkane distributions (e.g. Fig. 3a) have diverse ranges and carbon number maxima, generally from CH_4 to > C_{40}, with usual maxima in the mid-C_{20} region and no carbon number predominance (CPI, carbon preference index = 1.0) (Kawka and Simoneit, 1987). The generation of the complete suite of saturated (reduced) biomarkers (tracer compounds) from their biological precursors is additional evidence for the strongly reductive process operating during initial organic matter alteration. The major resolved peaks in the aromatic/naphthenic fraction are unsubstituted polynuclear aromatic hydrocarbons (PAH, Fig. 3c), a group of compounds uncommon in petroleums but ubiquitous in high temperature (> 250°C) pyrolysates (Geissman *et al.*, 1967; Blumer, 1975; Hunt, 1979). The dominant analogs are the pericondensed aromatic series (e.g. phenanthrene I, chemical structures are given in Fig. 4, pyrene II, chrysene III, etc.) (Kawka and Simoneit, 1990), and their pyrolytic origin is further supported by the presence of PAH with five-

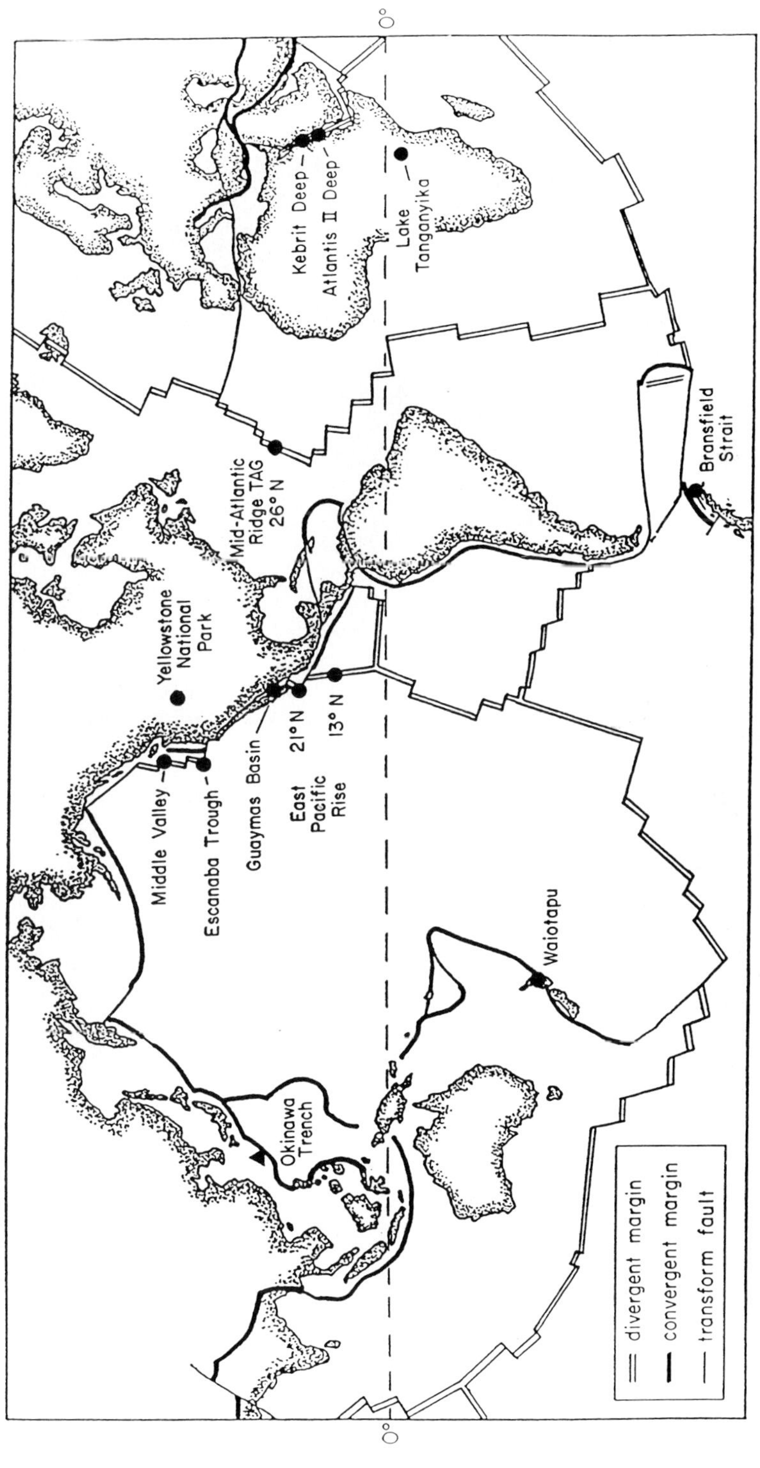

Fig. 1. General location map of the hydrothermal vent fields discussed here with the sketched global tectonics.

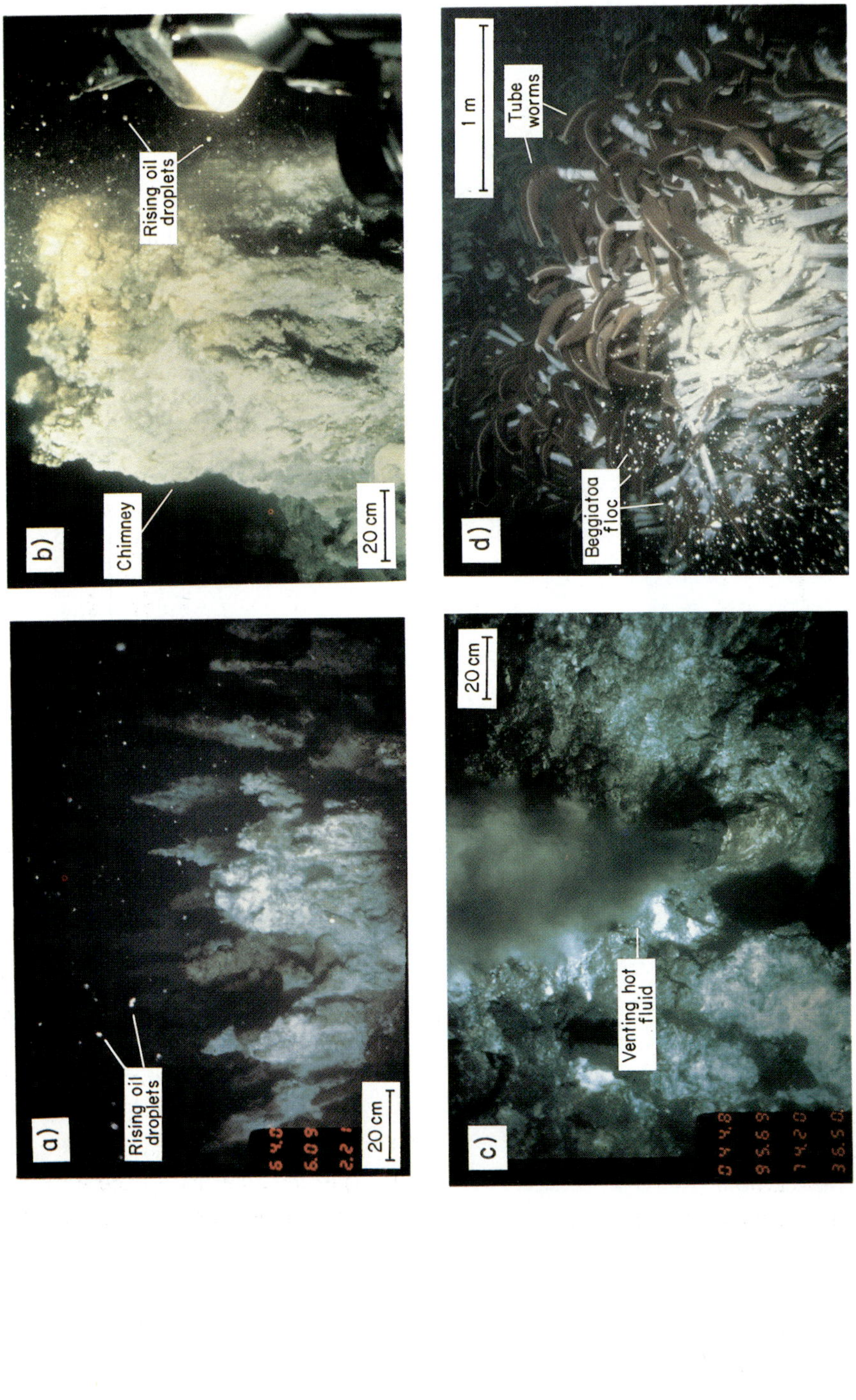

Fig. 2. Photographs of examples of Guaymas Basin vent systems: (a) petroleum rising from top of mound with small chimneys, (b) petroleum rising from base of chimney, (c) vent discharging hot fluid, (d) mound macrofauna (tube worms and beggiatoa floc).

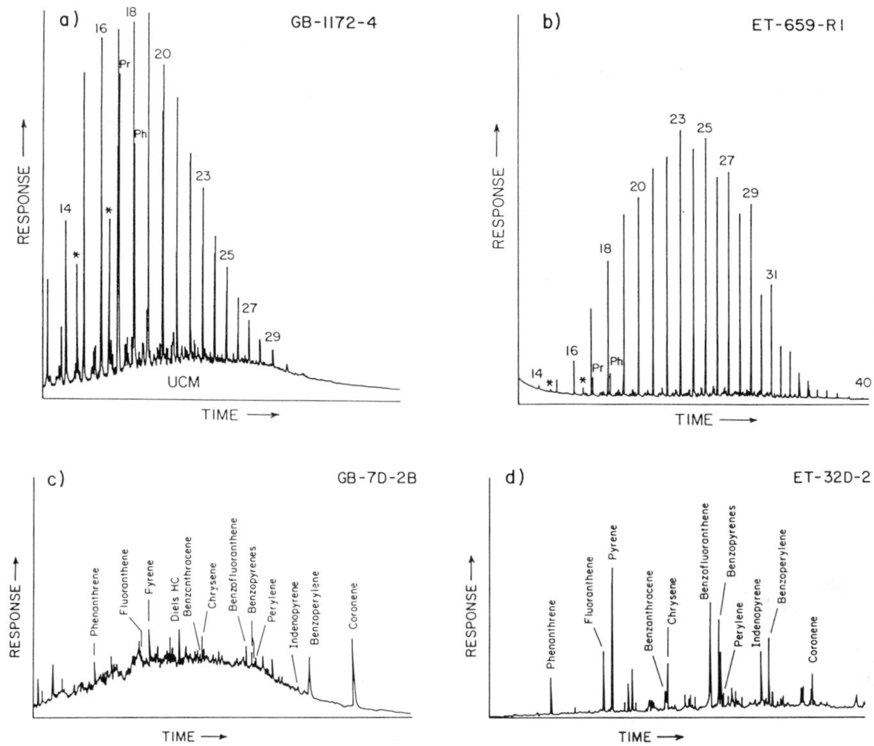

Fig. 3. Gas chromatograms of saturated (a, b) and aromatic (c, d) hydrocarbons in: (a-c) Guaymas Basin (GB) and (b, d) Escanaba Trough (ET) oils (Kvenvolden and Simoneit, 1990). (numbers refer to carbon chain length of *n*-alkanes, Pr = pristane, Ph = phytane, asterisk = other isoprenoids; UCM = unresolved complex mixture; PAH are labeled).

membered alicyclic rings (e.g. fluorene IV, methylenephenanthrene V, etc.), which are found in all pyrolysates from organic matter and once formed do not easily revert to the pericondensed PAH (Blumer, 1975, 1976; Scott, 1982). PAH become the dominant species due to their high thermal stability as well as enhanced solubility in near- and supercritical water (e.g. Sanders, 1986). The aromatic/naphthenic fractions of the Guaymas oils also contain significant amounts of N,S,O hetero-PAH (e.g. Gieskes *et al.*, 1988) and Diels' hydrocarbon, VI (Simoneit *et al.*, 1992). The chemical compositions of the aromatic fractions suggest an origin from oxidative alteration at high temperatures in the system (> 300°C).

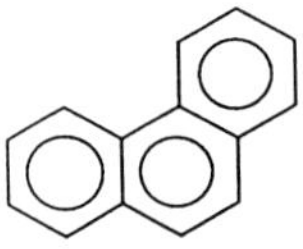

I. Phenanthrene

II. Pyrene

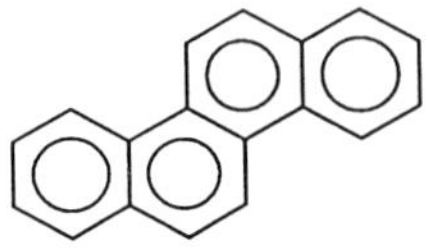

III. Chrysene

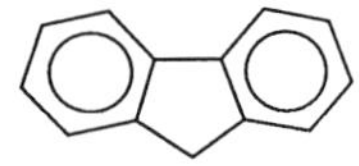

IV. Fluorene

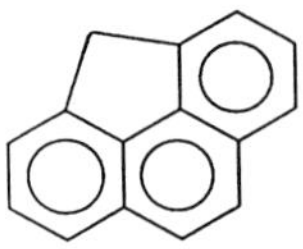

V. Methylene-phenanthrene

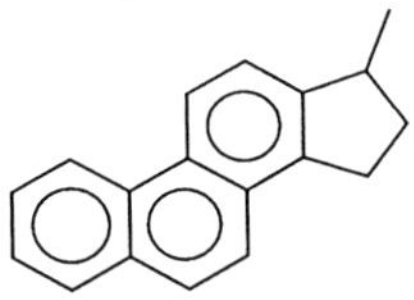

VI. Diels' Hydrocarbon

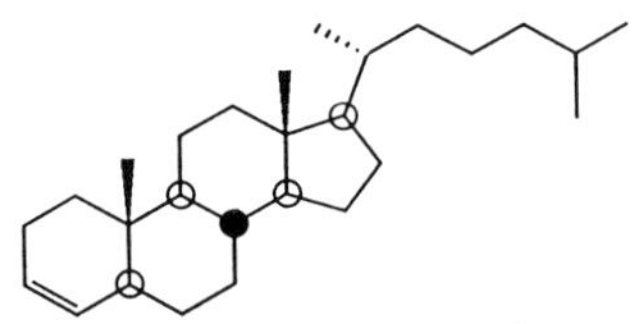

VII. Cholestene (e.g. Δ^3)

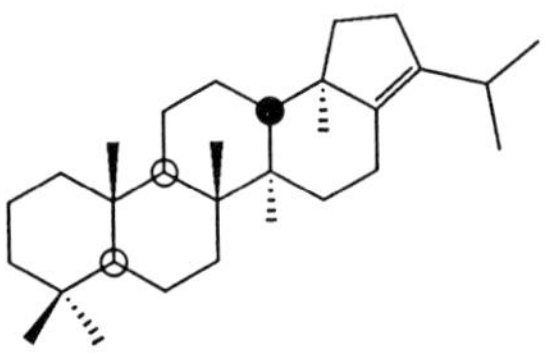

VIII. Hopene (e.g. $\Delta^{17,21}$)

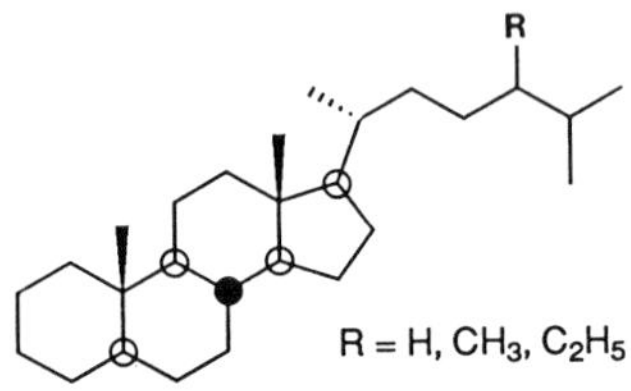

IX. Steranes (e.g. $\alpha\alpha\alpha$)

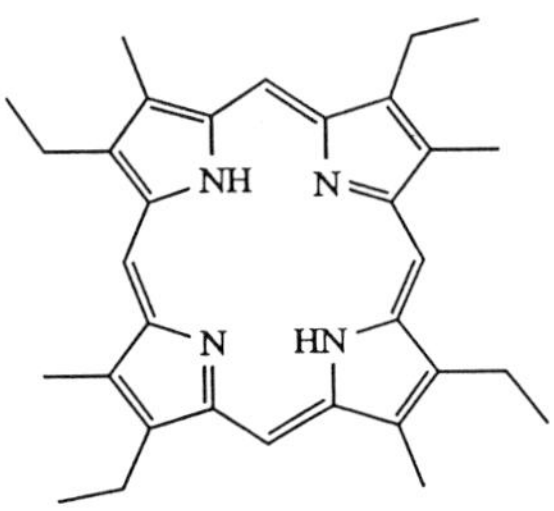

X. Porphyrin (free base, $C_{32}H_{38}N_4$)

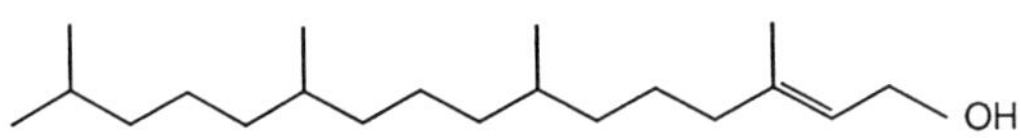

XI. Phytol

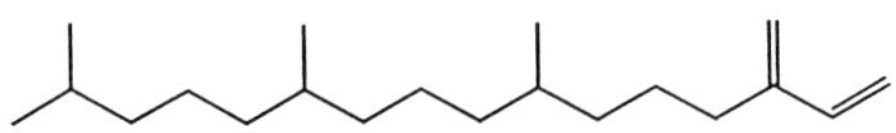

XII. Phytadiene (e.g. $\Delta^{1,3}$)

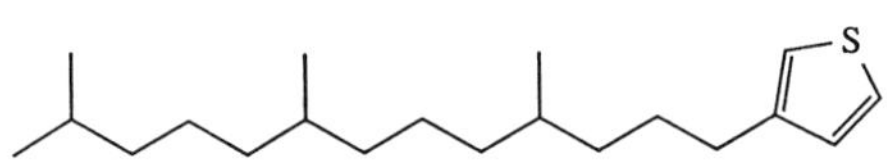

XIII. Phytylthiophene

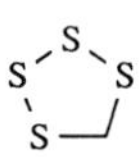

XIV. Tetrathiolane

XV. Pentathiane

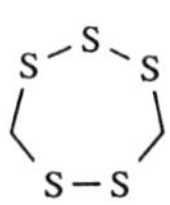

XVI. Pentathiepane

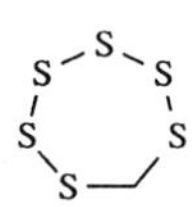

XVII. Hexathiepane

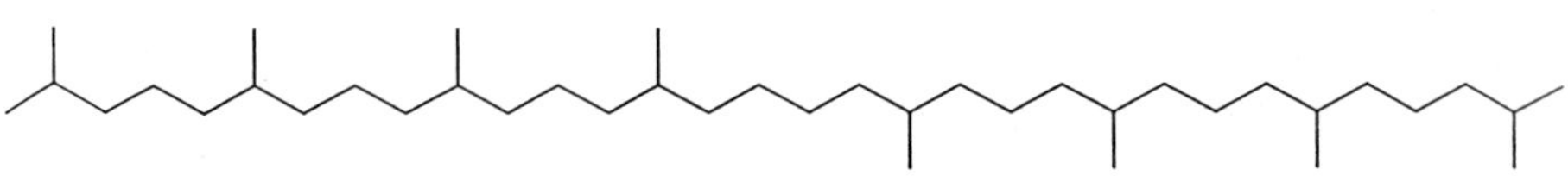

XVIII. Biphytane (lycopane), $C_{40}H_{82}$

Fig. 4. Chemical structures cited in the text.

The Escanaba Trough in the northeastern Pacific (Fig. 1) is an active oceanic spreading center about 300 km long and bounded on the north and south by the Blanco and Mendocino fracture zones, respectively. It is filled with up to 500 m of Quaternary turbidite sediments (Kvenvolden *et al.*, 1986). The petroleum which saturates the sediments and mineral ores is derived from hydrothermal alteration of sedimentary organic matter primarily from terrestrial sources (Kvenvolden *et al.*, 1986, 1990). Middle Valley is another sediment covered hydrothermal system in the northeastern Pacific (Fig. 1), with associated hydrothermal organic matter alteration. Examples of the compositions of the saturated and aromatic hydrocarbon fractions of a hydrothermal petroleum from Escanaba Trough are shown in Fig. 3b,d. The *n*-alkanes range from C_{14} to C_{40}, with a carbon number maximum at *n*-C_{27} and a significant odd carbon number predominance > *n*-C_{25} (CPI = 1.25), compared to conventional crude oils (CPI = 1.0). This is typical of a catagenetic (thermal alteration) product from organic matter of a terrestrial, higher plant origin (Kvenvolden *et al.*, 1990; Kvenvolden and Simoneit, 1990). The PAH are more concentrated relative to the UCM in the aromatic fraction when compared to the example from Guaymas Basin (Fig. 3c), although the relative yields are similar. This may be due to the deposition of more oxidized terrestrial organic matter and precursor biomarkers in Escanaba Trough than in Guaymas Basin.

The Bransfield Strait, Antarctica (Fig. 1) is a typical example of a back-arc rift, which is tectonically active with extensional features such as dip-slip faults and intrusives, and is also heavily sedimented (Whiticar *et al.*, 1985; Suess *et al.*, 1992). Hydrothermal activity is evidenced by mineral alteration and a slight petroliferous odor of the sediments. However, the bitumen compositions indicate only mild and localized heating from intrusions, which resulted in accelerated diagenesis and limited product migration (Brault and Simoneit, 1988).

The Atlantis II Deep (Fig. 1) contains stratified brine layers, the deepest of which is at a temperature of 62°C (Hartmann, 1980, 1985). Bulk organic matter and hydrocarbons have been analyzed in two sediment cores from the Deep and the results indicate mild hydrothermal alteration (Simoneit *et al.*, 1987). The reductive products, i.e. saturated hydrocarbons, are predominant and the oxidative products, i.e. PAH, are not detectable, confirming that hydrothermal alteration commences at low temperatures. Related data on hydrothermal petroleum from the Kebrit and Shaban Deeps of the Red Sea have also been reported; however, these systems appear to be at higher temperatures (Michaelis *et al.*, 1990).

Volatile compounds (mainly CH_4-C_{10} hydrocarbons) are not effectively retained with the heavy petroleum as it solidifies at the vents on the seafloor of Guaymas Basin. Upon exiting at the seabed the fluids are often saturated with a broad range of volatile hydrocarbons (CH_4 to *n*-C_{10}) as well as lower concentrations of heavy ends (> C_{15}) (Simoneit *et al.*, 1988). Interstitial gas in sediments of Deep Sea Drilling Project (DSDP) cores consists of biogenic methane (CH_4) overprinted by thermogenic CH_4 to C_5 hydrocarbons near the sills and, to a lesser extent, at increasing subbottom depths. These are of a similar composition as the venting volatile hydrocarbons (Simoneit *et al.*, 1988; Whelan *et al.*, 1988). Guaymas Basin vent water samples contain high amounts of light hydrocarbons, with CH_4 at corrected concentrations of about 150 cm^3 (STP)/kg (Welhan and Lupton, 1987). For comparison, the CH_4 concentrations in vent fluids from the East Pacific Rise at 21°N, a sediment-starved rift system, have been reported to be 1-2 cm^3 (STP)/kg (Welhan and Lupton, 1987). Sedimented hydrothermal systems generate higher amounts of natural gas. The headspace gases of a Guaymas Basin mound sample (1629-A3, Fig. 5a) can be compared with the hydrocarbon content of a 308°C vent water, which is highly enriched in the lower alkanes (< C_7, Fig. 5b; Simoneit *et al.*, 1988). The hot water has an enhanced content of aromatic (benzene, toluene, ethylbenzene and xylenes – i.e. more water soluble) versus aliphatic hydrocarbons (Fig. 5b). Hydrogen gas is a major

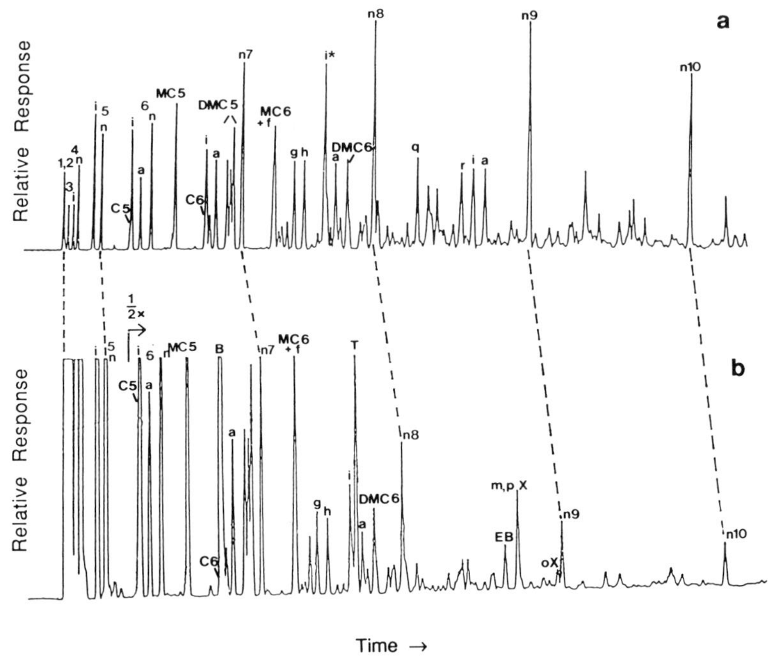

Fig. 5. Gas chromatograms of headspace analyses for comparison of the volatile hydrocarbons from: (a) hydrothermal crust, 1629-A3; (b) hot venting water, 1620-2C (T=308°C) (Simoneit *et al.*, 1988). Numbers refer to carbon chain length with n = normal, i =iso- (2-methyl-) and a = anteiso- (3-methyl-) of corresponding chain length. Other acyclic compounds are q = 2,6-dimethylheptane; r = 2,3-dimethylheptane; and s = 2,6-dimethyloctane. Cyclic compounds are the C = cyclo-, MC = methylcyclo-, DMC = dimethylcyclo- alkanes. The DMC5 triplet contains the c = *cis*-1,3-; d = *trans*-1,3-; e = *trans*-1,2- isomers. Other individual alkylcyclopentanes are f = 1,1,3-trimethyl-; g = 1,2,4-trimethyl-; and h = 1,2,3-trimethyl-. The aromatics are B = benzene, EB = ethylbenzene, T = toluene and X = xylenes (with (o) ortho-, (m) meta- and (p) para- isomers). The * indicates a coeluting unknown and the symbol 1/2x reflects a signal attenuation by a factor of 2.

component of the vent fluids in Guaymas Basin (Welhan and Lupton, 1987). Acetate and propionate ions are highly enriched in the hydrothermally altered sediments of Guaymas Basin relative to normal marine sediments (Martens, 1990). These volatile and water soluble acids are important as complexing agents for metals, substrates for heterotrophic microbial processes, buffers, and possibly in CO_2, CH_4 and C_2H_6 generation.

2.1.2. Marine non-sedimented systems

The examples of sediment-starved hydrothermal systems are located on basaltic rift areas as for example 26°N on the Mid-Atlantic Ridge and 13° and 21°N on the East Pacific Rise (Fig. 1). At 13°N (Hekinian *et al.*, 1983) aliphatic hydrocarbons have been analyzed in hydrothermal plumes and in metalliferous sediments near active vents and at the base of an inactive chimney (Brault *et al.*, 1985, 1988; Simoneit *et al.*, 1990). Hydrocarbons from metalliferous sediments have distributions characteristic of immature organic matter, which has recently been biosynthesized and microbiologically degraded, as might be expected from the low temperature environment in the surrounding talus of a vent system. Thermally mature compounds, such as *n*-alkanes with no carbon number predominance (CPI = 1.0) and biomarkers (17α(H)-hopanes, steranes), are present at trace levels in waters collected within ~1 km above the hydrothermal vents at 13°N. The hydrocarbon patterns of these waters are indicative in many cases of pyrolysis of bacterial matter entrained in ocean water mixed with the discharging high temperature fluids. Extensive hydrothermal activity occurs at the EPR in the region of 21°N (Fig. 1) in unsedimented oceanic crust with associated abundant faunal communities (Spiess *et al.*, 1980). Various samples of massive sulfides from vent chimneys have been analyzed for hydrocarbon contents, which are extremely low but from hydrothermal alteration (Brault *et al.*, 1989). All samples contain *n*-alkanes with no carbon number predominance (CPI = 1.0) and PAH, supporting evidence for an origin from high temperature alteration. A sample with pyritized tube worm residues also contains hydrothermally altered derivatives (e.g. cholestenes VII, hopenes VIII, Fig. 4) of biomarkers from this vent biota. However, these biomarkers may also have formed by acid-catalyzed, low temperature processes.

The Trans-Atlantic Geotraverse (TAG) hydrothermal field on the Mid-Atlantic Ridge crest at 26°N (Fig. 1) is an active vent system on a slow-spreading mid-oceanic ridge (Rona *et al.*, 1984). Various hydrothermal ores deposited directly on oceanic crust have been dredged from the area (TAG 1985-1) and four types of samples have been examined for lipid/bitumen content (Brault and Simoneit, 1989). A sample consisting of predominantly ferric oxide contained no hydrocarbons attributable to hydrothermal alteration of associated organic detritus. They were probably oxidized and lost. However, three other samples (consisting of mainly anhydrite, sphalerite and chalcopyrite, respectively) did contain minor amounts of lower molecular weight (C_{10}-C_{22}) hydrothermal petroleums, consisting of *n*-alkanes and PAH.

2.1.3. Continental systems

Continental systems are in volcanic or failed rift terranes as for example Yellowstone National Park, Lake Tanganyika and Waiotapu (Fig. 1). In most cases the thermal processes cause remobilization of organic matter in the form of bitumen as illustrated with the examples of oils from Yellowstone National Park (Love and Good, 1970; Clifton *et al.*, 1990). However, the following two cases are petroleums generated within continental hydrothermal systems. In the Waiotapu geothermal region of New Zealand small amounts of oil are presently being generated from volcanic sedimentary rocks of Lower Pleistocene age (Czochanska *et al.*, 1986). The source material is terrigenous organic matter present in vitric tuff which has been rapidly buried by volcanic overburden. The associated breccias serve as regional aquifers and surround the tuff with high temperature water. The generated oil, however, lacks the disequilibrium reaction products seen in the typical hydrothermal petroleums. The north Tanganyika trough in the East African Rift is another occurrence of massive sulfides and petroleum (Tiercelin *et al.*, 1989, 1992). Hydrothermal fluids pass through ~2 km of organic-rich lacustrine sediments (algal detritus), mobilizing asphaltic petroleum and venting with temperatures of 65-80°C at the lake bed in a water depth of ~20 m (the site described is in proximity to shore; vents at higher temperatures are suspected to occur in deeper water of the lake). The vent waters also contain thermogenic hydrocarbons (Tiercelin *et al.*, 1989). Hydrothermal activity can generate and migrate petroleum from continental source rocks, both lithified and unconsolidated. The invasion of hydrothermal fluids into mature source rocks will result in migration by remobilization with some alteration of the bitumen in the formation.

2.2. TEMPERATURE/PRESSURE CONSTRAINTS

Conventional petroleum formation is believed to occur in the temperature window of ~60-150°C and above that temperature the organic compounds are inferred to go to CH_4 and graphite (Hunt, 1979; Tissot and Welte, 1984). Geologically "instantaneous" (e.g., days-years) organic matter alteration in hydrothermal systems is a widespread process occurring over a temperature range from ~60°C to about 400°C. Formation of hydrothermal petroleums seems to commence in low temperature regions, generating products from weaker bonds and as the temperature regime rises, additional products are derived from more refractory organic matter and are even "reformed" (e.g. PAH). The products are continuously removed by fluid flow. The process progresses from reductive to more oxidative reactions of the residual organic matter as the temperature increases. At very high temperatures, organic matter is only partly destroyed, probably because the thermogenic products are soluble in the ambient fluid (Connolly, 1966; Sanders, 1986) and are thus rapidly removed from the hot zone by convection of fluids.

The aqueous solubility of petroleum and various hydrocarbon fractions has been determined experimentally (Price, 1976) and their solubility increased exponentially from

100°C to 180°C. It has also been demonstrated that CH_4 in the presence of water is an even better carrier for petroleum than water or methane alone (Price *et al.*, 1983). Both increases in pressure (to about 1800 bar) and temperature (to 250°C) raised the solubility of petroleum, and cosolubility was found at rather moderate conditions. Pure hydrocarbons, both aliphatic and aromatic, are soluble in all proportions in near critical water, as tested for low molecular weight compounds (Connolly, 1966), and for high molecular weight compounds (> 400 dalton, Sanders, 1986). Bitumen/tar solubility in supercritical water (400°C, 14–24.5 MPa) ranged as high as 90%, with little decomposition for Athabasca tar sand (Berkowitz and Calderon, 1990). All these laboratory experimental conditions are generally observed in the natural hydrothermal systems.

The thermal alteration products of organic matter in hydrothermal systems can be considered to be in a metastable equilibrium state (e.g. Shock, 1988, 1989, 1990b) during their brief formation and residence times at high temperatures. In this state not all of the stable equilibrium species are present due to kinetic constraints, but theoretical evaluations of the distributions of species at metastable equilibrium are analogous to those for stable equilibrium. Thus, Guaymas Basin vent fluids for example, concurrently contain reduced species (e.g. hydrogen, hydrogen sulfide and CH_4-C_{40} hydrocarbons), and oxidized species (e.g. CO_2, acetate, PAH).

The interactions of the fluid medium in terms of chemistry and solvent properties are not well understood. The dominant fluid is water, and in the example locales of Guaymas Basin and Escanaba Trough it is under temperatures approaching 350° and 400°C, respectively, with pressures exceeding 200 and 300 bar (20 and 30 MPa), respectively. Reduced density of hydrothermal fluids due to heating results in convective circulation, which in effect makes hydrothermal systems semi-open (a flow-through system) rather than closed as in most laboratory simulation experiments to date. These T and P conditions are in the near-critical domain of water (Fig. 6; Bischoff and Pitzer, 1989; Bischoff and Rosenbauer, 1984, 1988; Chen, 1981; Josephson, 1982; Pitzer, 1986). Supercritical water has enhanced solvent capacity for organic compounds and reduced solvation properties for ionic species due to its loss of aqueous hydrogen bonding (Fig. 6; Connolly, 1966; Shaw *et al.*, 1991; Tödheide, 1982). It is also a reactive medium for either reductive or oxidative reactions (Ross, 1984; Ross *et al.*, 1986) and some specific chemical reactions have been described by Ferris (this volume). Thus, the near-critical domain of water in hydrothermal systems is expected to aid reaction rates and enhance the solvation capacity for organic matter.

Fluids in hydrothermal systems also contain large concentrations of CH_4 and CO_2 (Sakai *et al.*, 1990; Simoneit and Galimov, 1984; Simoneit *et al.*, 1988; Welhan and Lupton, 1987). These gases, as well as many other possible trace components, are supercritical under the temperature and pressure conditions of the hydrothermal systems (Table 1) and their effects on the critical point of seawater are not known. Phase separation of CO_2 from water at reduced temperatures has been proposed for liquid CO_2 vents in a

TABLE 1

Critical Temperatures and Pressures for Possible Constituents of Hydrothermal Fluids[1]

Name	Composition	M.W.	Critical Temperature (°C)	Critical Pressure (bar)
Water	H_2O	18	374.2	218
Sea water	H_2O + 3.2% NaCl	18	407	299
Carbon dioxide	CO_2	44	31	73
Methane	CH_4	16	-82	46
Hydrogen sulfide	H_2S	34	100	89
Hydrogen	H_2	2	-240	13
Ammonia	NH_3	17	132	112
Nitrogen	N_2	28	-147	34
Ethane	C_2H_6	30	32	48
Ethylene	C_2H_4	28	10	51
Acetylene	C_2H_2	26	36	62
Propane	C_3H_8	44	97	42
Benzene	C_6H_6	78	289	49
Carbonyl sulfide	COS	60	105	65
Carbon monoxide	CO	28	-139	35
Carbon disulfide	CS_2	76	280	73
Hydrogen cyanide	HCN	27	184	49
Cyanogen	C_2N_2	52	127	
Sulfur dioxide	SO_2	64	158	78
Hydrogen chloride	HCl	36	51	82
Phosphine	PH_3	34	51	65
Nitrous oxide	N_2O	44	37	72
Nitric oxide	NO	30	-93	64
Nitrogen dioxide	NO_2	46	158	100
Helium	He	4	-268	2.3
Oxygen	O_2	32	-118	50
Sulfur	S	32	1040	116

[1]Source: CRC Handbook of Chemistry and Physics.

back-arc hydrothermal system (Sakai *et al.*, 1990). Carbon dioxide liquid is also an excellent solvent for organic compounds. Thus, hydrothermal fluids are efficient solvents for scavenging hydrothermal organic compounds (e.g. petroleum) from the source and migrating them away from the hot zone.

The reaction rates of organic matter alteration to petroleum in hydrothermal systems are rapid and fluid extraction is highly efficient. For example, carbon-14 dates have been obtained from hydrothermally-derived petroleum and calcite from the southern trough of Guaymas Basin (Peter *et al.*, 1991). The ages for petroleum range from 4240 to 5705 yBP (years before present, referenced to the year A.D. 1950 and using the ^{14}C half life of 5570 y). These are not true ages, but rather they reflect the age of carbon within these materials. Additional ^{14}C data on the aliphatic and aromatic hydrocarbon fractions of an

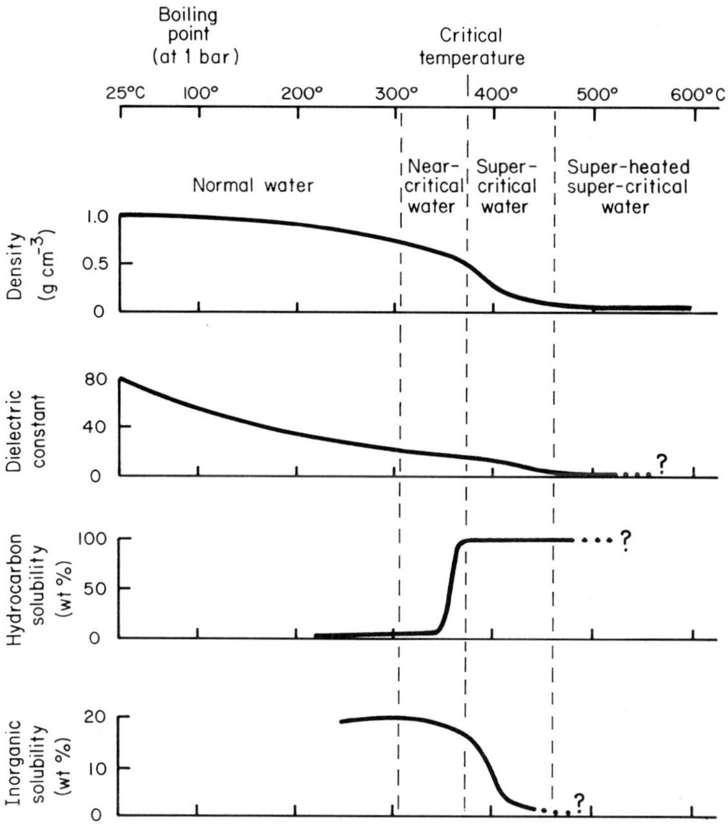

Fig. 6. Properties of water as a function of temperature at 200 to 300 bar pressure (adapted from Josephson, 1982).

oil sample from this area calculate for the same age (~4500 yBP), indicating that the PAH are generated from the same carbon pool as the saturated hydrocarbons at a subseafloor depth of ~12–30 m.

2.3. OXIDATIVE REACTIONS

Organic reactions occurring in hydrothermal systems under mild conditions of thermal stress (<150°C) have been shown to result in compound isomerizations and oxidations such as dehydration and deamination. Thus, amino acids racemize under prolonged hydrothermal stress (Silfer *et al.*, 1990) and are inferred to provide some of the ammonia in hydrothermal fluids of for example Guaymas Basin (Gieskes *et al.*, 1988). Dehydration of alcohols was inferred to occur near sill intrusions in the Guaymas Basin rift (Simoneit,

1984; Simoneit and Philp, 1982) and has been demonstrated with supercritical water in the laboratory yielding primary olefins (Ramayya *et al.*, 1987). Also, the thermally unaltered sediments of Guaymas Basin contain sterols (e.g. cholesterol) which are easily dehydrated and then both hydrogenated and dehydrogenated by hydrothermal fluids to sterane hydrocarbons (e.g. IX, Fig. 4) and Diels' hydrocarbon (VI), an aromatic compound characteristic of rapid thermal alteration (Kawka and Simoneit, 1987; Simoneit, 1985; Simoneit *et al.*, 1992). Deamination to hydrocarbons and loss of heteroatomic nitrogen to aromatic hydrocarbons have been demonstrated in supercritical water with model compounds (e.g. benzyl amine, quinolines) (Houser *et al.*, 1986; 1989).

In the intermediate temperature window (150-300°C) the alteration of organic matter in sediments under hydrothermal stress consists of cracking, reforming, sulfurization, decarboxylation and depolymerization reactions. Aliphatic biopolymers such as microbial membranes yield alkanes (e.g. van de Meent *et al.*, 1980); chlorophylls are converted to porphyrins (e.g. X, Fig. 4) with generally lower carbon number maxima (e.g. Didyk and Simoneit, 1990); phytol (XI) and/or phytadienes (XII), its dehydration products, are converted to phytylthiophenes (XIII) by reactions with H_2S, sulfur and/or $SO_4^=$ (Leif *et al.*, 1991); and carboxylic acids are reduced and/or decarboxylated to alkanes. Cracking and reforming products are saturated by the excess hydrogen present in the system (Welhan and Lupton, 1987) or by dimerization of the free radical fragments, yielding aliphatic products. It has been shown that the hydrogen in the reductive processes is derived from the water by using deuterium oxide in hydrous pyrolysis experiments (Hoering, 1984; Simoneit *et al.*, unpublished data). The hydrothermal fluids of Guaymas Basin contain major amounts of short chain organic acids as the ionic species, e.g. acetate and propionate (Martens, 1990), which may be derived from breakdown by cracking/oxidation of higher molecular weight lipids.

In the high temperature window (> 300°C) organic matter residues are heavily altered and the products consist primarily of low molecular weight species (e.g. methane, H_2S, benzene). Desulfurization of organic sulfur compounds occurs in this thermal region generating H_2S (Clark *et al.*, 1983; 1986). Polynuclear aromatic hydrocarbons in the higher molecular weight range (>228 dalton) are generated probably by reforming aromatic moieties into more stable configurations, but have not been formed in laboratory simulations up to 300°C (e.g. Simoneit, 1992). The residual organic matter ("spent" kerogen) after expulsion of all the products generated by the hydrothermal activity in Guaymas Basin sediments consists of activated amorphous carbon (Simoneit, 1982b; Curray *et al.*, 1982). In deeper sedimentary sections, where the temperature is higher, this carbonaceous kerogen residue is not found, indicating that it may have reacted (e.g. with S to yield CS_2) or migrated with the fluids.

3. Synthesis/Reductive Processes and Reactions

Evidence for synthesis of organic compounds in natural hydrothermal systems is sparse (also cf. Ferris, this volume) because such potential products would be masked by the excess of the contemporary pyrolysis products (e.g. hydrocarbons derived from organic detritus). Reductive alteration of organic compounds is the prevalent process in both sedimented and sediment-starved hydrothermal systems and this can readily be demonstrated with isotope labeling.

3.1. AQUEOUS ORGANIC CHEMISTRY

The reactivity of organic compounds in hot water (200-350°C) has been studied to model the geochemical and technological implications of this chemistry (see review by Siskin and Katritzky, 1991) and the potential of supercritical water as a medium for chemistry has also been reviewed (Shaw *et al.*, 1991; Towsend *et al.*, 1988). An extensive study of the hydrothermal alteration of some simple organic compounds has been published (Katritzky *et al.*, 1990a-o, 1991; Siskin *et al.*, 1990a-c, 1991). The results are summarized in Table 2. Four reaction categories have been defined, namely condensation, bond cleavage, hydrolysis and autocatalysis reactions. Reactions catalyzed by the presence of minerals are a related category. For organic compound synthesis in hydrothermal systems it is necessary to understand the chemistry of condensation, autocatalysis, catalysis and hydrolysis reactions.

3.2. SYNTHESIS

Preliminary evidence for the synthesis of heterocyclic sulfur compounds, as for example tetrathiolane (XIV, CH_2S_4, Fig. 4), pentathiane (XV, CH_2S_5), pentathiepane (XVI, $C_2H_4S_5$) and hexathiepane (XVII, CH_2S_6) has been reported (Kawka and Simoneit, 1987). These compounds are significant components in the Guaymas Basin samples and can be interpreted as being synthesized from formaldehyde and sulfur or HS_x^- in the hydrothermal fluids. Other products from potential synthesis reactions have not been defined in the natural systems but should be determined based on inferences from experimental interpretations. For example, more rigorous laboratory studies with model compounds should be carried out to test condensation reactions such as phytol (XI) yielding biphytane (lycopane, XVIII, Fig. 4) and dimerizations such as alkyl radicals or carbonium ions yielding extended alkanes. Since lycopane is a biomarker of bacterial origin (e.g. Albaiges *et al.*, 1985; Gillan and Johns, 1986) this experiment needs to be carried out with ^{13}C labels to test such a condensation reaction. The key indicator compounds can then be searched for in the natural hydrothermal environments. Fischer-Tropsch type synthesis also should first be demonstrated in the laboratory, because in the hydrothermal systems such products are masked by the overwhelming pyrolysates from organic matter alteration. The reaction is extensively utilized in the synthetic fuels industry. Similarly, the Strecker synthesis of amino acids is masked in hydrothermal systems by the extensive pyrolysis of proteinaceous residues from contemporary biota.

TABLE 2

Summary of organic compounds subjected to aqueous alteration at high temperatures.[1]

Reactant	Conditions[2]		Major Reactions and Products
	Temperature (°C)	Time	
Aliphatic compounds:			
1-decene	250	5.5 d	ionic (acid catalysis), isomerization, dimerization
1-decyne	250	5.5 d	ionic (acid catalysis), hydrogenation
1-decanal	250	1.5 d	aldol condensation, dimers
2-decanone	250	5.5 d	aldol condensation, dimers
decyl decanoate	250	1.5 d	hydrolysis
decanenitrile	250	2.5 d	hydrolysis
1-decanol	250	13.0 d	disproportionation, trace dehydration/hydrogenation
1-decylamine	250	5.5 d	condensation
methyl 1-nonyl sulfide	250	13.5 d	dialkyl sulfide
1-decanethiol	250	13.5 d	dialkyl sulfide
methyl 1-decyl ether	250	13.5 d	N.R.
decanoic acid	250	13.5 d	N.R.
Cyclohexyl phenyl compounds:			
cyclohexyl phenyl ketone	250	5.5 d	N.R.
cyclohexyl methylbenzene	250	5.5 d	N.R.
cyclohexylbenzene	250	5.5 d	N.R.
N-cyclohexylaniline	250	5.5 d	hydrolysis/rearrangement
cyclohexyl phenyl ether	250	5.5 d	hydrolysis/rearrangement
cyclohexyl phenyl sulfide	250	5.5 d	hydrolysis/rearrangement
Ethers and Esters:			
4-phenoxyphenol	250/343	5.5 d/2 h	hydrolysis/polymer
cyclohexyl phenyl ether	250/343	5.5 d/2 h	phenol, methylcyclopentene
benzyl phenyl ether	250/343	5.5 d/2 h	phenol, benzylphenol
	284-384*	2 h	phenol, benzyl alcohol, toluene
4-(benzyloxy)phenol	250/343	5.5 d/2 h	benzyl alcohol, hydroquinone, toluene

Table 2. continued

Reactant	Conditions[2]		Major Reactions and Products
	Temperature (°C)	Time	
methyl 1-naphthoate	250/343	5.5 d/2 h	hydrolysis/decarboxylation/isomerization, naphthalene
1-naphthoic acid	250/343	5.5 d/2 h	decarboxylation, naphthalene
dibenzyl ether	374-412*	1 h	benzyl alcohol, benzaldehyde, toluene
phenethylphenyl ether	375-413*	1.5 h	phenethyl alcohol, phenol, styrene
Benzenes:			
benzene	250/343	5.5 d/2 h	N.R.
toluene	250/343	5.5 d/2 h	N.R.
benzoic acid	250/343	5.5 d/2 h	trace decarboxylation
benzoic acid + formaldehyde	250/343	5.5 d/2 h	methyl benzoate
benzyl alcohol	250/343	5.5 d/2 h	dibenzyl ether, minor disproportionation
benzaldehyde	250/343	5.5 d/2 h	benzoic acid
benzaldehyde + phenol	250/343	5.5 d/2 h	polymerization
dibenzofuran	500*	1 h	N.R.
diphenylmethane	405*	1 h	N.R.
N-Heterocycles:			
pyridine	250/343	5.5 d/2 h	N.R.
3-methyl pyridine	250/343	5.5 d/2 h	N.R.
pyridine-3-carboxylic (nicotinic) acid	250/343	5.5 d/2 h	decarboxylation
pyridine-3-carboxaldehyde	250/343	5.5 d/2 h	decarbonylation
3-pyridyl methanol	250/343	5.5 d/2 h	3-methyl pyridine, pyridine-3-carboxaldehyde
2-methyl pyridine	250/343	5.5 d/2 h	N.R.
pyridine-2-carboxylic acid	250/343	5.5 d/2 h	rapid decarboxylation, pyridine
pyridine-2-carboxaldehyde	250/343	5.5 d/2 h	rapid decarbonylation, pyridine
2-pyridyl methanol	250/343	5.5 d/2 h	2-methyl pyridine, pyridine
4-methyl pyridine	250/343	5.5 d/2 h	N.R.
pyridine-4-carboxaldehyde	250/343	5.5 d/2 h	disproportionation
4-pyridyl methanol	250/343	5.5 d/2 h	4-methylpyridine, pyridine-4-carboxaldehyde
pyridine-4-carboxylic acid	250/343	5.5 d/2 h	decarboxylation
carbazole	450-550*	1 h	N.R.

Table 2. continued

Reactant	Conditions[2]		Major Reactions and Products
	Temperature (°C)	Time	
C_2-substituted benzenes:			
2-phenylethanol	250	5d	dehydration, styrene, ethylbenzene
phenylacetic acid	250	5d	N.R.
phenylacetaldehyde	250	5d	disporportionation/aldol condensation
phenylacetylene	250	5d	acetophenone, 1-phenylnaphthalene, 1,3,5-triphenylbenzene
styrene	250	5d	polystyrene
ethylbenzene	250	5d	acetophenone, toluene
1,2-diphenylethane	380-500*	3 h	N.R.
1,3-diphenylpropane	420*	1 h	N.R.
1-phenylethanol	250	5d	styrene, acetophenone
acetophenone	250	5d	traces ethylbenzene, benzaldehyde
phenylethanediol	200	6 h	condensation, phenylacetaldehyde
mandelic acid	250	1d	decarboxylation, disproportionation, condensation, dehydration
benzoylformic acid	200	1d	benzaldehyde, benzoic acid
Phenols:			
p-cresol (4-methylphenol) + $NaHSO_3$	250	3d	phenol, 2,7-dimethylxanthene
4-ethylphenol + $NaHSO_3$	250	56 h	phenol, 2-ethylphenol
4-*n*-propylphenol + $NaHSO_3$	250	56 h	phenol, dimers
4-*iso*-propylphenol + $NaHSO_3$	250	56 h	phenol, 2-*n*-propylphenol
4-*t*-butylphenol + $NaHSO_3$	250	56 h	phenol
4-*n*-dodecylphenol + $NaHSO_3$	250	1d	phenol, p-cresol
4-benzylphenol + $NaHSO_3$	250	57 h	phenol, 2-benzylphenol, benzoic acid
5,6,7,8-tetrahydro-2-naphthol + $NaHSO_3$	250	82 h	hydrogenation, 2-naphthol, 1,6-dihydroxynaphthalene
4-(hydroxymethyl)phenol + $NaHSO_3$	250	6 h	phenol
4-hydroxybenzaldehyde + $NaHSO_3$	250	44 h	phenol, 4-oxocyclohexanecarboxaldehyde
guaiacol	383*	45 min	o-cresol, methanol, methane

Table 2. continued

Reactant	Conditions[2]		Major Reactions and Products
	Temperature (°C)	Time	
4-hydroxybenzoic acid + $NaHSO_3$	250	44 h	decarboxylation
4-hydroxyacetophenone + $NaHSO_3$	250	1 d	phenol, p-cresol
o-cresol + $NaHSO_3$	250	11 d	phenol, 2,6-dimethylphenol
m-cresol + $NaHSO_3$	250	11 d	phenol (trace)
2-ethylphenol + $NaHSO_3$	250	4.6 d	phenol, o-cresol, furan
2-*iso*-propylphenol + $NaHSO_3$	250	3 d	phenol, o-cresol, 2-ethylphenol, 2,5-dimethylfuran
3,4,5-trimethylphenol + $NaHSO_3$	250	14 h	demethylation
2,4,6-trimethylphenol + $NaHSO_3$	250	13 h	demethylation
anisole	250	3 d	phenol, toluene
phenol + methanol	250	3 d	anisole, o- and p-cresol
diphenyl ether	250	3 d	N.R.
	405*	1 h	N.R.
n-butylphenyl ether	250	3 d	trace phenol
2,3-dihydrobenzofuran	250	3 d	phenol, 2-ethylphenol
Aryl amines:			
o-toluidine + H_3PO_4	250	3 d	deamination, o-cresol
p-toluidine + H_3PO_4	250	3 d	deamination, p-cresol
o- /p-toluidine + $NaHSO_3$	250	3 d	(aminophenyl)methylbenzothiazoles
4-ethylaniline	250	3 d	deamination/diphenylamine
4-*iso*-propylaniline	250	3 d	deamination/diphenylamine
N-methylaniline	250	3 d	disproportionation (aniline, N,N-dimethylaniline)
N-ethylaniline	250	3 d	aniline, N,2-diethylaniline, N,4-diethylaniline
N,N-dimethylaniline	250	3 d	N-methylaniline, aniline
N,N-diethylaniline	250	3 d	N-ethylaniline, aniline
1,2,3,4-tetrahydroquinoline	250	3 d	aromatization, dimers
2,4,6-trimethylaniline	250	3 d	N.R.
benzylphenyl amine	386*	1 h	aniline, benzyl alcohol

Table 2. continued

Reactant	Conditions[2]		Major Reactions and Products
	Temperature (°C)	Time	
Aryl nitriles, amides and amines:			
benzonitrile	250	5 d	hydrolysis, benzoic acid, benzamide
benzamide	250	5 d	benzoic acid, benzene
benzyl amine	250	5 d	toluene, benzaldehyde, benzyl alcohol
dibenzyl amine	250	5 d	toluene, 1-phenyl-*iso*-indole, benzyl amine
2-cyanopyridine	250	5 d	pyridine
pyridine-2-carboxamide	250	5 d	pyridine
2-pyridylmethylamine	250	5 d	2-methylpyridine, pyridine
3-cyanopyridine	250	5 d	nicotinic acid, pyridine, pyridine-3-carboxamide
pyridine-3-carboxamide	250	5 d	pyridine, nicotinic acid
3-pyridylmethylamine	250	5 d	3-methylpyridine, pyridine
4-cyanopyridine	250	5 d	pyridine-4-carboxylic acid, pyridine, pyridine-4-carboxamide
pyridine-4-carboxamide	250	5 d	pyridine-4-carboxylic acid, pyridine, pyridine-4-carboxamide
4-pyridylmethylamine	250	5 d	4-methylpyridine, dimers
Sulfides and disulfides:			
phenyl methyl sulfide	250	5 d	biphenyl, diphenyl sulfide
diphenyl sulfide	250	5 d	dibenzothiophene, thiophenol
benzyl phenyl sulfide	250	5 d	diphenyl disulfide, benzaldehyde, thiophenol
phenyl β-phenylethyl sulfide	250	5 d	ethylbenzene, thiophenol, diphenyl sulfide
benzyl methyl sulfide	250	5 d	benzaldehyde, toluene, dimethyl disulfide
dibenzyl sulfide	250	5 d	toluene, bibenzyl, benzoic acid, stilbene
diphenyl disulfide	250	5 d	diphenyl sulfide, thiophenol
dibenzyl disulfide	250	5 d	toluene, benzoic acid, bibenzyl, tetraphenylthiophene
di-2-pyridyl disulfide	250	5 d	2-pyridone, 2-pyridinethione
di-4-pyridyl disulfide	250	5 d	4-pyridinethione, 4-pyridone, di-4-pyridyl sulfide
dibenzothiphene	550*	3 h	N.R.

Table 2. continued

Reactant	Conditions[2]		Major Reactions and Products
	Temperature (°C)	Time	
Mercaptans and sulfonic acids:			
phenylmethanethiol	250	5 d	desulfurization, toluene
phenylethanethiol	250	5 d	desulfurization, ethylbenzene, toluene
phenylpropanethiol	250	5 d	3-phenylpropanol, n-propylbenzene
2-phenylethanesulfonic acid	250	5 d	N.R.
2-(2-pyridyl)ethanesulfonic acid	250	5 d	2-methylpyridine, pyridine
naphthalene-2-thiol	250	26 h	di-2-naphthyl sulfide, di-2-naphthyl disulfide
thiophenol	250	70 h	diphenyl sulfide, thianthrene, diphenyl disulfide
diphenyl sulfide + $NaHSO_3$	250	22 h	trace thiophenol
thioanisole + $NaHSO_3$	250	60 h	thiophenol, diphenyl sulfide
benzothiophene + $NaHSO_3$	250	82 h	N.R.
4-methylthiophenol + $NaHSO_3$	250	82 h	di-4-methylphenyl disulfide, di-4-methylphenyl sulfide

[1]Major examples from the open series of papers by Katritzky *et al.* (1990a-o, 1991) and Siskin *et al.* (1990a-c, 1991), and * = data from Townsend *et al.* (1988).
[2]Higher temperatures were used to increase reactivity. Solvents were water and simulated brine with addition of $NaHSO_3$ or H_3PO_4 for the aqueous tests and nonane, hexane, cyclohexane or decalin for non-aqueous thermal tests.
N.R. = no reaction.

3.3. REDUCTIVE REACTIONS

All hydrothermal systems examined to date yield organic products derived primarily from reductive alteration reactions. Dry pyrolysis generates equal amounts of alkenes (olefins) as alkanes from organic matter (e.g. van de Meent *et al.*, 1980). Alkenes are not detectable in the hydrothermal products because the precursor intermediates are hydrogenated yielding alkanes. The same is the case for other labile organic compounds as for example biomarkers such as sterols which are converted primarily to steranes (IX).

Deuterium labeling has shown that the hydrogen for the reductive reactions derives from the water medium (Hoering, 1984; Simoneit *et al.*, unpublished data). Hydrous pyrolysis of samples of an organic-matter-rich shale (Messel shale, Eocene age, Germany, extracted previously with organic solvents) in D_2O at 330°C for 3 days, generated a series of extensively deuterated, saturated hydrocarbons resembling petroleum (Hoering, 1984). A portion of the product was a series of *n*-alkanes in the range of C_{14} to C_{30}, with each alkane containing distributions of isomers having 0 to more than 14 deuterium atoms. The maximum deuterium substitution was in the D_4 to D_6 range and was very broad. This indicates extensive deuterium scrambling but confirms the D source from the water. However, further experiments are needed because these are enigmatic results. Furthermore, when *n*-octadec-1-ene, a terminal olefin, was subjected to these conditions, 60% of it was recovered as the corresponding deuterated alkane. This result confirms the presence of reducing chemistry, but the product contained significantly lower levels of deuterium than did the alkanes produced from the shale (Hoering, 1984). The deuterium distribution peaked at D_2 rather than at $D_{4\text{-}6}$, indicating possible matrix or catalytic effects and less scrambling. Nevertheless, such experiments carried out over a temperature/time series will permit the elucidation of the reaction pathways for hydrothermal alterations of organic species.

4. Examples of Industrial Applications

4.1. SUPERCRITICAL WATER

Supercritical fluid technology has been successfully applied in the processing industry in three major areas: 1) extraction, 2) coal liquefaction (hydrogenation), and 3) waste oxidation (Josephson, 1982). Some of the basic chemistry will have implications for hydrothermal systems and thus these chemical engineering research areas should be examined. There are numerous symposium volumes, reports, and reviews summarizing the state of the art in supercritical fluid technology (e.g. Franck *et al.*, 1983; Franck, 1987; Johnston and Penninger, 1989; Jones and Radding, 1980; Paulaitis *et al.*, 1983; Penninger *et al.*, 1985; Shaw *et al.*, 1991; Siskin and Katritzky, 1991; Squires and Paulaitis, 1987; Townsend *et al.*, 1988; Wu *et al.*, 1991). Additional information can be found in the chapter by Ferris (this volume). Reductive reactions with supercritical water have been applied primarily for upgrading shale oil and liquefying waste, biomass and coals (e.g. Antal *et al.*, 1985; Funazukuri *et al.*, 1987; Hong *et al.*, 1987, 1988; Köll *et al.*, 1983;

Paspek and Klein, 1990; Ross, 1984; Ross *et al.*, 1986). Oxidative reactions with supercritical water are utilized primarily for waste disposal and include the patented MODAR process (e.g. Hong *et al.*, 1987, 1988; Modell *et al.*, 1982; Staszak *et al.*, 1987). Various chemical reactions using organic compounds have been studied with both static and flow reactors using supercritical water (e.g. Helling and Tester, 1988; Houser *et al.*, 1986, 1989; Metzger *et al.*, 1983; Ramayya *et al.*, 1987; Shaw *et al.*, 1991). A more detailed perusal of the literature should uncover additional reports on specific organic reactions.

4.2. OTHER SUPERCRITICAL FLUIDS

Non-aqueous and aqueous/organic fluid mixtures have been applied in the supercritical state for various extraction processes (cf. reviews by Eckert *et al.*, 1986; Paulaitis *et al.*, 1982). Supercritical carbon dioxide is used extensively for extraction of polar organic compounds, e.g. for decaffeination and for extraction of oils from foods and seeds (Paulaitis *et al.*, 1982; Polak *et al.*, 1989). Supercritical hydrocarbon mixtures are utilized to extract thermally generated coal liquefaction products (Finn *et al.*, 1983). Phenols and carbohydrates can be extracted essentially quantitatively from wood by supercritical acetone or methanol (McDonald *et al.*, 1983). The phase behavior and critical properties of various fluid mixtures have been determined and reviewed (Franck *et al.*, 1983; Johnston *et al.*, 1982; Paulaitis *et al.*, 1982). These systems operate at lower temperatures and pressures, especially CO_2, and are generally excellent solvents for organic matter.

5. Acknowledgements

I thank the National Aeronautics and Space Administration (Grant NAGW-2833) for partial financial support during the preparation of this review.

CHAPTER 5
CHEMICAL ENVIRONMENTS OF SUBMARINE HYDROTHERMAL SYSTEMS

EVERETT L. SHOCK

Department Earth and Planetary Sciences and McDonnell Center for the Space Sciences,

Washington University, St. Louis, MO, 63130 U.S.A.

1. Hydrothermal Systems vs. Hydrothermal Vents

Perhaps because black-smoker chimneys make tremendous subjects for magazine covers, the proposal that submarine hydrothermal systems were involved in the origin of life has caused many investigators to focus on the eye-catching hydrothermal vents. In much the same way that tourists rush to watch the spectacular eruptions of Old Faithful geyser with little regard for the hydrology of the Yellowstone basin, attention is focused on the spectacular, high-temperature hydrothermal vents to the near exclusion of the enormous underlying hydrothermal systems. Nevertheless, the magnitude and complexity of geologic structures, heat flow, and hydrologic parameters which characterize the geyser basins at Yellowstone also characterize submarine hydrothermal systems. However, in the submarine systems the scale can be considerably more vast. Like Old Faithful, submarine hydrothermal vents have a spectacular quality, but they are only one fascinating aspect of enormous geologic systems operating at seafloor spreading centers throughout all of the ocean basins. A critical study of the possible role of hydrothermal processes in the origin of life should include the full spectrum of probable environments.

The goals of this chapter are to synthesize diverse information about the inorganic geochemistry of submarine hydrothermal systems, assemble a description of the fundamental physical and chemical attributes of these systems, and consider the implications of high-temperature, fluid-driven processes for organic synthesis. Information about submarine hydrothermal systems comes from many directions. Measurements made directly on venting fluids provide useful, but remarkably limited, clues about processes operating at depth. The oceanic crust has been drilled to ~ 2.0 km depth providing many other pieces of information, but drilling technology has not allowed the bore holes and core samples to reach the maximum depths to which aqueous fluids circulate in oceanic crust. Such determinations rely on studies of pieces of deep oceanic crust uplifted by tectonic forces such as along the Southwest Indian Ridge, or more complete sections of oceanic crust called ophiolite sequences which are presently exposed on continents owing to tectonic emplacement. Much of what is thought to happen in submarine hydrothermal systems is inferred from studies of ophiolite sequences, and especially from the better-exposed ophi-

Origins of Life and Evolution of the Biosphere **22**: 67–107, 1992.

olites in Oman, Cyprus and North America. The focus of much that follows is on a few general features: pressure, temperature, oxidation states, fluid composition and mineral alteration, because these features will control whether organic synthesis can occur in hydrothermal systems.

2. The Formation of Ocean Crust and Submarine Hydrothermal Systems

The presence of hydrothermal systems at mid-ocean ridges is a consequence of the injection of magma and its subsequent solidification and cooling. Understanding the variation and extent of submarine hydrothermal systems requires an understanding of the origin and structure of the oceanic crust. The hypothetical cross-section of one side of an active spreading center on an oceanic ridge in Fig. 1 helps to emphasize the scale of submarine hydrothermal systems. Note that the horizontal axis is compressed somewhat relative to the vertical. New oceanic crust is generated at the spreading axis which overlies a magma chamber at the divergent plate boundary. Magma erupts from the chamber at depth through a system of feeder channels to the seafloor. The basalt which erupts at the crust/seawater interface is rapidly cooled by the overlying seawater. The outer shell of the liquid basalt hardens into a glassy rind, and tends to form small mounds or 'pillows' as it is quenched. Underlying the zone of pillow basalts is a transition zone of a few hundred meters which is underlain by vertical dikes. These dikes are the solidified predecessors of the active feeder channels at the spreading center. Basalt from previous eruptions fills these older dikes which extend symmetrically away from the spreading center. Underlying the sheeted dikes are the layered gabbros which are thought to represent the solidified analogues of the magma chamber. New pillow basalts and sheeted dikes are constantly formed at the spreading center and new gabbros form at the margins of the magma chamber. It follows that the age of the crust increases away from the spreading center as indicated by the scale at the top of the diagram.

Even on a human time scale fluids exit quickly from black-smoker vents, where flow rates up to 10 l sec^{-1} are common (Hekinian *et al.*, 1983; Converse *et al.*, 1984; Little *et al.*, 1987). These rates are enormously rapid on a geologic timescale, as are the rates of generation of new crust at ridges and the spreading of crust across ocean basins. The oldest crust in the present ocean basins is not much more than 200×10^6 years old, implying that the forces driving plate tectonics have produced at least 20 or so times the present volume of oceanic crust over the lifetime of the Earth (this value depends, of course, on past seafloor spreading rates and on the timing of the initiation of plate tectonics on the early Earth). The enormous amount of new crustal material created at divergent plate boundaries is matched by an equally enormous amount of altered crustal material and sediments subducted back into the mantle and accreted on the continents.

As newly generated oceanic crust spreads away from ridge crests it is cooled by circulating fluids which flow in a pattern of forced and free convection cells (Figs. 1 and 2). These fluids move through the crust at rates which generally exceed those at which newly

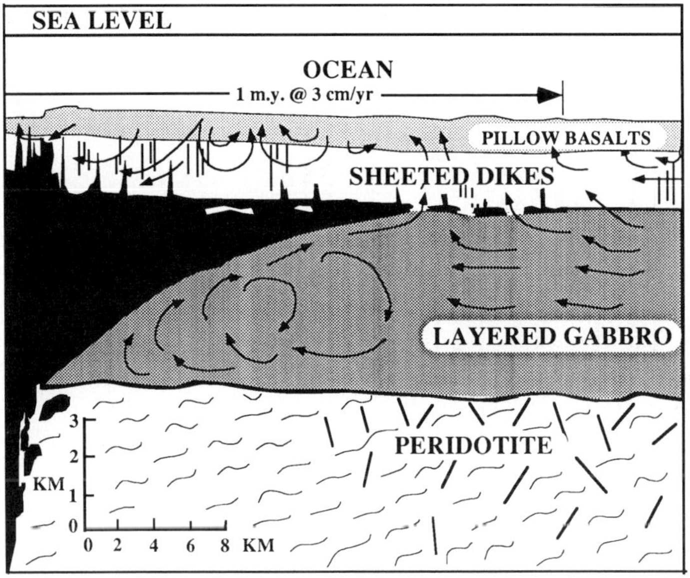

Fig. 1. (after Gregory and Taylor, 1981) Schematic cross section of oceanic crust at a spreading center based on petrologic and isotopic results from the Samail ophiolite in Oman. The dark region represents a hypothetical magma chamber which includes liquid and crystal-mush regions. Arrows indicate possible direction of fluid flow which penetrated to at least 6 km depth.

formed crust moves away from spreading centers. Depending on the rates of heat flow, local stress environments, chemical reactions and crustal structure, hydrothermal convection cells may remain more or less stationary in newly formed crust for thousands to millions of years. It should be kept in mind that oceanic crust moves laterally through these cells of convecting fluids over similar timescales. As a consequence, basalt which solidifies at the spreading center can be progressively altered by hydrothermal fluids for millions of years. The resulting geologic samples from ophiolite sequences or from drill holes in the seafloor contain the superimposed history of crustal formation and hydrothermal alteration.

The thermal and hydrologic histories which develop in these dynamic systems can be quite complex, and their accumulated products are often mixed and juxtaposed at any particular place in the altered oceanic crust. As an example, reference to Fig. 2 shows that basalt at approximately 1 km depth can pass through zones of high temperature (≥350°C) hydrothermal alteration, followed by low temperature (maybe ≤100°C) subseafloor weathering in zones of fluid recharge, and subsequent heating (temperatures up to ~250°C) in regions of fluid upwelling further off the axis of the spreading center. This pattern of oscillations can repeat in a somewhat dampened fashion until the fracture

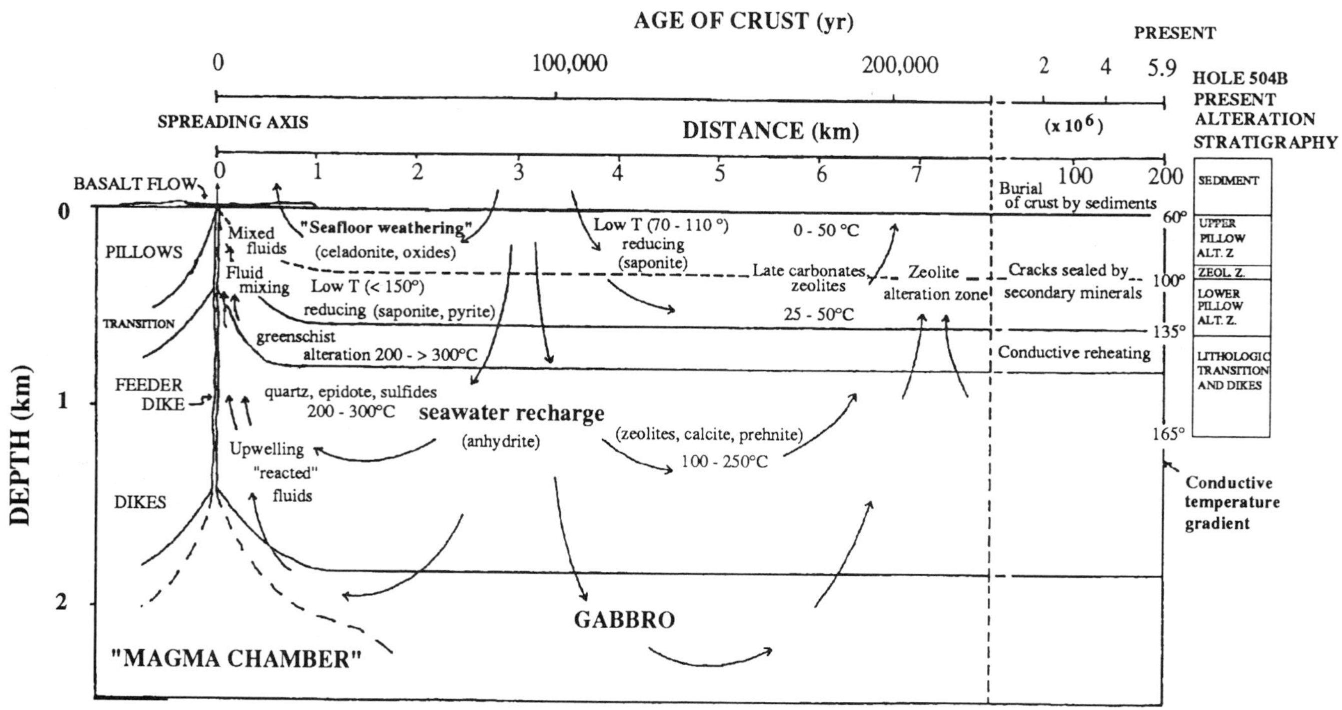

Fig. 2. (after Alt *et al.*, 1986a,b) Schematic cross sections of oceanic crust at a spreading center based on petrologic and isotopic results from DSDP site 504B. The structure of the oceanic crust, distribution of circulation cells and patterns of alteration are depicted. Stages 1 through 4 represent distinct fluid/rock alteration events which are eventually superimposed in the mineralogic record from the drill core. The arrows indicate fluid paths scaled to the model of Fehn *et al.* (1983).

network fills with secondary minerals and seals off. Although some crack sealing has occurred in core from Deep Sea Drilling Program (DSDP) hole 504B on the eastern flank of the Costa Rica Rift (Fig. 2), it is estimated that this process generally takes at least 60 to 80 million years to be completed based on the persistence of depressed heat flow in wide zones around the spreading centers (Anderson *et al.*, 1977). As fluid flow decreases, the crust will be more uniformly heated by conduction. Results from DSDP hole 504B indicate that temperatures in basalt at 1 km depth are ≥150°C even after nearly 6 million years of fluid circulation and alteration (Alt *et al.*, 1986a,b, 1989; Becker *et al.*, 1989).

3. Pressure and Temperature in Submarine Hydrothermal Systems

Hydrostatic pressures at the ocean/seafloor interface can be easily calculated from the depth of the overlying water column, the density of seawater, and the acceleration due to gravity. To the extent that fracture systems in seafloor sediments and basalt are open to the ocean/seafloor interface, pressures in fracture systems will also be hydrostatic. If geologic structures which serve as impermeable layers, such as fractures sealed with precipitated minerals, interfere with the permeability of geologic materials, the connectivity between aqueous fluids and the ocean/sediment interface can be blocked, and pressures can become lithostatic. In other words, the density of the overlying column of rock must be taken into account.

The fluid venting from black-smoker chimneys at mid-ocean ridges is connected through fracture systems to seawater and is at hydrostatic pressure. Fluids which are responsible for deep alteration of ophiolites (depths of 5 km or more) may have been at lithostatic pressures depending on the timing of fluid infiltration, the hydrodynamics of fracture systems, and the permeability of geologic materials (Gregory and Taylor, 1981). In either case, the fluid flows in response to buoyancy forces generated by the transfer of heat from the magma chamber to the hydrothermal fluid and the consequent fluctuation in fluid density (Norton, 1984). The rapid upward convection of these hot fluids can be thought to result from their relatively low density compared to the surrounding seawater.

Temperatures of aqueous fluids in submarine hydrothermal systems range from nearly 0°C at the seafloor to at least 700°C at depths of 3 km (Vanko, 1988). Higher temperature alteration (up to 900°C) has been documented in ophiolite sequences (Kimball, 1988), and is suggested in results from Ocean Drilling Program (ODP) site 735B (see below). Alteration of basalts, gabbros and other oceanic crustal rocks by hydrothermal fluids is generally more conspicuous at temperatures < 600°C than at higher temperatures. At the lower temperature range, the alteration occurs within the thermodynamic stability fields of several hydrous minerals (chlorite, talc, epidote, amphiboles, etc.) which are absent or uncommon in freshly crystallized oceanic igneous rocks. Although direct petrolgraphic evidence for hydrothermal alteration becomes more cryptic at higher temperatures (Norton and Taylor, 1979; Bird *et al.*, 1986, 1988; Schiffries and Skinner, 1987), oxygen isotopic data on samples from ophiolites indicate that pervasive hydrothermal alteration occurs to at least 6 km depths in the ocean crust (Gregory and Taylor, 1981; Cocker *et al.*, 1982; Ito and Clayton, 1983; Stakes *et al.*, 1983).

Studies of fluid inclusions, small fluid samples trapped within minerals which have precipitated during hydrothermal alteration, indicate that high-salinity brines (equivalent to 30 to 60 wt % NaCl) are generated deep in submarine hydrothermal systems (Kelley and Delaney, 1987; Cowan and Cann, 1988; Vanko, 1988; Kelley and Robinson, 1990; Nehlig, 1991). One explanation for the origin of these brines is phase separation above the consulate point in the H_2O-NaCl system at seawater salinities (Kelley and Delaney, 1987). Supporting evidence for phase separation in submarine hydrothermal systems comes from observations of Axial Seamount on the Juan de Fuca Ridge where fluids with dramatically different salinities, metal contents, and concentrations of dissolved gases vent from hot springs in close proximity (Massoth *et al.*, 1989; Butterfield *et al.*, 1990).

High temperatures in submarine hydrothermal systems are reached near the magma chamber during generation of new crustal material, but the duration of high-temperature alteration is likely to be shorter than lower-temperature alteration which continues for millions of years. Fluids circulate through a band of oceanic crust hundreds of kilometers wide around mid-ocean ridges (COSOD II, 1987), and at any moment the majority of these fluids are at temperatures considerably lower than those near the magma chamber or those flowing from hydrothermal vents. Recent estimates indicate that flow of warm water from the flanks of mid-ocean ridges is about 560 km^3 $year^{-1}$ ($\pm$ a factor of 3), or more than 20 times greater than flow from the ridge-crest hydrothermal vents (24 km^3 $year^{-1}$, COSOD II, 1987). The mass of fluid exiting the flanks is enormous and probably in the range 100 to 1000 x 10^{12} kg $year^{-1}$. If hydrothermal organic synthesis reactions can produce 1 ppm organic carbon in these solutions, the annual yield would be between 10^8 and 10^9 kg.

4. The Properties of H_2O at the Pressures and Temperatures of Submarine Hydrothermal Systems

Pressures from a few hundred bars to several kilobars and temperatures from 2°C (deep seawater) to at least 900°C can be encountered in submarine hydrothermal systems. These conditions are represented on the axes of Fig. 3 which contains a phase diagram for the system H_2O. The solid curve in Fig. 3 separates the liquid and vapor stability fields for H_2O. This curve terminates at the critical point of H_2O (373.917°C, 220.46 bar) beyond which H_2O is a single supercritical phase. At the critical point, the density of H_2O is 0.322778 gm cm^{-3} (Johnson and Norton, 1991). The dashed line extending from the critical point indicates the location of densities equal to this critical value (the critical isochore) which can be thought of as a boundary between the liquid-like and vapor-like regions of the single-phase, supercritical H_2O. The critical point of seawater, which should be referred to as a consulate point, is somewhat higher than that of pure H_2O owing to the

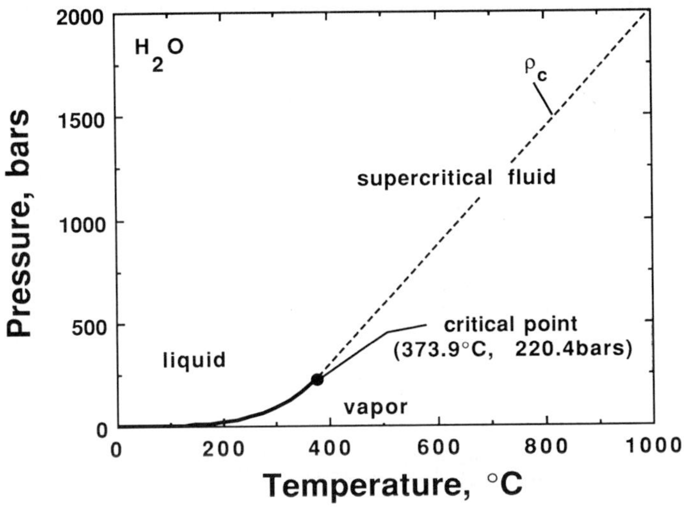

Fig. 3. Pressure/temperature phase diagram for the system H_2O showing the two phase (boiling) curve, critical point and location of the critical isochore (ρ_c). The triple point of H_2O and ice stability fields are omitted for clarity.

colligative properties of salt solutions. One recent estimate by Bischoff and Rosenbauer (1988) places the consulate point of seawater at 407°C and 298.5 bars. As in the case of the H_2O system, vapor-liquid phase separation occurs along the two-phase curve at conditions below this consulate point. Above the consulate point along the two-phase curve for the system H_2O-NaCl, phase separation yields a water-rich liquid and an NaCl-rich brine (Bischoff and Pitzer, 1989).

Although the density, and therefore the molar volume (V), of H_2O is finite at the critical point (see Fig. 4), the pressure and temperature derivatives of the density approach ∞. This can be illustrated by plotting values of the coefficient of isobaric expansivity, α, given by

$$\alpha \equiv \frac{1}{V}\left(\frac{\partial V}{\partial T}\right)_P \tag{1}$$

and the coefficient of isothermal compressibility, β, defined by

$$\beta \equiv -\frac{1}{V}\left(\frac{\partial V}{\partial P}\right)_T \tag{2}$$

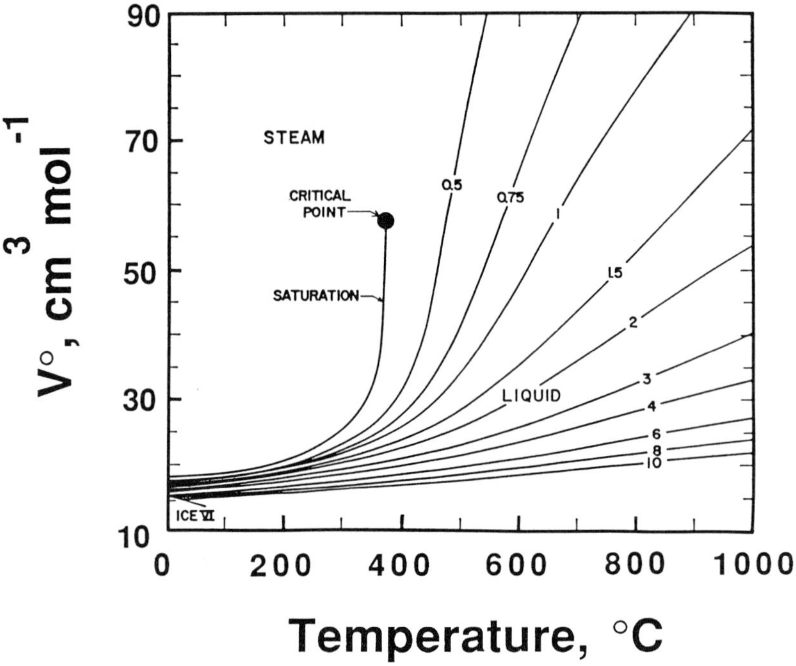

Fig. 4. (after Helgeson and Kirkham, 1974a) The molal volume of H_2O as a function of temperature at various pressures labeled in kilobars.

as functions of pressure and temperature. Plots of this type are shown in Figs. 5 and 6, where it can be seen that values of α and β increase dramatically with increasing temperature along the vapor-liquid saturation curve for H_2O. The isobaric molal heat capacity of H_2O also approaches ∞ at the critical point as inferred from the plot in Fig. 7.

Johnson and Norton (1991) have provided a unified equation of state for H_2O which allows accurate representation of these and many other properties in the region of pressure/temperature space around the critical point of H_2O. These authors also evaluate the extrema in dynamic viscosity, thermal conductivity, kinematic viscosity, thermal diffusivity and Prandtl numbers in the critical region, all of which are used to evaluate the physical properties of fluid flow. These extrema in transport and thermodynamic properties of H_2O in the critical region help to explain why hydrothermal circulation is particularly vigorous and effective in the critical region (Norton and Knight, 1977; Norton, 1984, 1990; Taylor, 1990). These effects are summarized by Taylor (1990) who states that in the critical region, "the buoyancy and heat transport properties of the fluid are maximized and drag forces are minimized", and that "it is very likely that these physical properties of

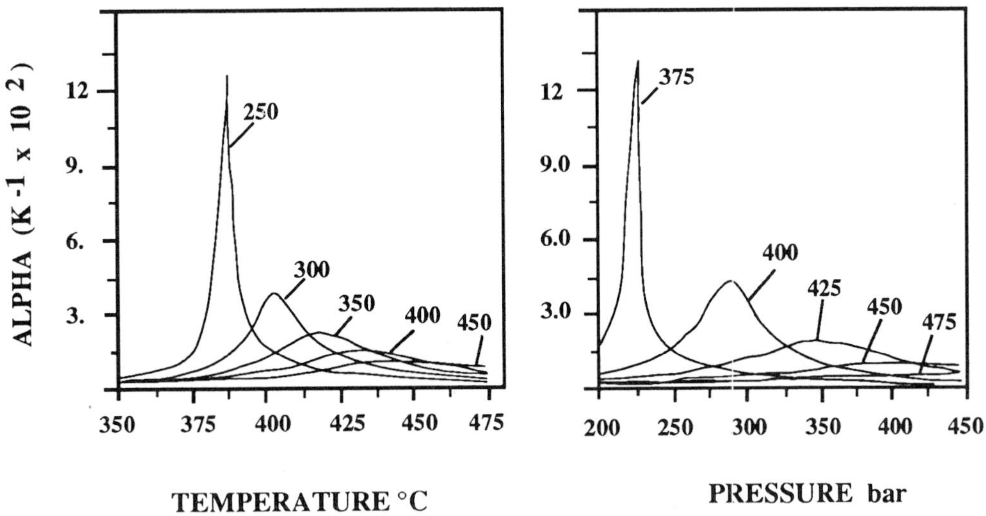

Fig. 5. (after Johnson and Norton, 1991) Isobaric expansivity of H_2O as a function of temperature and pressure in the vicinity of the critical point.

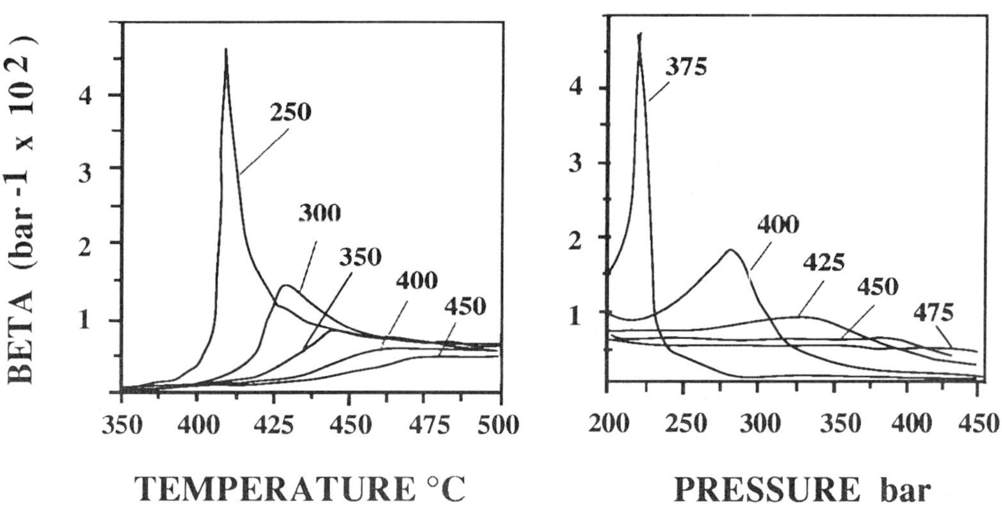

Fig. 6. (after Johnson and Norton, 1991) Isothermal compressibility of H_2O as a function of temperature and pressure in the vicinity of the critical point.

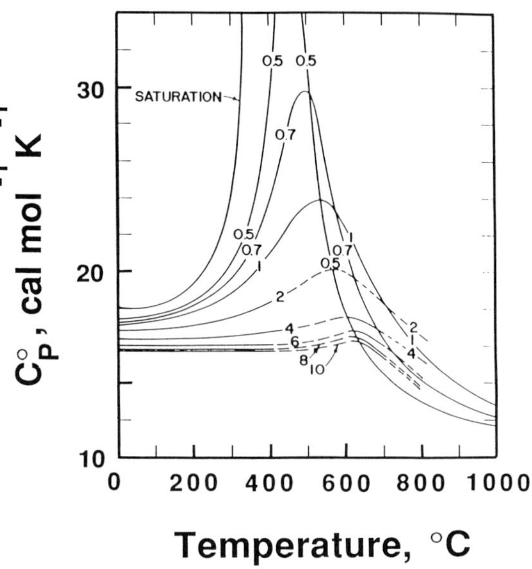

Fig. 7. (after Helgeson and Kirkham, 1974a) The standard molal isobaric heat capacity of H_2O as a function of temperature a various pressures labeled in kilobars.

H_2O control the observed upper limit of 350° to 400°C observed during surface venting of modern hydrothermal systems."

The effects of hydrothermal fluids on the basalts and layered gabbros of the oceanic crust are recorded by assemblages of alteration minerals. The ability of hydrothermal fluids to transport ions and other aqueous species into and away from zones of alteration is strongly correlated with changes in the electrostatic properties of the fluid. As shown in Fig. 8, the dielectric constant of H_2O decreases steadily from a value of ~ 80 at 25°C and 1 bar to ~ 7 at the critical point. It can also be seen in Fig. 8 that pressure has a mild effect on the dielectric constant, and that values less than the critical value are attained in the supercritical region. Also shown along the right-hand vertical axis of Fig. 8 are values of the dielectric constant for several organic solvents at 25°C and 1 bar, emphasizing the point that H_2O is an enormously different solvent at hydrothermal conditions than it is at typical laboratory conditions. This observation may be particularly important for considerations of H_2O-organic compound reactions at hydrothermal conditions (Shock and Helgeson, 1990; Siskin and Katritzky, 1991).

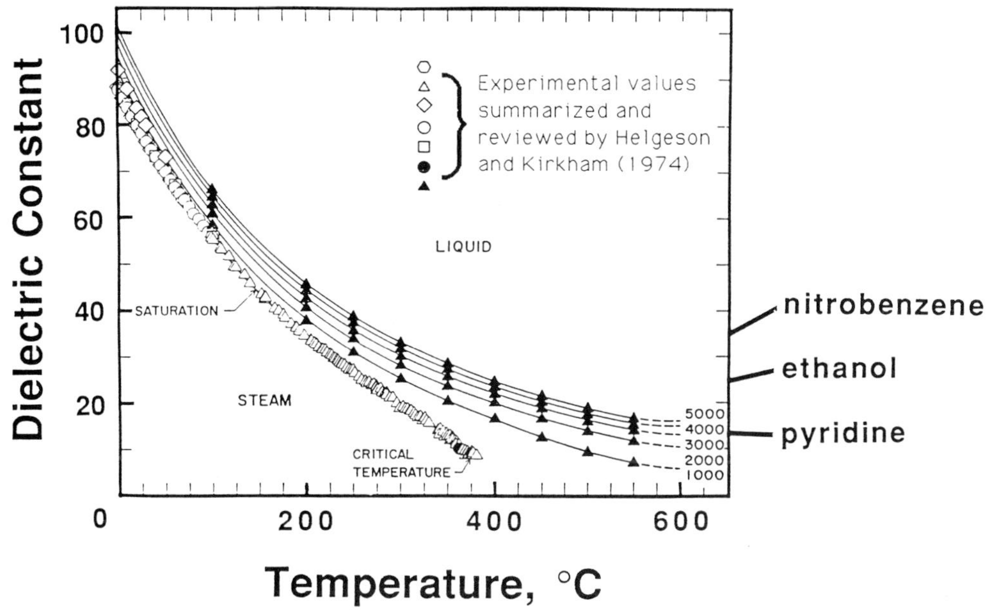

Fig. 8. (after Helgeson and Kirkham, 1974a) The dielectric constant of H_2O as a function of temperature at various pressures labeled in bars.

5. Thermodynamic Properties of Aqueous Species and Reactions at High Pressures and Temperatures

The fluctuations in the thermodynamic, transport and electrostatic properties of H_2O described above can be used to evaluate thermodynamic and transport properties of aqueous species, as well as thermodynamic properties of reactions among aqueous species and between aqueous species, minerals, organic compounds and gases. One approach to evaluating thermodynamic properties of aqueous species at elevated pressures and temperatures which is widely applicable in the study of hydrothermal systems involves the Helgeson-Kirkham-Flowers equations of state and their recent revisions (Helgeson and Kirkham, 1976; Helgeson, Kirkham and Flowers, 1981; Tanger and Helgeson, 1988; Shock *et al.*, 1992). Thermodynamic data and equation-of-state parameters for hundreds of inorganic and organic species are presently available (Shock and Helgeson, 1988, 1990; Shock *et al.*, 1989; Sassani and Shock, 1990a, 1992; Sverjensky *et al.*, 1992; Shock, 1992a,c,d,e). These data and parameters allow calculations to 1000°C and 5 kb. The accuracy and generality of these equations, data and parameters are demonstrated by the comparisons between calculations (solid curves) and experimental data (symbols) in

Figs. 9-11. The examples selected include many aqueous species of interest to the study of organic synthesis in hydrothermal systems.

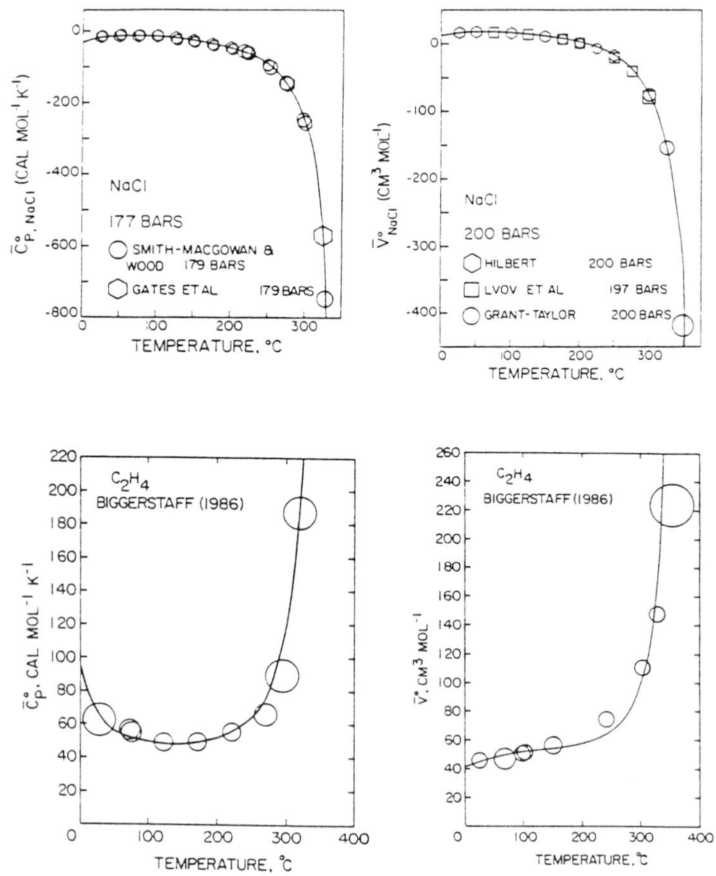

Fig. 9. (after Shock *et al.*, 1989, 1992) Plots of the standard partial molal heat capacity ($\overline{C}_p^\circ$) and volume ($\overline{V}^\circ$) of NaCl and C_2H_4 in aqueous solution as functions of temperature at constant pressure. Symbols represent experimental data from the references indicated, and curves correspond to regression results with the revised HKF equations of state (see text).

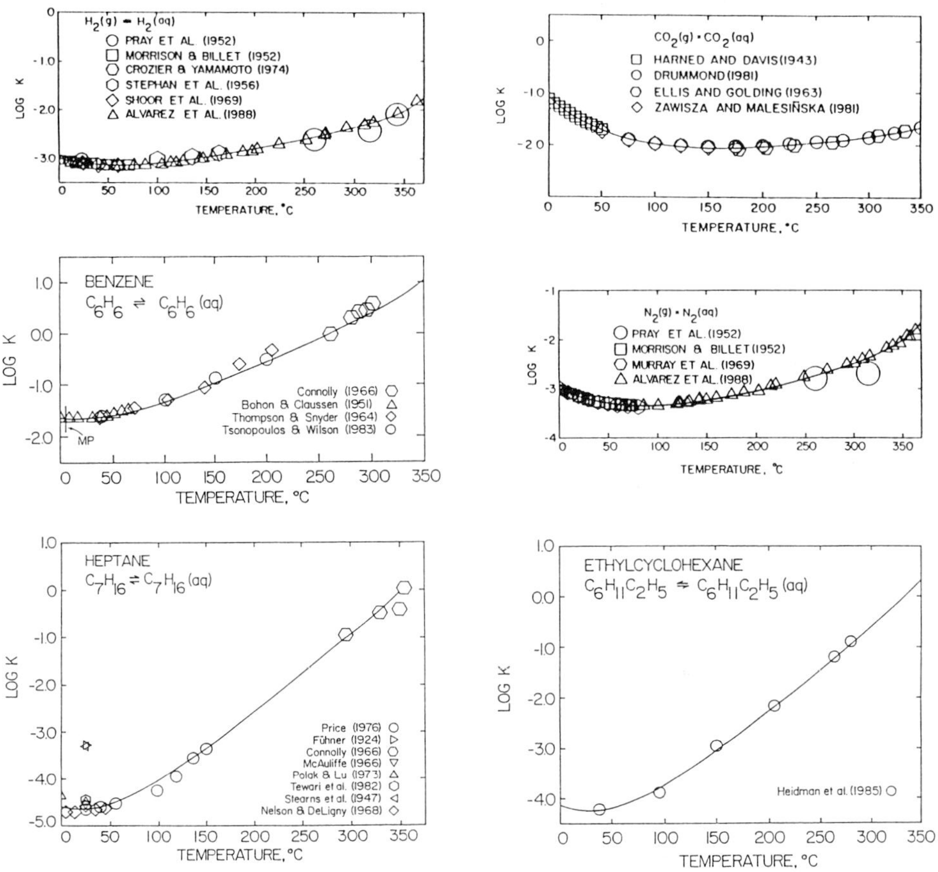

Fig. 10. (after Shock *et al.*, 1989; Shock, 1992f) Plots of standard state equilibrium constants for solubility reactions for H_2, CO_2, N_2, benzene, heptane and ethylcyclohexane. Symbols represent experimental data from the references indicated, and curves correspond to values calculated with the revised HKF equations of state (see text).

5.1. INORGANIC AND ORGANIC IONS AND ELECTROLYTES

Standard partial molal volumes ($\bar{V}°$), isobaric heat capacities ($\bar{C}p°$), and entropies ($\bar{S}°$), as well as apparent standard partial molal enthalpies of formation ($\Delta\bar{H}_f°$) of aqueous ions and electrolytes approach $-\infty$ at the critical point of H_2O. This behavior is a direct consequence of ion solvation which in turn reflects the temperature and pressure dependence of the dielectric constant of H_2O (Helgeson and Kirkham, 1974b, 1976). Experimental evidence of this behavior is indicated by the symbols shown in Fig. 9 for $\bar{V}°$ and $\bar{C}p°$ of the electrolyte NaCl. The curves in Fig. 9 are calculated with the revised

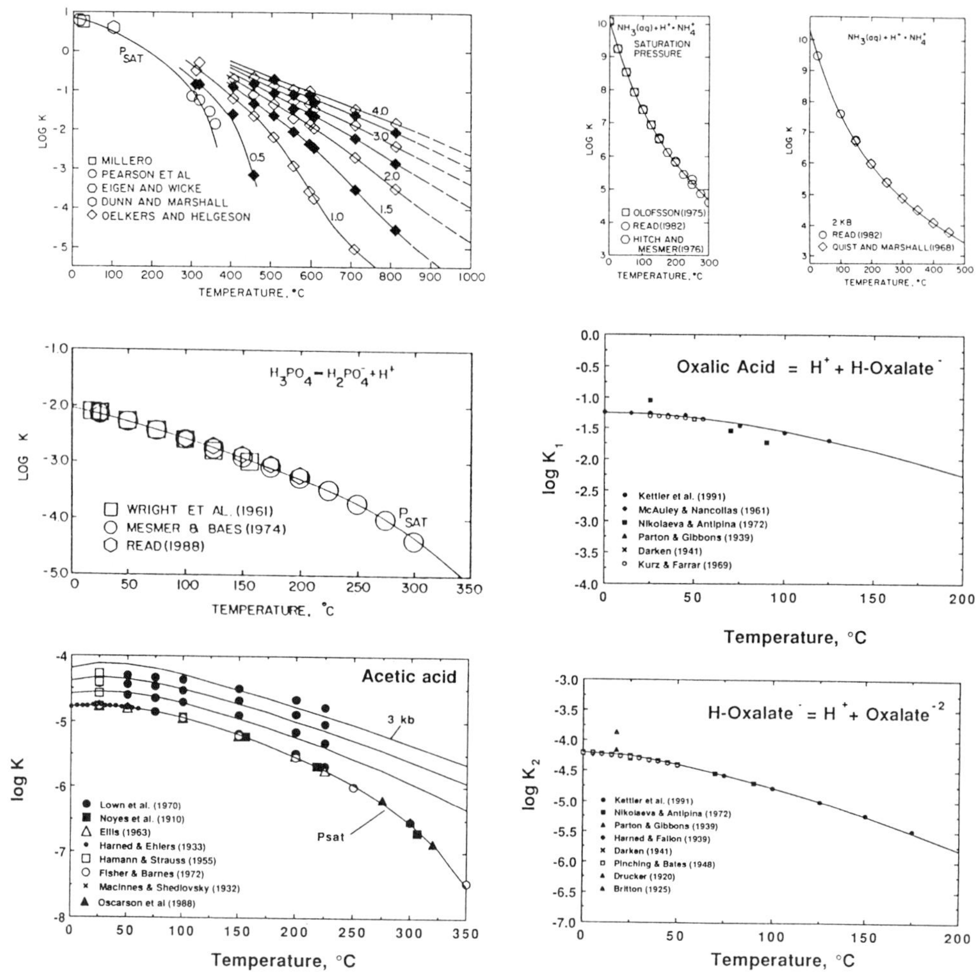

Fig. 11. (after Shock *et al.*, 1989, 1992; Shock, 1992d) Plots of equilibrium constants for complex and acid dissociation reactions of NaCl, H_3PO_4, oxalic and acetic acids, as well as for the protonation of NH_3. All log K values are plotted as functions of temperature at the saturation pressure or at higher pressures labeled in kilobars. Symbols represent experimental data from the references indicated, and the curves are calculated with the revised HKF equations of state (see text).

Helgeson-Kirkham-Flowers (HKF) equations of state (Helgeson *et al.*, 1981; Tanger and Helgeson, 1988; Shock *et al.*, 1992). Parameters in these equations are obtained by regression of experimental data and by application of correlation algorithms, and are available for hundreds of inorganic and organic ions (Shock and Helgeson, 1988, 1990; Sassani and Shock, 1990a; Shock, 1992d). These parameters, together with standard state thermodynamic data for the ions at 25°C and 1 bar allow calculation of the corresponding thermodynamic properties of ions to 1000°C and 5kb.

5.2. INORGANIC AND ORGANIC NEUTRAL SPECIES

In contrast to aqueous ions and electrolytes, $\bar{V}°$, $\bar{C}p°$, $\bar{S}°$, and $\Delta\bar{H}_f°$ of many aqueous neutral species approach ∞ at the critical point of H_2O. Experimental evidence of this behavior is shown by the symbols representing $\bar{V}°$ and $\bar{C}p°$ for aqueous ethylene in Fig. 9. The curves in Fig. 9 were generated with the revised HKF equations of state (Shock and Helgeson, 1990; Shock *et al.*, 1992). The approach of standard partial molal properties of neutral species to ∞ is typical behavior for those species which in pure form exist as gases or supercritical fluids at the critical point of H_2O. Neutral species which have solid-phase analogs at the critical point of H_2O, such as aqueous silica for which the pure analog is quartz, will exhibit behavior like aqueous ions. Regression of gas solubility and quartz solubility data (Shock *et al.*, 1989) provides standard state data and equation-of-state parameters for numerous neutral aqueous species including many organic compounds (Shock and Helgeson, 1990; Shock, 1992f). Comparisons of experimental gas and hydrocarbon solubility data with calculations performed with the revised HKF equations of state are depicted in Fig. 10. The close agreement between experimental and calculated values of log K for solubility reactions provides the framework for regressing dissociation constant data for acids and aqueous complexes such as NaCl°, which are also neutral aqueous species.

5.3. COMPLEX ASSOCIATION AND ACID DISSOCIATION

Dissociation constants for the reaction

$$NaCl° = Na^+ + Cl^- \quad (3)$$

are shown in Fig. 11 where the symbols designate experimental values and the curves correspond to calculations with the revised HKF equations of state (Shock *et al.*, 1992). Data for numerous other neutral and charged inorganic aqueous complex dissociation reactions were regressed by Sverjensky *et al.* (1992), who provide a general framework for predicting dissociation constants at elevated temperatures and pressures which have not been measured experimentally. As a consequence, it is now possible to make a detailed accounting of the speciation of major and trace elements in hydrothermal fluids to 1000°C and 5kb (see for example: Molling and Sverjensky, 1989; Sassani and Shock, 1990a,b; Sverjensky, 1991; Willis and Shock, 1991). Speciation calculations allow evaluation of total concentrations of elements in hydrothermal fluids from ion activity ratios obtained from consideration of mineral/fluid reactions.

Acid dissociation reactions are treated in a manner analogous to complexes. Equilibrium constants obtained from experiments are regressed with the revised HKF equations of state to obtain the equation-of-state parameters that are used to construct correlations which, in turn, allow estimates for the many acids which have not been studied experimentally (Shock *et al.*, 1989; Shock and Helgeson, 1990; Shock, 1992d). Close agreement between experimental and calculated values of acid dissociation constants is demonstrated in Fig. 11 for acetic acid, oxalic acid, and H_3PO_4, as well as for the protonation of aqueous NH_3.

5.4. OXIDATION/REDUCTION REACTIONS

Thermodynamic data and equation of state parameters for aqueous species provide the means of evaluating the magnitude of the driving force for organic synthesis at hydrothermal conditions (Shock, 1990a,b). As an example, let us consider the potential synthesis of acetic acid in submarine hydrothermal systems. The predominant inorganic form of volatile carbon leaving the mantle is likely to be CO_2 (see below). Synthesis of acetic acid will require reduction of CO_2 through an overall reaction of the type

$$2\,CO_2\,(g) + 4\,H_2(g) = CH_3COOH(aq) + 2\,H_2O\ . \tag{4}$$

We can write the expression of the law of mass action for this reaction as

$$K_4 = \frac{a\,(CH_3CHOOH)\,a^2(H_2O)}{f^2(CO_2)\,f^4(H_2)}\ , \tag{5}$$

where K_4 designates the equilibrium constant for reaction (4), a stands for activity and f represents fugacity. Activities of H_2O rarely stray far from unity in hydrothermal fluids except at extremes of concentration (Helgeson, 1979, 1985). Taking the logarithm of Eqn. (5) with this in mind allows us to write

$$\log fCO_2 = 2 \log fH_2 + \frac{1}{2}(\log a\ CH_3COOH - \log K_4)\ , \tag{6}$$

which corresponds to an equation of a line on a plot of $\log fCO_2$ vs $\log fH_2$. Such a plot is shown in Fig. 12 constructed for 150°C and 500 bars, where $\log K_4 = 0.18$ (Shock, 1992c).

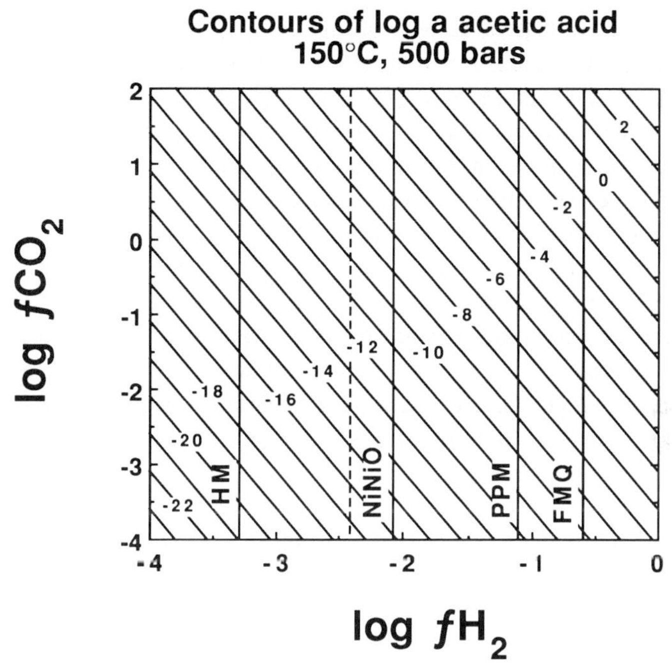

Fig. 12. Fugacity diagram showing contours of log a acetic acid in fH_2-fCO_2 space at 150°C and 500 bars. Vertical lines show positions of various mineral buffers of fH_2 (reactions 7 - 10) at the same temperature and pressure.

Contours of log *a* CH_3COOH are shown in Fig. 12 together with the locations of mineral redox buffer assemblages corresponding to the hematite-magnetite (HM), nickel-nickel oxide (NiNiO), pyrite-pyrrhotite-magnetite (PPM) and fayalite-magnetite-quartz (FMQ) reactions given by

$$\underset{\text{magnetite}}{2\,Fe_3O_4} + H_2O = \underset{\text{hematite}}{3\,Fe_2O_3} + H_2(g) \quad , \tag{7}$$

$$Ni + H_2O = NiO + H_2(g) \quad , \tag{8}$$

$$\underset{\text{pyrrhotite}}{1.5\,FeS} + H_2O = \underset{\text{pyrite}}{0.75\,FeS_2} + \underset{\text{magnetite}}{0.25\,Fe_3O_4} + H_2(g) \quad , \tag{9}$$

and

$$\underset{\text{fayalite}}{1.5\,Fe_2SiO_4} + H_2O = \underset{\text{magnetite}}{Fe_3O_4} + \underset{\text{quartz}}{1.5\,SiO_2} + H_2(g) \quad . \tag{10}$$

It can be seen in Fig. 12 that the activity of acetic acid is ~10^{-2} at equilibrium with 10 bars of CO_2 and the log fH_2 values set by the PPM assemblage at 150°C and 500 bars. The activity of acetic acid is high because CO_2 is unstable at this fH_2 value as indicated by the location of the dashed line which represents the fH_2 value corresponding to equal fugacities of CO_2 and CH_4 if stable equilibrium attains. If stable equilibrium between CO_2 and CH_4 is blocked by kinetic barriers in hydrothermal systems, then metastable organic synthesis may prevail (Shock, 1990b, see below).

Results of speciation calculations conducted to test the possibility of hydrothermal organic synthesis are described below. In contrast to the calculations used to construct Fig. 12, speciation calculations account for mass-balance constraints together with consideration of simultaneous metastable equilibrium constraints for several possible products of reactions like Eqn. (4). Any attempt to constrain such calculations, or experiments which simulate submarine hydrothermal systems, must take into account the buffering capacity of fluid/rock reactions.

6. Fluid/Rock Reactions in Submarine Hydrothermal Systems

Taken together, studies of hydrothermal vent fluids, ophiolites, volcanogenic sulfide ore deposits, and samples from DSDP and ODP sites like 504B and 735B yield a detailed picture of the products of fluid/rock reactions in submarine hydrothermal systems (Bischoff and Seyfried, 1978; Edmond *et al.*, 1979a,b, 1982, 1987; Zierenberg *et al.*, 1984, 1988; Merlivat *et al.*, 1987; Shanks and Seyfried, 1987; Richardson *et al.*, 1987; Kadko and Moore, 1988; Kawahata and Shikazono, 1988; Woodruff and Shanks, 1988; Campbell *et al.*, 1988a,b; Harper *et al.*, 1988; Alt *et al.* 1989; Becker *et al.*, 1989; Hayman, 1989; Campbell and Edmond, 1989; Berndt and Seyfried, 1990; Von Damm, 1990; Dick *et al.* 1991; among many others). These observations can be coupled with experimental studies (Seyfried and Bischoff, 1977, 1979, 1981; Mottl and Seyfried, 1980; Hajash and Chandler, 1981; Seyfried and Mottl, 1982; Janecky and Seyfried, 1986; Shiraki *et al.*, 1987; Seyfried, 1987; Cole *et al.*, 1987; Bischoff and Rosenbauer, 1983, 1987; Thornton and Seyfried, 1987; Seyfried *et al.*, 1987, 1988; Berndt *et al.*, 1988, 1989; Seewald and Seyfried, 1990; among others), and theoretical calculations of the dynamics of fluid flow and chemical mass transfer to reveal many of the hydrologic, thermodynamic and kinetic factors which control mineral alteration and compositional changes in fluids (Fehn *et al.*, 1983; Haymon, 1983; Mottl, 1983; Reed, 1983; Janecky and Seyfried, 1984; Bowers and Taylor, 1985; Seyfried and Janecky, 1985; Bowers *et al.*, 1985, 1988; Von Damm, 1988; Janecky and Shanks, 1988; Goldfarb and Delaney, 1988; Cann and Strens, 1989; Brikowski and Norton, 1989; Bowers, 1989; Fox, 1990; Norton, 1990; Lowell, 1990, 1991; Fisher *et al.*, 1990; Tivey and McDuff 1990; Fisher and Becker, 1991; Fisher and Narasimhan, 1991; Lowell and Burnell, 1991; among many others). These same factors are likely to control, or strongly influence, possible organic synthesis in submarine hydrothermal systems. Evidence is reviewed in this section for controls on pH, oxidation state and major element composition in these systems.

6.1. MINERAL ALTERATION

The original igneous mineral assemblages of oceanic crust are replaced to varying extents during hydrothermal alteration. Evidence for alteration of oceanic crust at temperatures from 0° to 900°C comes from drill core, dredge samples and ophiolite sequences (Coish, 1977; Humphris and Thompson, 1978; Böhlke *et al.*, 1981; Oudin and Constantinou, 1984; Mevel, 1987, 1988; Schiffman *et al.*, 1987; Delaney *et al.*, 1987; Vanko, 1988; Schiffman and Smith, 1988; Silant'yev *et al.*, 1988; Schöps and Herzig, 1990; Dick *et al.*, 1991; among many others). Fig. 13 contains a stratigraphic section through the Josephine ophiolite of California and Oregon together with the occurrence of various minerals (Harper, 1986; Harper *et al.*, 1988). Mineral formulas are given in Table 1. Note that most of the original igneous minerals are replaced, especially throughout the middle depths. These petrogenetic relations are quite similar to those recorded in a detailed study of core samples from DSDP site 504B. Stratigraphic data and assemblages of secondary minerals reported by Alt *et al.* (1986a, 1989) from 504B are shown in Fig. 14. Comparison of the lists and distribution of minerals in Fig. 13 and 14 indicates that chlorite, albite, epidote, actinolite, magnetite and quartz are among the most common alteration products in the deeper portions of the hydrothermally altered sequences. Note that pyrite and magnetite coexist through many hundred meters of oceanic crust as shown in Fig. 14. At shallower depths, many low-temperature alteration products are found including pumpellyite, aragonite, phillipsite, saponite, celadonite, and Fe-hydroxides.

Recently, the Ocean Drilling Program was successful in drilling 500 meters into oceanic layered gabbros located on the Southwest Indian Ridge at Site 735B (Robinson *et al.*, 1991; Dick *et al.*, 1991). Evidence for high-temperature metamorphism (Cannat *et al.*, 1991) as well as hydrothermal vein assemblages (Stakes *et al.*, 1991) attest to the presence of aqueous fluids at temperatures to at least 700°C, and that they may have been present at 900°C (Stakes *et al.*, 1991; Dick *et al.*, 1991). Isotopic data on oxygen and hydrogen (Stakes *et al.*, 1991; Kempton *et al.*, 1991), strontium (Kempton *et al.*, 1991), and sulfur (Alt and Anderson, 1991) also indicate pervasive high-temperature alteration. In addition, fluid inclusions in plagioclase, hornblende, diopside, apatite and quartz are present in these samples (Vanko and Stakes, 1991). The combined evidence from assemblages of alteration minerals and fluid inclusions indicates that early development of hornblende in the metagabbro occurred at ~720°C, followed by hornblende-plagioclase vein formation at 500° to 600°C, and retrograde greenschist alteration at temperatures below 400°C. Veins containing plagioclase and diopside appear to have formed at ~500°C and 1kb (Vanko and Stakes, 1991).

At the higher temperatures and pressures at which submarine hydrothermal fluids equilibrate with layered gabbros, peridotites and other mafic and ultramafic rocks, consid-

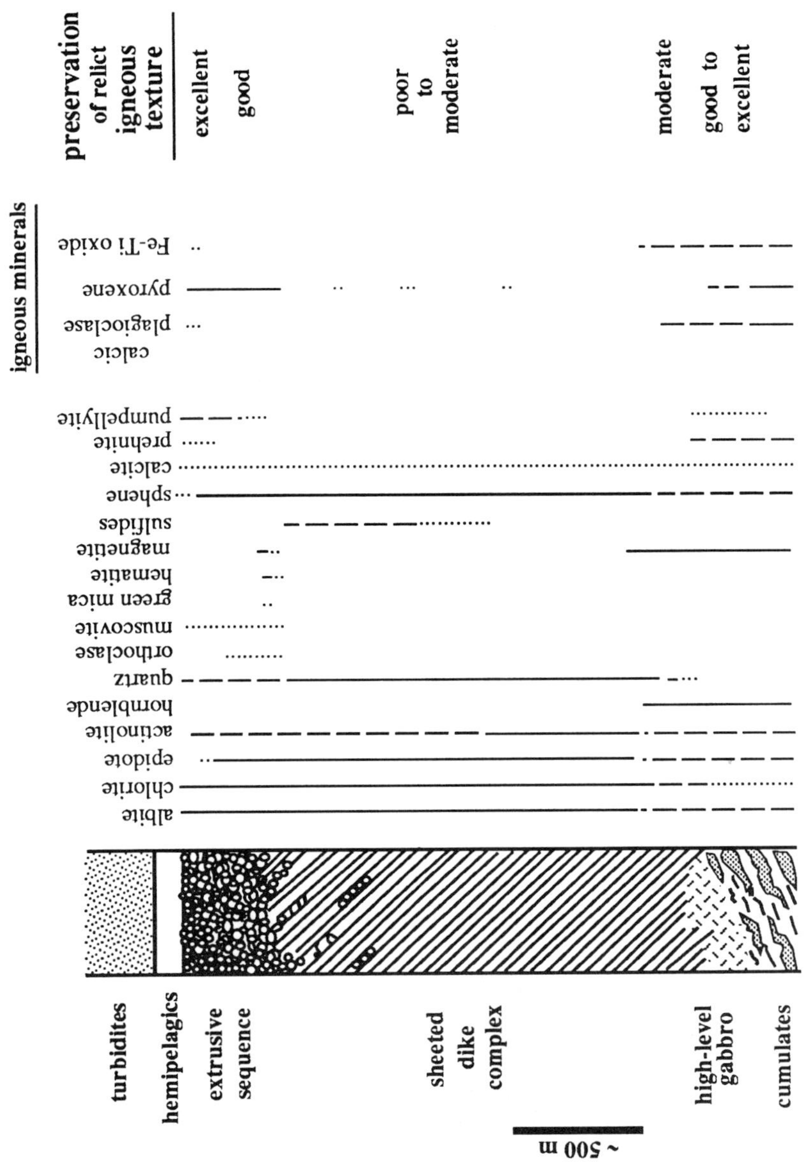

Fig. 13. (after Harper *et al.*, 1988) Lithostratigraphy and mineralogy vs. depth in the Josephine ophiolite. A solid line indicates that a mineral is characteristically present, a dashed line indicates that a mineral is commonly present and a dotted line indicates a minor or sporadic phase.

TABLE 1

Minerals in submarine hydrothermal systems refered to in the text and/or Figs. 13 and 14 and their formulas

Mineral Name	Formula
albite	$NaAlSi_3O_8$
analcite	$NaAlSi_2O_6 \bullet H_2O$
anhydrite	$CaSO_4$
apatite	$Ca_5(PO_4)_3(OH,F,Cl)$
apophyllite	$KCa_4Si_8O_{20}F \bullet 8\ H_2O$
aragonite	$CaCO_3$
celadonite	$KMgFeSi_4O_{10}(OH)_2$
chabazite	$CaAl_2Si_4O_{12} \bullet 6\ H_2O$
chalcopyrite	$CuFeS_2$
diopside	$CaMg(SiO_3)_2$
epidote	$Ca_2FeAl_2Si_3O_{12}(OH)$
fayalite	Fe_2SiO_4
galena	PbS
gyrolite	$Na(Ca, Mg, Fe)_{16}(Si, Al)_{24}O_{60}(OH) \bullet 15H_2O$
hematite	Fe_2O_3
magnetite	Fe_3O_4
mesolite	$Na_2Ca_2(Al_2Si_3O_{10})_3 \bullet 8\ H_2O$
muscovite	$KAl_2(AlSi_3)O_{10}(OH)_2$
natrolite	$Na_2Al_2Si_3O_{10} \bullet 2\ H_2O$
orthoclase	$KAlSi_3O_8$
prehnite	$CaAl(AlSi_3)O_{10}(OH)_2$
pyrite	FeS_2
pyrrhotite	FeS
quartz	SiO_2
sphalerite	ZnS
sphene	$CaTiSiO_5$
stilbite	$(Ca, Na_2, K_2)Al_2Si_7O_{18} \bullet 7\ H_2O$
talc	$Mg_3Si_4O_{10}(OH)_2$
thompsonite	$NaCa_2[(Al, Si)_5O_{10}]_2 \bullet 6\ H_2O$
tremolite	$Ca_2Mg_5Si_8O_{22}(OH)_2$
Mineral Group Names	Composition
actinolite	Ca, Mg, Fe amphibole
aegerine-augite	$(Na, Ca)(Fe, Mg)Si_2O_6$
chlorite	Mg, Fe, Al sheet silicate
hornblende	Na, Ca, Mg, Fe, Al amphibole
melanite	titanium-bearing garnet
phillopsite	Na, Ca, K zeolite
plagioclase	Na, Ca feldspar
pumpellyite	hydrous Ca, Mg, Fe, Al silicate
pyroxene	$(Mg, Fe)SiO_3$
saponite	Mg, Fe clay mineral

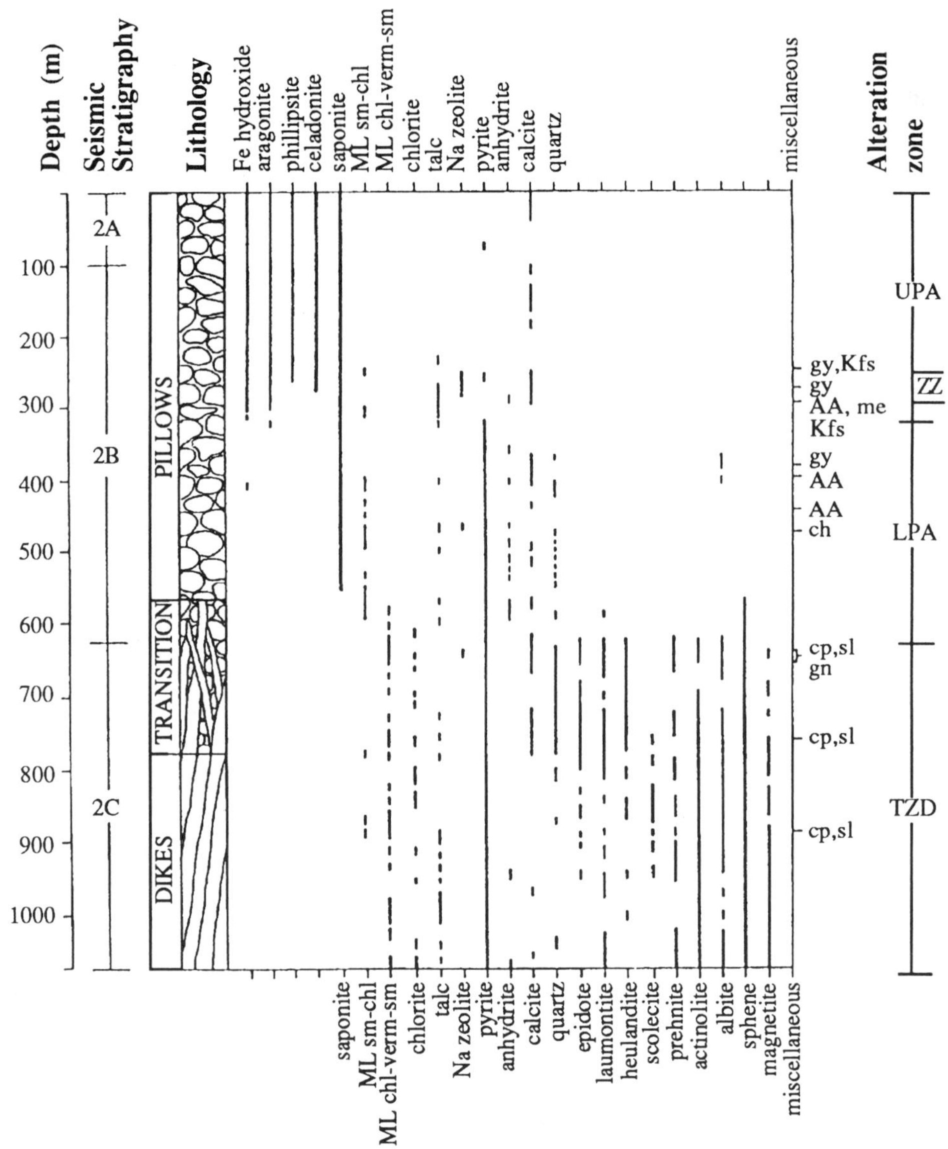

Fig. 14. (after Alt *et al.*, 1986a,b) Lithostratigraphy and distribution of alteration minerals in drill core from DSDP site 504B. Alteration zones correspond to those shown in Fig. 2. Depths given are below the overlying 274.5 m of sediment. ML sm-chl, mixed layer smectite-chlorite; ML chl-verm-sm, mixed layer chlorite-vermiculite-smectite; gy, gyrolite; AA, aegerine-augite; me, melanite; Kfs, K-feldspar; gn, galena; sl, sphalerite; cp, chalcopyrite; ch, chabazite. Na-zeolite includes analcite, stilbite, natrolite, mesolite, thomsonite, and apophyllite.

erable mass transfer is likely to occur. Preliminary mass-transfer calculations for reaction of 1.0 molal NaCl solutions initially at quartz saturation, with various mafic assemblages at 2kb and 400° to 750°C are reported by Sassani and Shock (1990b). In these calculations, pH values hover around neutrality, and the activities of the major rock-forming elements fluctuate over several orders of magnitude as a function of reaction progress. In addition, the resulting equilibrium mineral assemblages range from those containing ample talc, tremolite and other hydrous phases at lower temperatures (~500°C), to those consisting of strictly anhydrous, mafic minerals at higher temperatures (≥ 700°C). The latter differ from the mineral assemblages of the unaltered rocks, but might be (incorrectly) identified as being of strictly igneous origin since they lack obvious alteration phases. This type of cryptic alteration is probably common in rocks at great depths in the oceanic crust and upper mantle, and is consistent with oxygen isotopic measurements for ophiolites (Gregory and Taylor, 1981; Harper *et al.*, 1988). Cryptic alteration may already be documented in the samples from ODP site 735B.

6.2. MAJOR ION COMPOSITION OF SUBMARINE HYDROTHERMAL FLUIDS

The most thoroughly studied submarine hydrothermal fluids are those venting from hot springs on the ridge crest. Von Damm (1990) summarizes results from many investigators on 32 hot-spring samples with temperatures up to 380°C. Relative to seawater concentrations, Li, K, Rb, Cs, Be, Ca, B, Al, As, Se, Fe, Mn, Cu, Zn, Cd, Co, Ag, Pb, CO_2, CH_4, H_2S, SiO_2, ^3He, ^4He, and H_2 are generally enriched in hydrothermal vent fluids. Concentrations of Na and Cl show both depletions and enrichments relative to seawater values, and Mg, F, and sulfate appear to be greatly depleted. An overriding assumption in reports of vent fluid composition is that the Mg concentration is essentially zero in the vent fluids, and that any Mg in the sample results from contamination by ambient seawater during sampling. This assumption is based on results of basalt/seawater experiments at vent temperatures which indicate an enormous depletion of Mg at temperatures above 150°C (Seyfried and Bischoff, 1979). As a result, most analyses are referenced to zero Mg concentrations by subtracting the appropriate fraction of seawater. This approach can introduce difficulties owing to the possibility that not all the Mg is removed from fluids which vent at lower temperatures, and in identifying evidence of low and high temperature fluid mixing (Von Damm, 1990). In addition, it sometimes leads to reports of negative concentrations for sulfate in the hydrothermal fluids.

Theoretical mass-transfer and speciation calculations for hydrothermal vent fluids indicate that their composition can be explained by reactions with mineral assemblages containing epidote, plagioclase, chlorite, and quartz (Bowers *et al.*, 1985, 1988). This assemblage compares closely with the observations from the deeper position of DSDP site 504B (Fig. 14), suggesting that the venting fluids react with mineral assemblages at depth in the crust and move rapidly to the surface through ridge-crest hydrothermal systems.

6.3. pH BUFFERING DURING FLUID/ROCK REACTIONS

Measured values of pH for vent fluids at 25°C and 1 bar are in the range 3 to 4 for fluids venting from ridge crests without sediments (Michard *et al.*, 1983, 1984; Von Damm *et al.*, 1985a; Von Damm and Bischoff, 1987; Campbell *et al.*, 1988; Butterfield *et al.*, 1990; among others). In the case of fluids from hot springs in the Guaymas Basin, where there are ~500 meters of terrigenous and marine sediments above the ridge-crest basalt, the pH at 25°C and 1 bar is 5.9 (Von Damm et al., 1985b). It must be emphasized that these pH values may be quite different from those of the fluids at the temperatures and pressures of the vent environment which can not be measured directly.

Values of pH can be calculated by taking account of fluid speciation at the appropriate temperature and pressure. These calculations require equilibrium constants for the numerous major element complexes which may be present, as well as activity coefficients at the appropriate salinity. As described above, many of the thermodynamic data required for these calculations have been determined from experiments and many more can be estimated with a variety of correlation algorithms based on regression of data from those experiments. Calculated pH values for fluids from ridge crests without sediments are in the range 4.1 to 4.7 at the elevated temperatures of the hydrothermal vents (Bowers *et al.*, 1988). As shown in Fig. 15, neutral pH changes as a function of temperature and pressure, and it can be seen that the pH values of hydrothermal fluids are acidic by about 1 to 2 units at 350°C. Somewhat acidic conditions are likely to prevail at lower temperatures as well, depending on the intensity of hydrous alteration of basalt. If the early oceanic crust was more enriched in Mg and Fe than present day oceanic basalt, which may have been the case given the presence of komatiites early in Earth history (Nisbet, 1987), pH values may have been somewhat basic in the coexisting hydrothermal fluids (Janecky and Seyfried, 1986).

6.4. REDOX BUFFERING DURING FLUID/ROCK REACTIONS

Oxidation states of hydrothermal fluids are seldom determined by direct measurement, but are often calculated by assuming equilibrium between fluids and mineral assemblages. In the case of submarine hydrothermal vent fluids, dissolved H_2 has been measured and can be compared to values consistent with mineral buffer assemblages such as those represented by reactions (7-10). As an example, Evans *et al.* (1988) report concentrations of H_2 in hydrothermal vent fluids collected at the southern Juan de Fuca ridge. The vent fields are at 2300m depth (Von Damm, 1990) and the samples were collected at approximately 285°C. Concentrations of H_2 in the fluid samples range from 148 to 313μm. Using the zero Mg assumption, Evans *et al.* (1988) take account of the relative contributions of hydrothermal fluids and sea water to each sample and estimate that the concentrations of H_2 in the pure hydrothermal fluids range from 270 to 527μm.

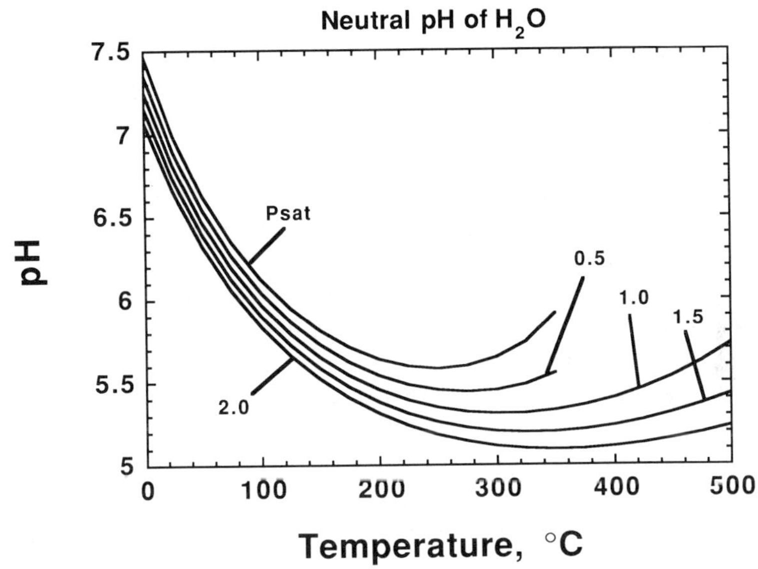

Fig. 15. Plot of neutral pH of pure H_2O as a function of temperature at the saturation pressure and higher pressures labeled in kilobars.

Fugacities of H_2 corresponding to these concentrations can be evaluated from calculated values of log K for the reaction

$$H_2(g) \quad = \quad H_2(aq) \tag{11}$$

using the computer code SUPCRT92 (Johnson *et al.*, 1992) which employs data, equations and parameters for $H_2(aq)$ from Shock *et al.* (1989). Activity coefficients for $H_2(aq)$ in hydrothermal fluids at these conditions have not been measured, so as a first approximation we can take molality and activity to be equivalent. The calculated log K for reaction 11 at 285°C and 230 bars is -2.52, and the log activities of $H_2(aq)$ calculated from the samples range from -3.57 to -3.28. These values yield log fugacities of H_2 from -1.05 to -0.76, and the more positive of these values is in close agreement with the value of log fH_2 = -0.72 set by the PPM assemblage at 285°C and 230 bars (also calculated with SUPCRT92). This result is consistent with the presence of these minerals in the hydrothermal vent chimney, plume and mineral deposits. Similar calculations for East Pacific Rise fluids by Janecky and Seyfried (1984) are also closely consistent with buffering by PPM as proposed by Hayman and Kastner (1981). Recent calculations by Roberts and Norton (1992) indicate that many measured vent-fluid compositions are consistent with the coexistence of pyrite and magnetite but not pyrrhotite at 250 bars and 350°C. These results indicate that the fluids may be more oxidized than the PPM assemblage, consistent with the more negative values of log fH_2 calculated above. As discussed in the next section, interplay between the temperature and the oxidation states of hydrothermal fluids are the factors which control whether metastable organic synthesis can occur in submarine hydrothermal systems.

7. The Potential for Organic Synthesis in Submarine Hydrothermal Systems

As emphasized elsewhere (Shock, 1990b), the crucial condition for organic synthesis in hydrothermal systems is the presence of kinetic barriers which block the attainment of stable thermodynamic equilibrium in the C-H-O-N system. If stable equilibrium is attained, the predominant species will be H_2O, H_2, CO_2, CH_4, graphite, N_2 and NH_3, and their relative abundances will be controlled by temperature, pressure and bulk composition (Dayhoff *et al.*, 1967; Holloway, 1984). At stable equilibrium in the C-H-O-N system the activities of organic compounds of interest to the study of the origin of life are minuscule. However, organic synthesis may proceed if kinetic barriers block stable equilibrium but allow metastable equilibrium states.

Evidence supporting the presence of kinetic barriers to stable equilibrium in the C-H-O-N system at hydrothermal conditions comes from isotopic studies of CO_2, CO, CH_4, N_2 and NH_3 (see: Shock, 1990b for a review). These data indicate that the lowest temperatures at which CO_2 and CH_4 equilibrate isotopically in submarine hydrothermal fluids is > 500°C (Welhan, 1988). At all lower temperatures there are kinetic barriers which prohibit equilibration of CO_2 and CH_4. Therefore, stable equilibrium is blocked in these systems.

Ample evidence from studies of organic acid concentrations in sedimentary basin brines, as well as consideration of reactions between organic acids, carbonate minerals and petroleum, indicates that metastable states among organic compounds and CO_2 are reached in oil reservoirs (Shock, 1988, 1989, 1992c; Helgeson and Shock, 1988; Helgeson *et al.*, 1991; Helgeson, 1991). These same calculations demonstrate that CH_4 and other light hydrocarbons are excluded from these metastable states. Therefore, it appears that *unlike* reactions between CO_2 and CH_4, those between CO_2, organic acids and petroleum hydrocarbons may *not* be hindered by kinetic barriers at temperatures from ~ 80° to 200°C or higher. As a consequence, the present understanding of reactions in the C-H-O-N system, based on a thermodynamic analysis of natural systems, is consistent with the possibility of organic synthesis from CO_2 and N_2 at hydrothermal conditions.

7.1. SOURCES OF CARBON AND NITROGEN

Identifying the inorganic forms in which carbon and nitrogen occur is essential in order to critically evaluate the potential for organic synthesis in submarine hydrothermal systems. So far, little is known about nitrogen in present-day hydrothermal vent fluids except those in the Guaymas Basin which contain from 10 to 15 mmol kg^{-1} NH_4^+. In this location, the origin of the ammonium is thought to be from organic matter in the 500 meters of sediments overlying the ridge (Von Damm, 1990).

Considerably more is known about volatile forms of inorganic carbon in present-day hydrothermal vent fluids. Those from the southern Juan de Fuca ridge contain ~ 4 mmol kg^{-1} CO_2, those at 21°N ~ 6 mmol kg^{-1}, those at the Galapagos spreading center ~ 10 mmol kg^{-1} and those at 11 - 13 °N on the East Pacific Rise ~ 15 mmol kg^{-1} (Von Damm, 1990). In contrast, fluids venting from the Virgin Mound on the Axial Volcano on the Juan de Fuca ridge contain ~ 285 mmol kg^{-1} CO_2 (Butterfield *et al.*, 1990). Methane and short-chain hydrocarbon concentrations are much lower than CO_2 concentrations at spreading centers without sediments such as the Southern Juan de Fuca ridge (Evans *et al.*, 1988; Welhan, 1988). Methane and other light hydrocarbons are abundant in hydrothermal fluids from sedimented ridges (Welhan and Lupton, 1987), as well as in the hydrothermal petroleum which is characteristic of these systems (Simoneit, 1988, 1990). The present-day fluids which provide constraints on organic synthesis in submarine hydrothermal systems are those which are free from contamination by sediments. Therefore, it appears that CO_2 is the predominant form of volatile carbon in these fluids, at least at high temperatures.

An explanation why CO_2 and not CH_4 is the dominant form of volatile carbon in present day hydrothermal vent fluids can be found in a consideration of stable thermodynamic equilibrium in the C-H-O system at the oxidation states set by mineral assemblages deep in the oceanic crust and upper mantle where the appropriate fluid/rock reactions occur. As mentioned above, isotopic data indicate that CO_2/CH_4 equilibration only occurs at temperatures > 500°C which indicates that the reaction

$$CO_2 + 4\,H_2 = CH_4 + 2\,H_2O \tag{12}$$

is reversible only at high temperatures. This observation is consistent with laboratory experiments designed to study this reaction and carbon isotope exchange (see: Shock, 1990b, for a review). Assuming unit activity for H_2O, rearrangement of the logarithmic form of the law of mass action for reaction (12) yields

$$\log f\mathrm{H_2} = \frac{1}{4}\left[\log\left(\frac{f\mathrm{CH_4}}{f\mathrm{CO_2}}\right) - \log K_{12}\right]. \tag{13}$$

Values of log K_{12} can be used to evaluate the fugacity ratio of CO_2 to CH_4 consistent with various mineral buffers as shown in Fig. 16 where log fH_2 is plotted against log (fCH_4/fCO_2) at 600°C and 500 bars, conditions where stable equilibrium should prevail. The solid line in Fig. 16 is consistent with Eqn. (13) and log K_{12} = -6.17 at 600 °C and 500 bars (calculated with SUPCRT92). The dashed vertical lines indicate the values of log $f\,H_2$ set by the mineral buffer assemblages represented by reactions (7-10) at the same temperature and pressure. The points at which the dashed lines intersect the solid line give the fugacity ratios of CH_4 to CO_2 for each buffer assemblage. Note that these values are all less than 1. It follows that CO_2 should be the predominant form of carbon at the high temperatures and pressures encountered deep in hydrothermal systems, and that the high CO_2 to CH_4 ratios in present-day submarine hydrothermal vent fluids are likely to be indicative of high-temperature stable equilibration.

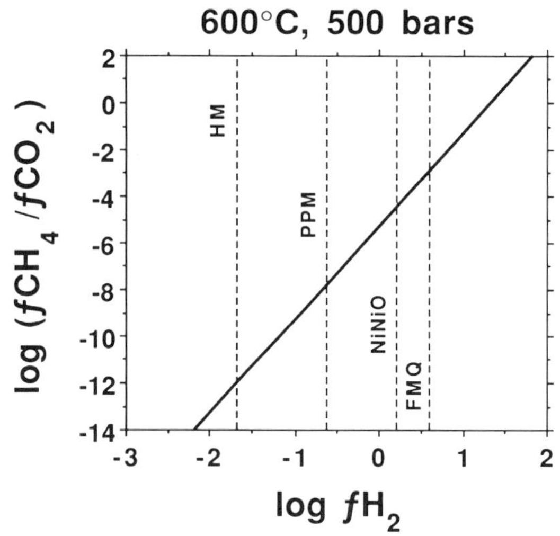

Fig. 16. The ratio of the fugacities of CH_4 and CO_2 as a function of fH_2 at 600°C and 500 bars (solid line). Dashed vertical lines show values of fH_2 buffered by various mineral assemblages (reactions 7 - 10) at the same temperature and pressure.

Calculations of the type shown in Fig. 16 indicate that CO_2 is likely to be the predominant form of volatile carbon presently leaving the Earth's mantle through submarine hydrothermal systems unless the oxidation state of the mantle is considerably more reduced than generally thought (Wood and Virgo, 1989; Bryndzia *et al.*, 1989). These results also suggest that CO_2 would be the predominant product of alteration of carbon-bearing compounds introduced into the high-temperature regions of submarine hydrothermal systems from sources other than the outgassing mantle. Therefore, the extent of high-temperature hydrothermal alteration of the earliest oceanic crust material and the composition of that crust place constraints on the extent to which organic compounds from accreted meteorites, comets or interplanetary dust particles (Anders, 1989; Chyba *et al.*, 1990; Chyba and Sagan, 1992), or those generated by impacts (Fegley *et al.*, 1986; Fegley and Prinn, 1989; Chyba *et al.*, 1990), would have been altered in early submarine hydrothermal systems. If those systems were generally analogous to present-day systems, then the pervasive high-temperature alteration of oceanic crust near ridge crests would have also altered the accreted carbon-bearing compounds and generated considerable CO_2 (Shock, 1991a). Combined with the rapid annihilation of reduced gases by ultraviolet radiation and hydroxylradical reactions in the atmosphere (Kuhn and Atreya, 1979; Levine *et al.*, 1982; Levine, 1985), high temperature (>500°C) alteration deep in submarine hydrothermal systems may have purged reduced species (such as CH_4, NH_3, HCN, etc.) from the early Earth.

7.2. ORGANIC SYNTHESIS AND BASALT ALTERATION AS COUPLED IRREVERSIBLE PROCESSES

If atmospheric photochemistry and deep-crustal hydrothermal alteration were successful in converting CH_4 and NH_3 to CO_2 and N_2 on the early Earth, then the central dogma of the predominant origin of life theories, chemical evolution from reduced starting materials, is highly doubtful. Under such circumstances, are there environments in which organic synthesis from CO_2 and N_2 can proceed? As mentioned above, these conditions are likely to exist in the upper part of the oceanic crust where hydrothermal systems operate at lower temperatures and at favorable oxidation states set by mineral assemblages. Attainable oxidation states in altered oceanic crust are likely to be spanned by those set by the FMQ and HM buffer assemblages. Is it possible that the same mineral assemblages which would drive the conversion of carbon to CO_2 at stable equilibrium, which would prevail at high temperatures, might also permit organic synthesis at the lower temperatures where metastable states have been shown to persist? The answer to this question is found in calculations similar to those conducted to construct Fig. 16.

Equilibrium constants for reactions (7-10) allow calculation of $\log fH_2$ values set by the mineral buffer assemblages in altered basalt as functions of temperature and pressure. Values of fH_2 at 500 bars can be compared in Fig. 17. Also shown in Fig. 17 are three contours of $\log (fCO_2/fCH_4)$ as a function of temperature at 500 bars calculated from the equilibrium constants for reaction (12). These contours are calculated for stable equilibrium, and are plotted here to help delineate conditions where metastable states may exist. As described above, stable equilibrium between CO_2 and CH_4 is unlikely to occur below 500°C in submarine hydrothermal systems. Because organic synthesis requires a departure from stable equilibrium, the dashed vertical line at 500°C indicates an attempt to place an upper temperature boundary on hydrothermal organic synthesis. Note that with decreasing temperature mineral buffers move to lower values of $\log (fCO_2/fCH_4)$.

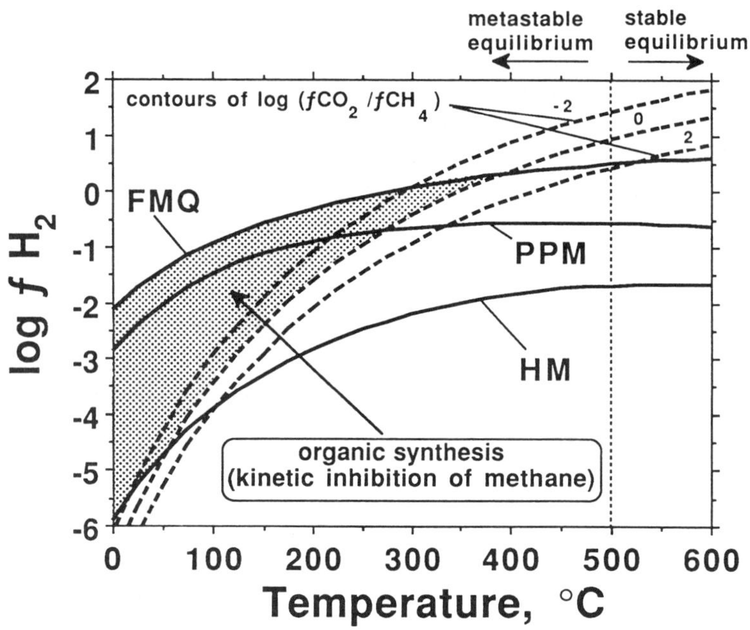

Fig. 17. Plot of log fH_2 against temperature. Solid curves show values of log fH_2 buffered by the FMQ, PPM and HM mineral assemblages as functions of temperature. Dashed curves correspond to contours of log (fCO_2/fCH_4) equal to 2, 0 and -2 as functions of temperature. Dotted vertical line at 500°C separates range of temperature where stable equilibrium in the C-H-O system is attained in submarine hydrothermal systems (T > 500°C), from that at which CO_2 reduction to CH_4 is kinetically inhibited, and where metastable equilibrium states between CO_2 and aqueous organic compounds may prevail. Stippled area corresponds to the region where synthesis of aqueous organic compounds in metastable states may be most easily detected (see text).

Therefore, CO_2 would persist metastably at lower temperatures along a mineral buffer owing to the kinetic inhibition of CH_4 formation. At some point along a mineral-buffered trajectory in Fig. 17, CO_2 would be unstable relative to many organic compounds. Organic synthesis could occur by reduction of CO_2 if such reactions are allowed by reaction kinetics. Although the mechanisms of these reactions are not yet known, the evidence from oil reservoirs cited above indicates that kinetic pathways between CO_2 and organic compounds appear to exist and permit the establishment of metastable equilibrium states in the natural system.

These considerations allow the designation of the field labelled "organic synthesis" in Fig. 17. This field is bounded in log fH_2 by the FMQ and HM curves, which delineate the range of possible fH_2 values owing to the composition of the oceanic crust, and more arbitrarily by the log (fCO_2/fCH_4) = 0 contour. Organic synthesis along a mineral buffer curve in Fig. 17 may begin at lower values of log (fCO_2/fCH_4), but it should be detectable at the zero boundary. With reference to plots of this type, we can follow likely trajectories taken by fluids as they circulate through the oceanic crust and evaluate whether they encounter the field of possible organic synthesis in hydrothermal systems.

Two schematic fluid circulation paths are shown in the cross section in Fig. 18. Path A follows a trajectory which takes fluids deep into the sheeted dike complex, into the vicinity of the magma chamber and out through ridge-crest black smokers. Path B follows a lower temperature and somewhat shallower trajectory through the flanks of the ridge system. The dotted lines in Fig. 18 indicate isotherms similar to those calculated in many theoretical studies of fluid circulation in the oceanic crust. The numbers along paths A and B indicate specific points along these trajectories where speciation calculations were conducted in the present study. These points are located in $\log fH_2$ vs. temperature space in Fig. 19, and are listed in Table 2. Because temperature and oxidation state are defined for each of these points, we can test the relative effectiveness of these positions in the oceanic crust for organic synthesis from CO_2 and N_2.

General calculations of the type described below are presented for > 80 aqueous organic compounds by Shock (1990b). For the purpose of illustration, and to evaluate the potential of organic synthesis in various regions in the oceanic crust, the calculations described here are for a shortened list of simple organic compounds. The compounds considered in the following calculations are listed in Table 3 and include compounds containing 1 or 2 carbon atoms and/or 1 or 2 nitrogen atoms per molecule. Thermodynamic data and equation-of-state parameters for these aqueous species are taken from Shock et al. (1989), Shock and Helgeson (1990), Shock (1992d), and Schulte and Shock (1992). In the metastable states considered in these calculations, methane, ethane, ethene and ethyne are excluded in order to maintain consistency with evidence from sedimentary basins. In contrast, NH_3 is included in the calculations even though kinetic barriers to its formation from N_2 may exist. Insufficient analytical data from natural systems precludes a quantitative assessment of this possibility. Therefore, without the necessary data to exclude NH_3, the decision was made to include it. If a kinetic barrier to NH_3 formation exists, then calculated activities of N-bearing organic compounds would be higher than those obtained in this study. All of the calculations conducted in this study are for 500 bars total pressure, initial fugacities of CO_2 and N_2 equal to 10 bars and 1 bar, respectively, and unit activity of H_2O. These high values for CO_2 and N_2 are used to simulate either a rapid outgassing of the mantle or efficient cycling of volatiles from comets and/or meteorites through the deepest portions of the oceanic crust.

Calculation of the distribution of species in the metastable states described above requires mass-action and mass-balance constraints. Mass-action constraints for the fourteen product compounds listed in Table 3 are given by the equilibrium constants for reactions of the type

$$a\,CO_2 + b\,N_2 + c\,H_2 = d\,\text{product} + e\,H_2O \qquad (14)$$

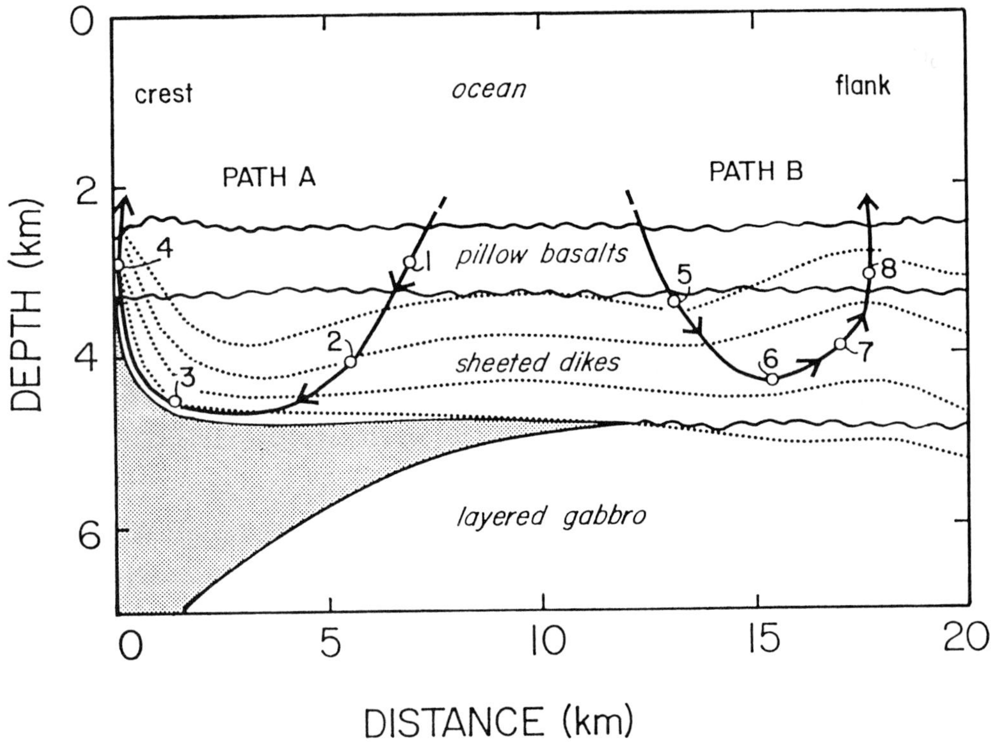

Fig. 18. Schematic cross section of oceanic crust at a spreading center showing two possible flow paths for hydrothermal fluids. Dotted curves represent isotherms consistent with the model of Fehn *et al.* (1983). Path A follows fluids from a seawater recharge zone to depths in excess of 4 km in oceanic crust where they are heated near the magma chamber before exiting at the ridge-crest in black smoker vents. Path B follows fluids from a seawater recharge zone through the flanks of the ridge system. Estimates of fluid flow through the oceanic crust indicate that flow through paths like B exceeds the flow through paths like A by more than a factor of 20. The stippled area indicates a magma chamber consisting of liquid and crystal-mush regions. Locations numbered 1-8 correspond to points along the flow paths which are likely to be characterized by the temperature and mineral assemblages listed in Table 2, and are used for reference in Figs. 19-24.

where a-e represent stoichiometric reaction coefficients. The set of equilibrium constants differs at each temperature considered. Fourteen reactions of the type given by reaction (14) are combined with four mass-balance constraints for C, N, H and O to construct a matrix of eighteen equations solved simultaneously. One example of the mass-balance constraints is that all of the carbon redistributed into products, as well as that remaining as CO_2, must sum to equal the initial 10 bars of CO_2. Results of these calculations at points 1-8 defined in Table 2 are listed in Table 4. This set of results is depicted in various sub-groupings in Figs. 20-24.

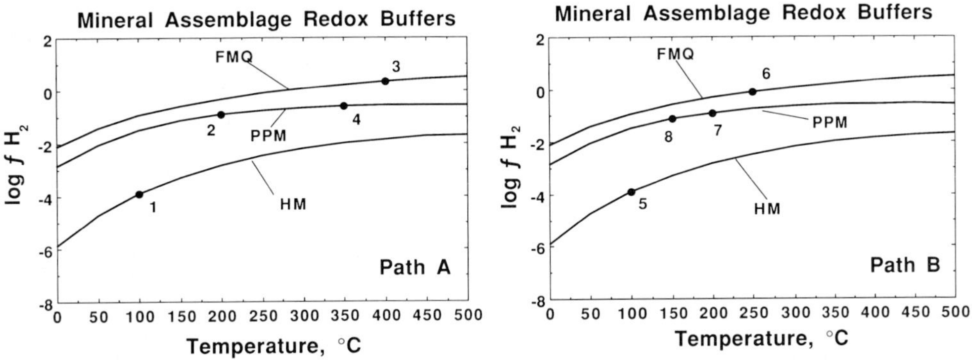

Fig. 19. Plots of log fH_2 against temperature (in °C) showing locations of points on the flow paths in Fig. 18 and listed in Table 2.

TABLE 2

Temperatures and fH_2 buffer assemblages at points along paths A and B in Fig. 18

Point	Temperature, °C	fH_2 buffer
1	100	HM
2	200	PPM
3	400	FMQ
4	350	PPM
5	100	HM
6	250	FMQ
7	200	PPM
8	150	PPM

TABLE 3
Species involved in the distribution of species calculations depicted in Figs. 20-24

Reactants:	
CO_2, N_2, H_2, H_2O	
Products:	
CO, NH_3, HCN,	
formaldehyde	H_2CO
acetaldehyde	CH_3HCO
methanol	CH_3OH
ethanol	C_2H_5OH
formic acid	HCOOH
acetic acid	CH_3COOH
methanamine	CH_3NH_2
ethanamine	$C_2H_5NH_2$
urea	$CO(NH_2)_2$
oxalic acid	$(COOH)_2$
glycine	$C_2H_5NO_2$

Results from Table 4 for the fourteen aqueous organic species which are reaction products in the metastable equilibrium state examined by the distribution of species calculations are shown in the following groupings: compounds with one carbon atom in Fig. 20, compounds with two carbon atoms in Fig. 21, compounds with one oxygen atom in Fig. 22, compounds containing nitrogen in Fig. 23, and organic acids in Fig. 24. These groupings allow many species to be represented more than once which facilitates comparisons among the results. A general trend can be observed on each pair of figures showing results along paths A and B. In each case, the calculated activities of aqueous organic species are higher at the final point along path B than at the end of path A. This indicates that organic synthesis in submarine hydrothermal systems is more likely in fluids circulating through the flanks rather than those circulating through the ridge-crust and venting at black smokers (see Fig. 18), a point made previously by Shock (1990b, 1991b).

TABLE 4

Log activities of aqueous species from speciation calculations for points along paths A and B in Fig. 18 using temperatures and fH_2 values consistent with mineral buffers as listed in Table 2, and initial fugacities of CO_2 and N_2 set to 10 bars and 1 bar, respectively

Products	#1 (also #5)	#2 (also#7)	#3	#4	#6	#8
CO	-10.00	-6.56	-3.55	-5.05	-5.41	-7.05
NH_3	-4.19	-1.99	-2.50	-3.47	-1.58	-1.33
HCN	-18.69	-12.74	-10.58	-12.82	-11.15	-12.68
formaldehyde	-14.25	-9.02	-6.84	-8.78	-7.63	-9.15
acetaldehyde	-16.50	-8.84	-9.71	-13.19	-7.29	-6.78
methanol	-12.55	-7.18	-6.79	-9.04	-5.96	-6.29
ethanol	-16.25	-7.93	-9.92	-13.87	-6.36	-5.09
formic acid	-8.11	-5.91	-5.09	-6.07	-5.35	-5.80
acetic acid	-9.22	-4.66	-7.11	-9.56	-3.99	-2.33
methanamine	-15.32	-8.59	-9.77	-12.77	-7.28	-6.67
ethanamine	-20.56	-10.65	-14.04	-18.76	-8.93	-6.88
urea	-10.37	-7.46	-10.08	-11.78	-7.19	-5.47
oxalic acid	-17.79	-15.49	-14.66	-15.63	-14.91	-15.41
glycine	-16.78	-12.12	-16.15	-18.56	-11.65	-9.24

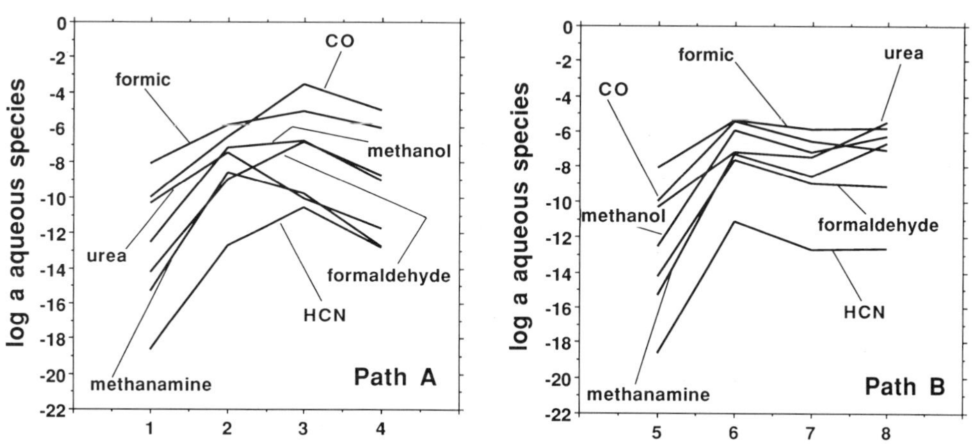

Fig. 20. Plots of log activity of aqueous species containing one atom of carbon per molecule taken from Table 4 as functions of the points along the flow paths indicated in Figs. 18 and 19 and Table 2.

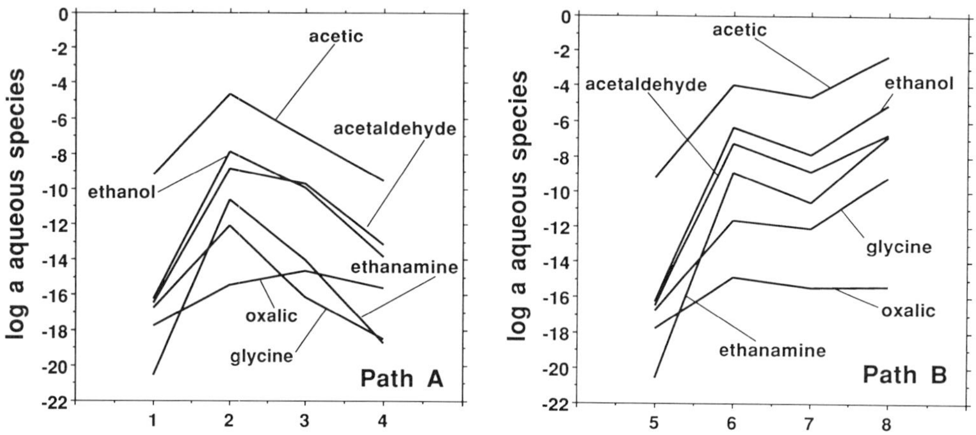

Fig. 21. Plots of the log activity of aqueous species containing two carbon atoms per molecule taken from Table 4 as functions of the points along the flow paths indicted in Figs. 18 and 19 and Table 2.

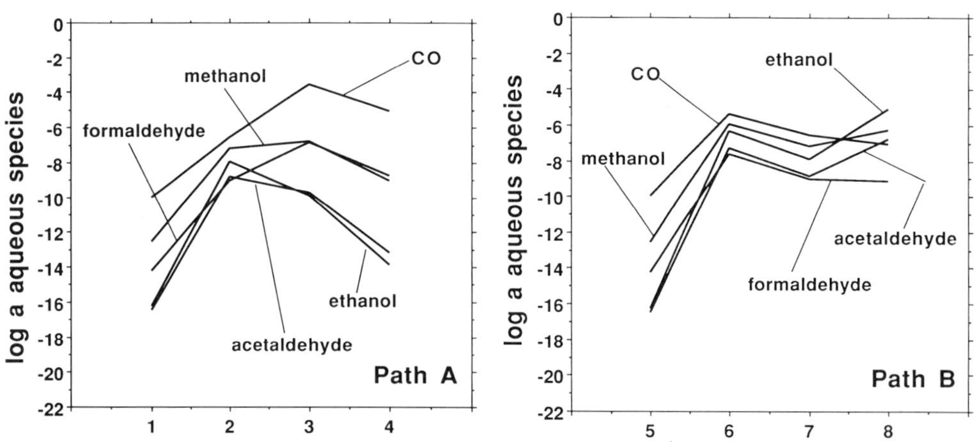

Fig. 22. Plots of the log activity of aqueous species containing one oxygen atom per molecule taken from Table 4 as functions of the points along the flow paths indicted in Figs. 18 and 19 and Table 2.

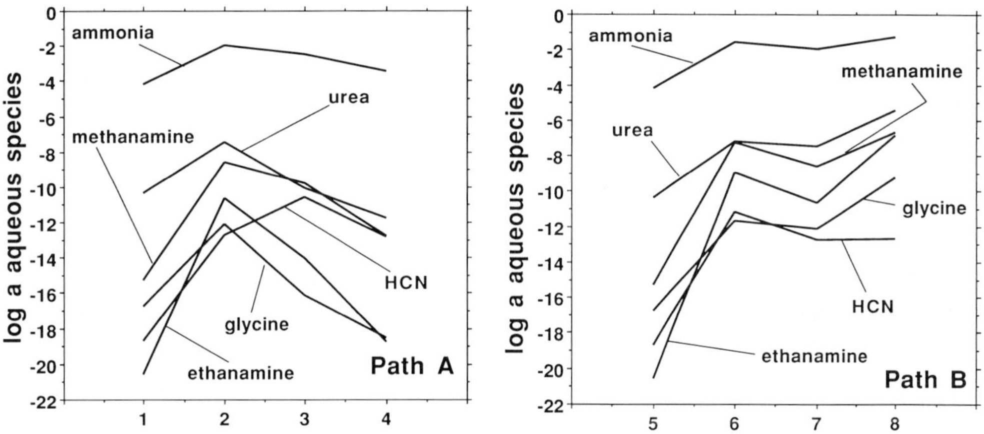

Fig. 23. Plots of the log activity of aqueous species containing nitrogen taken from Table 4 as functions of the points along the flow paths indicted in Figs. 18 and 19 and Table 2.

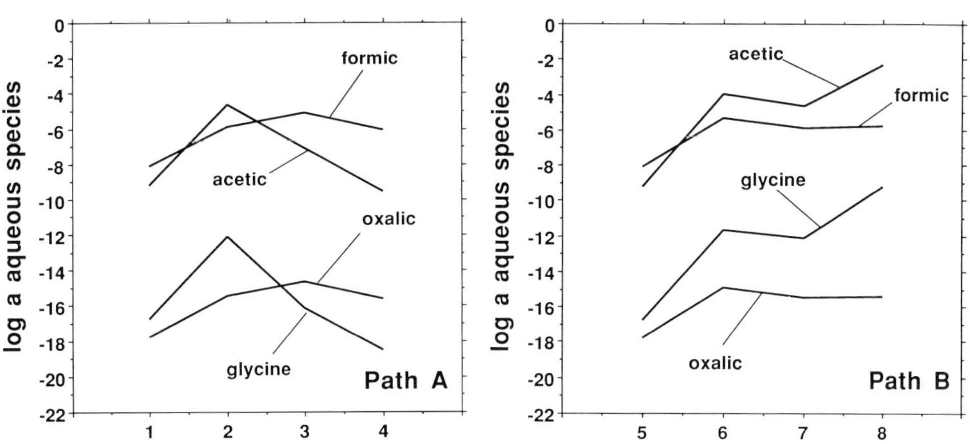

Fig. 24. Plots of the log activity of aqueous organic acids taken from Table 4 as functions of the points along the flow paths indicted in Figs. 18 and 19 and Table 2.

Specific comparisons of the results in Figs. 20-24 may help to focus experimental investigations on the most likely environments in submarine hydrothermal systems for organic synthesis. Comparison of Figs. 20A and B shows that CO is the predominant 1-carbon species at points 3 and 4 along path A, and second only to formic acid at points 1 and 2. In contrast, the activity of CO is lower than those of urea, formic acid, methanol and methanamine at the final point along path B. For the sake of comparison, let us consider activities = 10^{-8} as a threshold above which the activities of aqueous species become significant in the speciation of hydrothermal fluids. Along path A, CO, formic acid, methanol, urea and formaldehyde cross this threshold as shown in Fig. 20A. At point 4, which represents the possible final metastable equilibration before venting at black smokers, only CO and formic acid are across this threshold. In contrast, calculated activities of all of the 1-carbon species except HCN cross this 10^{-8} threshold along path B as shown in Fig. 20B, and all but HCN and formaldehyde are present at activities > 10^{-8} at point 8, a possible final metastable equilibrium state achieved before venting.

Differences between paths A and B are more dramatic for the 2-carbon species shown in Figs. 21A and B. Note that acetic acid is the only 2-carbon species to cross the 10^{-8} threshold at points along path A (Fig. 21A), but that acetic acid, ethanol, acetaldehyde and ethanamine cross this threshold at points along path B (Fig. 21B). Comparison of points 4 and 8 shows that no 2-carbon species have calculated activities > 10^{-8} at black smoker conditions, but that four 2-carbon species have activities > 10^{-8} in calculations appropriate for flank hot spring environments. Note that among these aqueous organic species, the calculated activity of acetic acid approaches 10^{-2} at point 8 along path B. As a comparison, activities of acetic acid in sedimentary basin brines are in the range 10^{-4} to 10^{-2} (Shock 1988, 1989, 1990c).

Among the 1-oxygen compounds shown in Figs. 22A and B, CO is predominant at all points along all of path A and most of path B. Calculated activities of methanol, formaldehyde and CO cross the 10^{-8} threshold along path A, but only CO remains at point 4. In contrast, calculated activities of all of the 1-oxygen species cross this threshold along path B, and all but formaldehyde remain at point 8. Note that the calculated activities of ethanol, methanol and acetaldehyde exceed that of CO at point 8 and that the calculated activity of ethanol approaches 10^{-5}. As in the cases of the 1- and 2-carbon species shown in Figs. 20 and 21, the results shown in Fig. 22 suggest that fluids circulating through the flanks of ridge systems have the greater potential for organic synthesis.

The calculated activities of nitrogen-bearing species shown in Fig. 23 are also consistent with the hypothesis that flank environments may be more suitable for organic synthesis than crest environments. Comparison of Figs. 23A and B shows that ammonia is the predominant product at all points along both paths. Along path A, the calculated activity of urea crosses the 10^{-8} threshold at point 2 but plunges well below this value, like the other N-bearing organic compounds, at points 3 and 4. In contrast, urea, methanamine and ethanamine have calculated activities which exceed 10^{-8} at points along path B. Note that all of the calculated activities of N-bearing organic compounds increase along path B, and that the calculated activity of urea is ~ $10^{-5.5}$ at point 8. In contrast, the calculated activities of N-bearing organic compounds increase from point 1 to point 2 along path A, corresponding to downward flow of hydrothermal fluids near ridge crests, but decrease dramatically along the upward flow path to the black smokers as represented by points 3 and 4. As mentioned above, kinetic barriers to the formation of NH_3 from N_2 at low temperatures would allow greater calculated activities of N-bearing aqueous organic compounds.

The results shown in Figs. 24A and B are also shown in the other figures, but are isolated in this figure in order to compare results for the organic acids considered in these calculations. Results for formic and oxalic acids are remarkably similar along paths A and B as shown in Fig. 24. In contrast, those for acetic acid and glycine differ dramatically between the two paths. Along path A, the calculated activity of formic acid is generally greater that for acetic acid. However, the opposite is the case along path B. The predominance of acetic acid over formic acid along path B resembles the relative abundances in sedimentary basin brines in which acetic acid is abundant and formic acid is seldom detected (Carothers and Kharaka, 1978). In addition, the calculated activity of glycine, which decreases to nearly 10^{-19} at point 4 along path A, increases by eight orders of magnitude along path B and approaches 10^{-9} at point 8. These results, together with those summarized in Figs. 20-23, as well as those presented by Shock (1990b), can be used to identify environments in submarine hydrothermal systems where organic synthesis may be favorable. They also serve as predictions which can be used to guide experimental studies. In addition, these results can be used to estimate the possible contributions of submarine hydrothermal synthesis as the budget of organic carbon on the early Earth.

7.3. SUBMARINE HYDROTHERMAL SYSTEMS AS SOURCES OF ORGANIC COMPOUNDS

The results summarized above indicate that calculated activities of aqueous organic compounds in metastable equilibrium states are greater along flow paths through the flanks of mid-ocean ridge systems than they are along flow paths through the ridge crest which exit the oceanic crust at black-smoker vents. For example, the calculated activity of acetic acid along path B reaches $10^{-2.33}$. It should be kept in mind that the calculations described above are for an initial CO_2 fugacity of 10 bars which may be high, and that the corresponding activities of aqueous organic species may represent maximum values. Therefore, if we take 10^{-5} as an attainable activity of acetic acid at ~200°C in ridge-flank fluids, we can place some constraints on the carbon-fixing potential of submarine hydrothermal systems.

The molecular weight of acetic acid is 60.053 g mol^{-1}, and assuming that molalities and activities are equivalent for highly dilute species in hydrothermal fluids the corresponding concentration of acetic acid would be 6 x 10^{-4} g kg^{-1} or 0.6 ppm by weight.

Acetic acid is ~40% (by weight) carbon, so the corresponding concentration of organic carbon would be 0.24 ppm. As described above, estimates of fluid flux through present-day ridge-flank hydrothermal systems yield 10^{14} to 10^{15} kg yr^{-1}. Therefore the potential organic carbon productivity of submarine hydrothermal systems would be ~2×10^7 to 2×10^8 kg yr^{-1} at present. Fluid flow and the potential for organic synthesis may have been considerably greater on the early Earth which was probably considerably warmer with higher heat flow and more rapid seafloor spreading rates.

TABLE 5

Comparison of various potential sources of organic compounds on the early Earth

Source[a]	Organic production reduced atmosphere (kg yr^{-1})	Organic production neutral atmosphere (kg yr^{-1})
Lightning	3×10^9	3×10^7
Coronal discharge	2×10^8	2×10^6
Ultraviolet light[b]	2×10^{11}	3×10^8
Atmospheric shocks from meteors	1×10^9	3×10^1
Atmospheric shocks from post-impact plumes	2×10^{10}	4×10^2
Interplanetary dust particles	6×10^7	6×10^7
Hydrothermal synthesis	2×10^8	2×10^8

[a] values for all sources other than hydrothermal synthesis are taken from Chyba and Sagan (1992).

[b] values for ultraviolet light represent the sum of productivity from various wavelengths (see Chyba and Sagan, 1992).

Part of the organic production in the flank systems would be offset by destruction in ridge-crest hydrothermal systems as advocated by Miller and Bada (1988, 1991a,b), and consistent with results described above. However, the destructive flow through the ridge-crests is estimated to be 5% of the productive flow through the flanks (COSOD II, 1987, see above). It follows that destruction of aqueous organic species at ridge-crests is nearly insignificant and has only the slightest impact on the potential of submarine hydrothermal systems for organic synthesis. Given the potential for greater productivity on a warmer early Earth than at present, 2×10^8 kg yr^{-1} is likely to be a reasonable estimate of the potential productivity of hydrothermal systems.

To place hydrothermal organic synthesis in perspective, comparisons should be made of the rates of organic productivity obtained above with those from other sources both terrestrial and extraterrestrial. Chyba and Sagan (1992) provide a summary of their best estimates of various sources of organic compounds on the early Earth other than hydrothermal systems. Rates of organic production form Chyba and Sagan (1992) are listed in Table 5, together with that obtained in this study for hydrothermal synthesis. Chyba and Sagan (1992) perform calculations for a reduced atmosphere rich in CH_4 with H_2O and either N_2 or NH_3, and a neutral atmosphere rich in CO_2, with H_2O and N_2. For reasons described above, the results for the reduced atmosphere are unlikely to be applicable to the early Earth. Comparison of hydrothermal synthesis and all other sources can be made with the values in the right-hand column of Table 5. The sum of organic contributions and production from all sources other than hydrothermal synthesis is $\sim 4 \times 10^8$ kg yr^{-1} for a neutral atmosphere. This value is only a factor of two greater than the potential contribution from hydrothermal synthesis as evaluated in the present study. This suggests that hydrothermal synthesis of organic compounds should receive considerable experimental and theoretical attention in order for progress to be made in the study of the origin of life.

Acknowledgements

The research described above benefitted from discussions with Dimitri Sverjensky, Mitch Schulte, Harold Helgeson, Dave Sassani, Nils Holm, Bruce Fegley, Sherwood Chang, Mike Engel, Bernd Simoneit, Steve Macko and Tom McCollom. Technical assistance from Carla Koretsky, Patty DuBois and Allison Shock is greatly appreciated, as are reviews from John Kerridge and Sherwood Chang. This work was funded in part by NSF grant EAR-9018468 and NASA grant NAGW-2818.

CHAPTER 6

CHEMICAL MARKERS OF PREBIOTIC CHEMISTRY IN HYDROTHERMAL SYSTEMS

JAMES P. FERRIS

Department of Chemistry, Rensselaer Polytechnic Institute, Troy, NY 12180, U.S.A.

1. Introduction

The goal of this chapter is to suggest some organic compounds which may be indicative of prebiotic processes in hydrothermal systems or laboratory simulations of them. While the exact processes which led to the origins of life are not known, studies of life's origins of the past forty years have uncovered a plethora of potential precursor molecules. Some of these same molecules were probably present in hydrothermal systems if chemical processes there had a role in the origins of life.

The types of molecules formed in primitive Earth simulation experiments and observed in the interstellar medium, on comets and meteorites will be reviewed in Section 2 of this chapter. Some reactions involving these molecules which may have been important in prebiotic syntheses will be outlined. Since near- to supercritical water is found in hydrothermal systems, its properties and aspects of organic chemistry in supercritical water at high temperature and pressure will be discussed in Section 3. Fischer-Tropsch type (FTT) reactions, which are a potential source of the building blocks of biological molecules in hydrothermal systems, are discussed in Section 4. In the concluding section, Section 5, the possible formation in hydrothermal systems of organic molecules that are believed to have been important for the origins of life is discussed.

2. Molecules Formed in Prebiotic Simulation Experiments and Observed in Extraterrestrial Environments

The molecules leading to the origins of life may have either been formed in the hydrothermal systems or transported to them after being formed elsewhere on earth or after being delivered to the earth by impacting carbonaceous meteorites or comets. This section gives a brief overview of the range of organics that may have been delivered to or possibly formed in hydrothermal systems. Although every attempt has been made to provide a comprehensive compilation of the organics that may have been present on the primitive earth, no attempt has been made to report on all the reactions that have been described in the literature for these compounds.

Origins of Life and Evolution of the Biosphere **22**: 109–134, 1992.

2.1. HYDROGEN CYANIDE (HCN) AND SIMPLE HCN ADDUCTS

2.1.1. HCN

HCN is likely to have been present in prebiotic hydrothermal systems because it is formed by a variety of processes driven by thermal energy (Ferris and Hagan, 1984). Indeed, HCN has been observed in volcanic emissions and has been postulated to be present in hydrothermal systems (Mukhin, 1974). Most, if not all of this HCN is probably formed by the pyrolysis of contemporary biomolecules and not prebiotic precursors (Ferris and Hagan, 1984; Johnson and Kang, 1971; Morikawa, 1978). HCN is the product of passing an electric discharge through gaseous mixtures which contain carbon at the oxidation levels of CH_4, CO and CO_2, nitrogen as NH_3 and N_2, hydrogen gas and water (Lemmon, 1970; Ferris and Hagan, 1984). Shock heating of comparable mixtures also generates HCN (Bar-Nun and Shaviv, 1975). Simple pyrolysis of CH_4-NH_3 at 1200°C is a preparative source of HCN (Koberstein, 1973). HCN and nitriles are formed by the reaction of CO, NH_3 and H_2 (or the corresponding deuterated derivatives) in a FTT reaction at around 500°C (Hayatsu *et al.*, 1968, 1972). HCN is one of the products of the pyrolysis of formamide and nitriles (Yasuda, 1967; Ashworth, 1971; Brunmuller *et al.*, 1986). Photolysis of mixtures of CH_4 and NH_3 (Ferris and Chen, 1975a) or CO and NH_3 (Ferris *et al.*, 1974) also generates HCN. A model of chemical processes in the atmosphere of the primitive Earth is consistent with HCN formation if CH_4 and NH_3 were present (Zahnle, 1986) but an alternative model with carbon in the CO or CO_2 oxidation state and nitrogen as N_2 is more widely accepted (see Chapter 1; Cogley and Henderson-Sellers, 1984; Kasting, 1988). HCN has also been detected in the atmospheres of Titan and Jupiter, in comets, and in the interstellar medium. Thus the ease of formation of HCN under a variety of conditions strongly suggests that it, or adducts of it with other organic molecules (e.g., the formation of $HOCH_2CN$ by reaction with CH_2O), were present in hydrothermal systems on the primitive earth. There it may have been complexed with iron as $Fe(CN_6)^{4-}$ or $Fe(CN)_6{}^{3-}$ which is bound to insoluble minerals (Arrhenius *et al.*, 1989; Kuma *et al.*, 1989). Alternatively the HCN may have been combined with formaldehyde as the cyanohydrin (Section 2.1.3).

2.1.2. Diaminomaleonitrile (DAMN)

The self condensation of HCN in mildly basic solution (Sanchez *et al.*, 1967) results in the formation of DAMN, a central intermediate in the formation of the purine ring (Ferris and Orgel, 1966) and HCN oligomers (Ferris and Edelson, 1978; Ferris and Hagan, 1984). DAMN will be formed if the concentration of HCN exceeds 0.01 M and the pH is in the 7-10 range (Sanchez *et al.*, 1967). HCN is hydrolyzed to formate faster than it is converted to DAMN if the HCN concentration is lower than 10^{-2} M. The effect of inorganic and organic substances on its formation has been investigated (Sanchez *et al.*, 1967).

DAMN formation is accelerated by the presence of formaldehyde (Schwartz and Goverde, 1982).

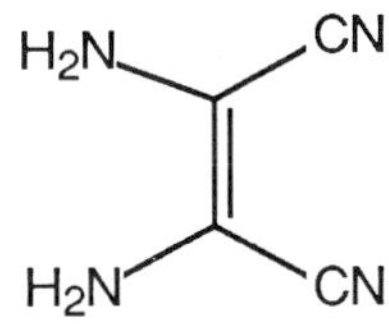

DAMN

2.1.3. Glycolonitrile and Other Cyanohydrins

HCN undergoes a facile addition to formaldehyde (CH_2O) (equation 1) and other aldehydes to form glycolonitrile (Walker, 1964). The reaction proceeds at cyanide concentrations varying between 10^{-2} to 10^{-4} M depending on the reactivity of the aldehyde used. Aminonitriles (e.g., glycine nitrile) are formed if the free NH_3 concentration is greater than 10^{-2} M (equation 2) (Miller, 1957).

Glycolonitrile dimerizes to 2-hydroxymethyl-5-aminooxazole at 0°C and pH 8.0 and forms polymers at 25° and pH 8 (equation 3) (Arrhenius, 1990, pers. communication).

$$HCN + CH_2O \longrightarrow HOCH_2CN \qquad (1)$$

$$CH_2O + NH_3 \longrightarrow \longrightarrow CH_2{=}NH \xrightarrow{HCN} NH_2CH_2CN \qquad (2)$$

glycinenitrile

$$2\ HOCH_2CN \longrightarrow$$ 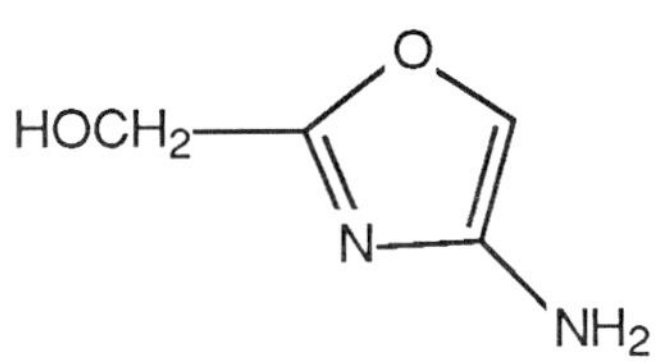 (3)

2-hydroxymethyl-5-aminooxazole

2.1.4. Simple Nitriles, Urea and Guanidine

2.1.4.1. Acetonitrile (CH_3CN)

CH_3CN is detected in comets (Donn, 1982), the interstellar medium (Mann and Williams, 1980) and is formed in FTT reactions from CO, NH_3 and H_2 (Hayatsu *et al.*, 1968, 1972), by the photolysis of mixtures of acetylene and NH_3 (Ferris and Ishikawa, 1988) and the photolysis of acetamide (CH_3CONH_2) (Spall and Steacie, 1957). No studies suggest that CH_3CN has a role in prebiotic synthesis so its presence is not indicative of processes leading to the origins of life but it may be formed in hydrothermal systems.

2.1.4.2. Cyanoacetylene (HC≡CCN)

HC≡CCN is formed when an electric discharge is passed through a CH_4/N_2 mixture (Sanchez *et al.*, 1966). It is present in the interstellar medium (Mann and Williams, 1980) and the atmosphere of Titan (Kunde *et al.*, 1981). HC≡CCN is a starting compound in a prebiotic synthesis of the pyrimidine ring system (Ferris *et al.*, 1968, 1974) and aspartic acid (Ferris *et al.*, 1968). It is readily converted to polymers in thermal and photochemical reactions (Ferris and Guillemin, 1990). Although cyanoacetylene may have had an important role in the origins of life, its high reactivity to nucleophilic addition suggests that it will be rapidly converted to a phosphate or other adducts in hydrothermal systems (Ferris *et al.*, 1970, 1974).

2.1.4.3. Cyanogen $(CN)_2$

$(CN)_2$ is formed from HCN photochemically (Airey and Dainton, 1966; Lohrmann, 1972), thermally (Harada and Fox, 1965), in electric discharges (Schavo and Winkler, 1959) and in FTT syntheses (Hayatsu *et al.*, 1968, 1972). It has been detected in the atmosphere of Titan (Kunde, 1981). It is also a very reactive compound and is not likely to be detected in hydrothermal systems but its formation may be inferred if cyanate ($N{=}C{=}O^-$) or urea ($H_2N\overset{\overset{\displaystyle O}{\|}}{C}NH_2$) are observed (Naumann, 1910; Lohrmann, 1972).

2.1.4.4. Cyanamide (NH_2CN)

Cyanamide, a product of the photolysis of ammonium cyanide (Lohrmann, 1972) or the electron irradiation of CH_4, NH_3 and H_2O (Schimpl *et al.*, 1965), has also been observed in the interstellar medium (Mann and Williams, 1980). It dimerizes to dicyanamide (cyanoguanidine) ($H_2N\overset{NH}{\overset{\|}{C}}NHCN$) on heating in the dry state or in solution and is hydrolyzed to urea in aqueous acid and base (Prager *et al.*, 1910). It may have effected the condensation of amino acids to oligopeptides or nucleotides to oligonucleotides on the primitive earth (Ibanez *et al.*, 1971).

2.1.4.5. Urea and Guanidine

The first laboratory synthesis of a biological compound from simple inorganic materials was accomplished by Wöhler (1828) when he formed urea by heating ammonium cyanate. Urea has been obtained in a number of prebiotic studies using a spark discharge, UV light, β-rays and X-rays as the energy sources. Urea is also formed from HCN, nitriles and cyanamide (NH_2CN). A recent review of the prebiotic formation of urea is given by Navarro-Gonzalez *et al.* (1989). Guanidine is formed by the thermal (Lowe *et al.*, 1963; Labadie *et al.*, 1968) and photochemical reactions (Lohrmann, 1972) of NH_4CN and by the reaction of cyanamide with NH_3 (Blair and Braham, 1924).

Gamma irradiation of aqueous urea generates CO_2, NH_3 and H_2 as the main products along with smaller amounts of oxalic [$(CO_2H)_2$] and malonic [$(CH_2(CO_2H)_2)$] acids (Navarro-Gonzalez *et al.*, 1989). Thermolysis of urea at temperatures as low as 120°C gives the triazine, cyanuric acid, via biuret (equation 4) (Smolin and Rapoport, 1959a). The yield increases as the temperature is raised to 279°C (Smolin and Rapoport, 1959a; Okazaki *et al.*, 1974). Pyrolysis of urea at 350-600°C yields isocyanic acid, presumably via cyanuric acid (equation 4) (Takahasi and Ozako, 1974). Pyrolysis of urea in the presence of ammonia at 310°C (equation 5), pyrolysis of guanidine or cyanamide yields melamine (equations 6 and 7) (Smolin and Rapoport, 1959b).

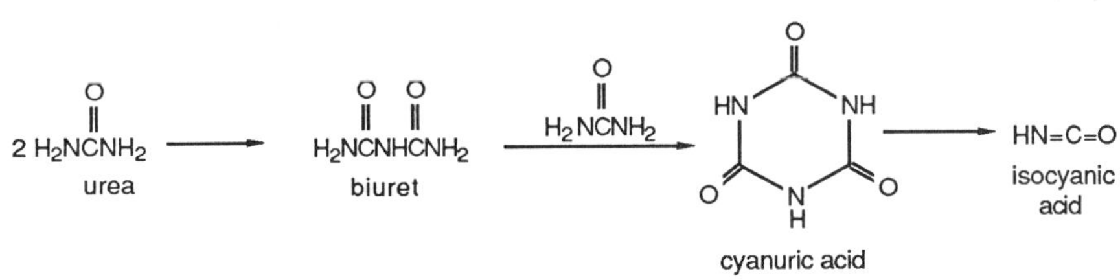

(4)

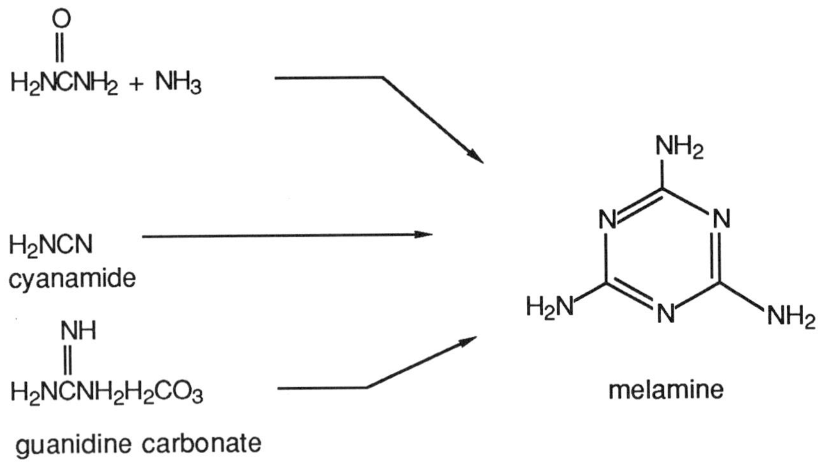

(5)

(6)

(7)

Diglycine and N-carbamylglycylglycine (equation 8) are formed when a mixture of urea and glycine are heated at 105°C in aqueous solution containing a mixture of metal ions believed to have been present in the oceans of the primitive earth. Some of the same products are formed on heating mixtures of urea and formaldehyde under the same reaction conditions (Sakurai and Yanagawa, 1984).

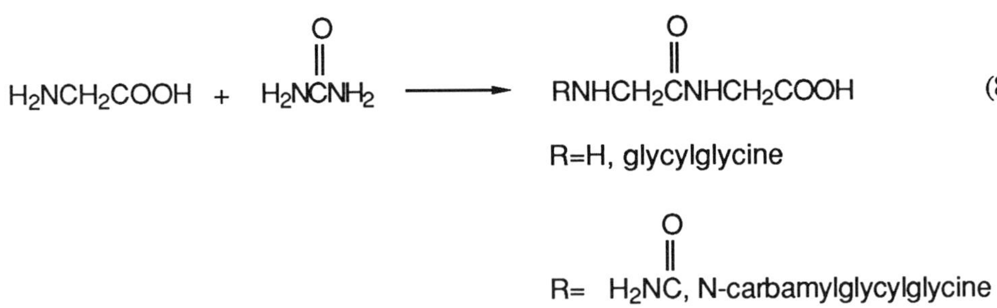

(8)

2.2. FORMALDEHYDE AND OTHER ALDEHYDES

Formaldehyde (CH_2O) is a major product resulting from passing an electric discharge through mixtures of gases which simulate primitive atmospheres (Schlesinger and Miller, 1983). The yields of higher aldehydes decrease approximately in the order of their increasing complexity. Helium atom irradiation of mixtures of CO_2, H_2, Fe^{2+} in a cyclotron yields traces of CH_2O and formic acid (Garrison *et al.*, 1951).

CH_2O and higher aldehydes are also formed by the photolysis of H_2O in the presence of CH_4 (Ferris and Chen, 1975b; Bar-Nun and Chang, 1983), the photolysis of CO-H_2O mixtures (Park and Getoff, 1988) and the photolysis of methanol (Allamandola *et al.*,

1988). It has been tentatively identified as a product of the photolysis of ferrous carbonate (Joe *et al.*, 1986) and β-irradiation of calcium carbonate (Albarran *et al.*, 1987). Formaldehyde has been detected in the interstellar medium (Mann and Williams, 1980) and the presence of polyoxymethylene, a polymeric form of formaldehyde, on Comet Halley has been inferred, but not proven, from mass spectral data (Huebner, 1987).

Aldehydes are reactive compounds and they may form adducts with other substances in hydrothermal systems. The reaction with HCN to form cyanohydrins (equation 1) and aminonitriles (equation 2) has already been discussed. The self condensation of formaldehyde proceeds in the presence of divalent metal ions to yield glycolaldehyde and eventually a mixture of sugars from trioses to hexoses (equation 9) (Miller and Orgel, 1974). A concentration of 0.01 M CH_2O or greater is required for the detection of sugars when alumina, kaolinite (Gabel and Ponnamperuma, 1967) and carbonate-apatite (Reid and Orgel, 1967) are used as catalysts. A 0.001 M formaldehyde solution apparently yielded only glycolaldehyde and no higher molecular weight compounds (Gabel and Ponnamperuma, 1967).

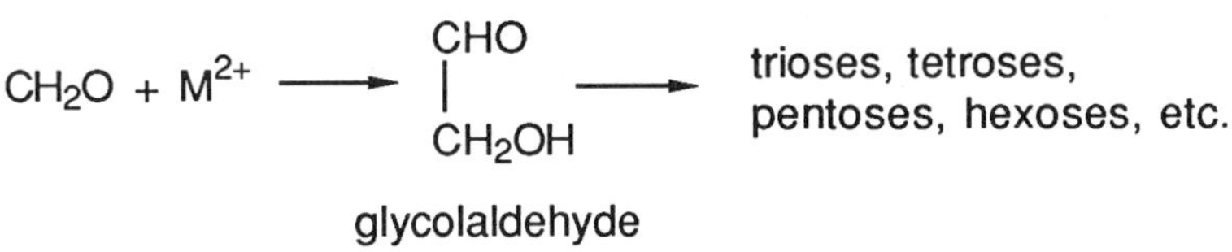

(9)

The complex mixture of sugars formed from formaldehyde condensation reactions was considered to be a major flaw in the proposed prebiotic syntheses of RNA (Reid and Orgel, 1967; Ferris, 1987; Joyce *et al.*, 1987). Recent results from Eschenmoser's group (Müller *et al.*, 1990) have demonstrated that a very limited array of sugars, with ribose as the major product, are obtained in the condensation reaction of glycolaldehyde phosphate and formaldehyde in strongly alkaline solution (equation 10).

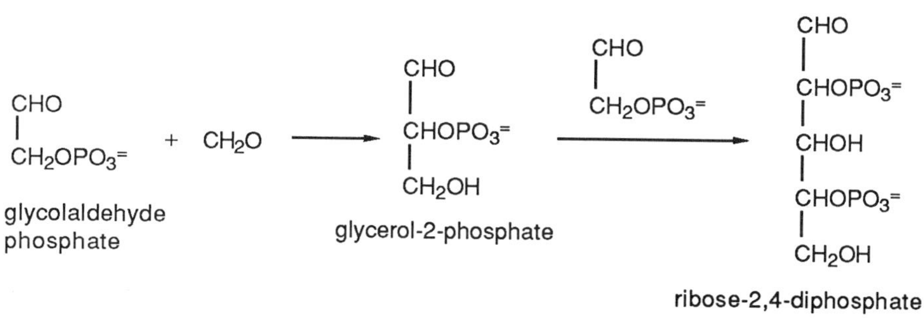

(10)

2.3. FORMIC, ACETIC AND OTHER CARBOXYLIC ACIDS

There are a host of different ways carboxylic acids may have accumulated on the primitive Earth. They have been reported to be the products of passing an electric discharge through CH_4 and H_2O (Yuen *et al.*, 1981; Allen and Ponnamperuma, 1967), photolysis of CH_4-H_2O mixtures (Ferris and Chen, 1975b) and photolysis of CO, H_2O, NH_3 under conditions which simulate reactions on grains in the interstellar medium (Agarwal *et al.*, 1985). Carboxylic acids and hydrocarbons are formed in the FTT reaction of CO and D_2 over a catalyst of meteoritic iron with either carbonaceous chondrite or ferric oxide and potassium carbonate as a promoter at 300-450°C (Nooner *et al.*, 1976; Leach *et al.*, 1978). Fatty acids of carbon number 6-18 were detected with a maximum yield at C_9 to C_{10}. There was no preference for odd versus an even number of carbon atoms but the linear fatty acids predominated over the branched-chain fatty acids by factors ranging between 1.7-5.0. Autoradiolysis of ^{14}C-$CaCO_3$ (-β) yields formic (HCO_2H), oxalic, glyoxylic (HO_2CCHO), acetic and glycolic acids (HO_2CCH_2OH) (Albarran *et al.*, 1987). The identity of the acids requires confirmation by mass spectrometry or related techniques. Gamma irradiation of aqueous urea yields oxalic and malonic acids (Navarro-Gonzalez *et al.*, 1989) and oxalic acid is formed in processes leading to HCN oligomers from HCN (Ferris *et al.*, 1978). The corresponding acids are formed by hydrolysis of amides and nitriles; for example, HCN is hydrolyzed to formic acid at neutral pH (Sanchez *et al.*, 1967). Aliphatic, carboxylic acids (C_1-C_8); hydroxy acids (C_1-C_5) and dicarboxylic acids (C_2-C_9) were detected in the Murchison meteorite (Cronin *et al.*, 1988) and formic acid has been detected in the interstellar medium (Mann and Williams, 1980).

Pyrolysis of acids at 250-550°C in the absence of water and the presence of ThO_2, Ga_2O_3, In_2O_3, MnO, or mordenite (a clay), generates ketones (Fuki *et al.*, 1974; Cornejo *et al.*, 1979; Sosnina, 1977). Heating formic acid at 282-415°C in the presence of Al_2O_3, SiO_2 and other inorganics results in the formation of CO, CO_2, H_2 and H_2O (Schwab and Schwab-Agallidis, 1949). This decomposition proceeds on other mineral surfaces at temperatures as low as 150°C-200°C (Hinshelwood and Topley, 1923). Gamma irradiation of acetic acid (CH_3CO_2H) yields a mixture of dicarboxylic and tricarboxylic acids (Negron-Mendoza and Ponnamperuma, 1982). The main photochemical reaction of carboxylic acids is the formation of alkane and CO_2 (equation 11) (Wilkerson and Guillory, 1977; Ferris and Joshi, 1979).

$$RCO_2H \xrightarrow{h\nu} RH + CO_2 \tag{11}$$

2.4. AMINO ACIDS AND AMINONITRILES

A voluminous literature developed on the prebiotic synthesis of amino acids in the wake of Miller's (1953) pioneering study. Amino acids are formed by the action of electric discharges on almost any gaseous mixture of carbon, hydrogen and oxygen compounds so long as it is more reduced than CO_2, N_2 and H_2O (Schlesinger and Miller, 1983). Aminonitriles appear to be intermediates in the formation of some of the amino acids (Ponnamperuma and Woeller, 1967). Amino acids are produced when mixtures of simple carbon, nitrogen and oxygen compounds are subjected to shock heating (Bar-Nun *et al.*, 1970), ultraviolet radiation, ionizing radiation, high temperatures (Yanagawa *et al.*, 1984a; Harada and Fox, 1964; Lawless and Boynton, 1973; Lemmon, 1970) or proton irradiation (Kobayashi *et al.*, 1990). Glycine has been detected as a product of the photolysis of CO, H_2O, NH_3 at 10 K in a reaction designed to simulate synthesis in the interstellar medium (Briggs *et al.*, 1991). The self condensation of HCN in aqueous solution yields oligomeric products which release amino acids upon hydrolysis with 1 N HCl or at pH 8.5 (Oró and Kamat, 1961; Lowe *et al.*, 1963; Ferris *et al.*, 1978). These oligomers are not polypeptides since the amino acids formed constitute a small percent of the starting oligomers (Ferris *et al.*, 1978). Amino acids are also formed by the pyrolysis of formamide (Harada, 1967) and in reaction sequences starting from the sugars glucose or erythrose or from malic acid (Shen *et al.*, 1987, 1990). Amino acids are produced in small amounts in FTT reactions of CO, D_2 and ND_3 over meteoritic iron or iron oxide or montmorillonite and alumina at 200-1000°C (Yoshino *et al.*, 1971; Hayatsu *et al.*, 1971). Tyrosine and histidine were detected as reaction products in addition to many of the same protein amino acids found in the electric discharge studies. Tentative identification of some non-protein amino acids was noted but no GC/MS confirmation was reported. Small amounts of amino acids have been detected in the Murchison and other meteorites. Seventy-four amino acids extracted from the Murchison meteorite have been conclusively identified (Cronin *et al.*, 1988). So far, amino acids have not been identified in the interstellar medium, however, the formation of glycine in simulation studies is suggestive of their presence (Briggs *et al.*, 1991).

Pyrolysis of aminonitriles at 240°-300°C in the absence of water gives HCN and the corresponding imine (Guillemin and Denis, 1985, 1988). Pyrolysis of amino acid mixtures at 160°-200°C for 4-6 h gives UV absorbing compounds, some of which have been identified as pteridines (Heinz *et al.*, 1979). Diketopiperazine formation and decarboxylation was observed on heating amino acids in the presence of silica (Basiuk *et al.*, 1991). When glycine is heated with a mixture of $MnCO_3 \cdot Mn(OH)_2$ and Al_2O_3 at 240°C under nitrogen and the product is hydrolyzed with 6 N HCl, ten protein and non-protein amino acids were separated and identified by nuclear magnetic resonance (Ivanov and Slavcheva, 1977). When glycine and alanine were heated between 310°-340°C HCN is formed. None was detected when leucine, phenylalanine and tyrosine were heated in this temperature range (Heyns and Pavel, 1957). Pyrolysis of sixteen amino acids at 700-1000°C (Johnson and Kang, 1971) or heating nitrogen-containing materials such as nylon at temperatures greater than 600°C generates HCN (Morikawa, 1978).

2.5 PURINES, PYRIMIDINES AND OTHER HETEROCYCLIC COMPOUNDS

The synthesis of adenine* by the electron bombardment of a mixture of CH_4, NH_3, H_2O and H_2 was reported by Ponnamperuma *et al.* (1963). The formation of purine and pyrimidine bases by passage of an electric discharge through CH_4, N_2 and H_2O was reported in a meeting abstract (Kobayashi *et al.*, 1986) but the experimental details have not appeared in a refereed journal. Proton irradiation of a mixture of CO, N_2 and H_2O yields imidazole along with amino acids (Kobayashi *et al.*, 1990). The FTT reaction of CO, D_2 and ND_3 at about 500°C, in the presence of a mixture of powdered iron meteorite and alumina or ferric hydroxide and alumina, yields an array of biologically important heterocycles including adenine, guanine, xanthine, thymine and uracil (Hayatsu *et al.*, 1972, 1968). In addition, the formation of dimethyluracil, the triazine melamine, pyridine, pteridines and pyrroles was also observed. The formation of the triazines ammeline and cyanuric acid was suggested by chromatography but it was not possible to confirm these findings by mass spectrometry.

Adenine and substituted imidazoles are formed by the self condensation of mildly alkaline solutions of 0.1-15 M NH_4CN (Oró and Kimball, 1961, 1962). No free adenine was detected in the condensation of 0.1 M HCN but it was released on hydrolysis of the oligomeric reaction products (Ferris *et al.*, 1978; Ferris and Hagan, 1984). It has been suggested that the 8-cyano or 8-carbamyl derivative of adenine is formed first and this undergoes hydrolysis and decarboxylation to adenine (Voet and Schwartz, 1983; Schwartz *et al.*, 1984). The formation of 8-hydroxymethyladenine is observed if the condensation of HCN is performed in the presence of formaldehyde (Schwartz and Bakker, 1989). The pyrimidines 4,5-dihydroxypyrimidine, 5-hydroxyuracil, orotic acid (Ferris *et al.*, 1978) and uracil (Voet and Schwartz, 1982) are also formed on hydrolysis of products of the self-condensation of HCN.

Diaminomaleonitrile is a readily detectable intermediate in the self condensation of HCN (Section 1.2). It undergoes a facile photochemical rearrangement to 4-aminoimidazole-5-carbonitrile (equation 12) (Ferris and Orgel, 1966), a compound which reacts with HCN or simple derivatives of HCN to yield adenine, guanine, diaminopurine and hypoxanthine (Sanchez *et al.*, 1968).

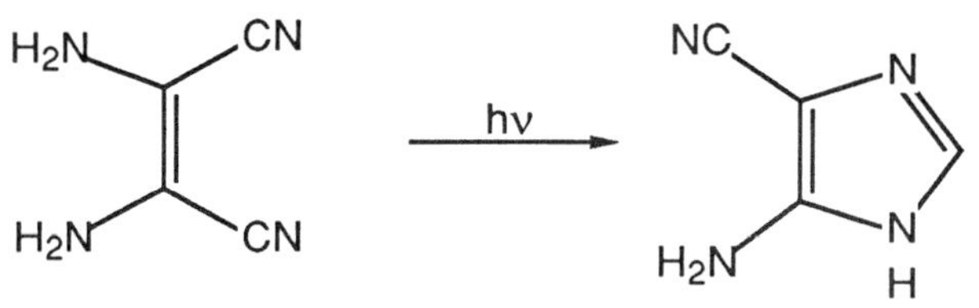

(12)

4-aminoimidazole-5-carbonitrile

*The structures of the heterocyclic compounds discussed in this section are given in Table 1.

TABLE 1
Potential prebiotic heterocyclic compounds

Compound	Structure	Compound	Structure	Compound	Structure
Adenine	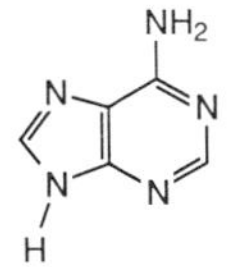	1,3-Dimethyluracil	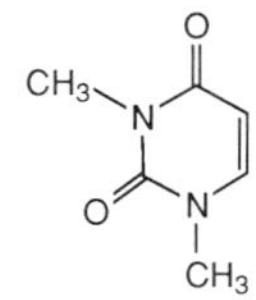	Pteridine	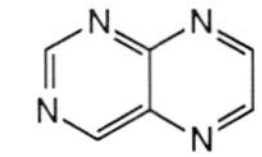
4-Aminoimidazole-5-carbonitrile	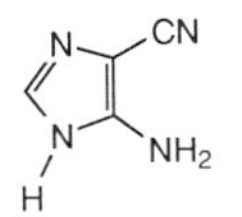	5,6-Dimethyluracil	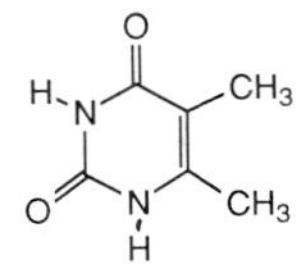	Pyridine	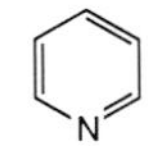
4-Aminoimidazole-5-carboxamide	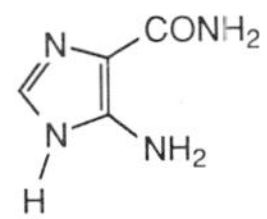	Guanine	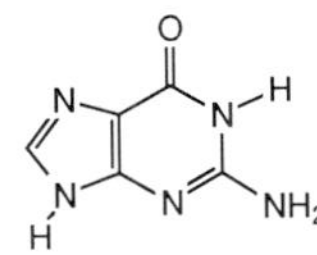	Pyrrole	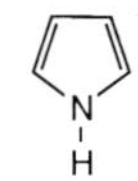
8-Carbamyladenine	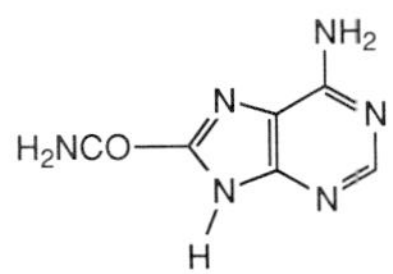	8-Hydroxymethyladenine	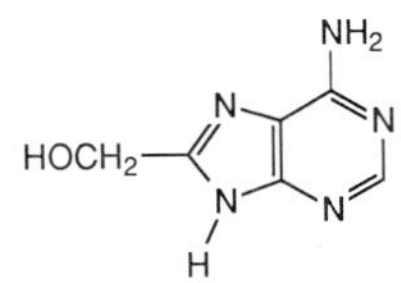	Thymine	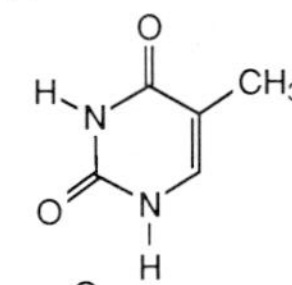
Cytosine	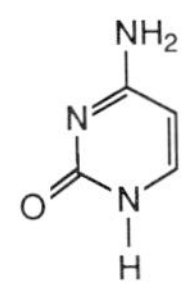	5-Hydroxyuracil	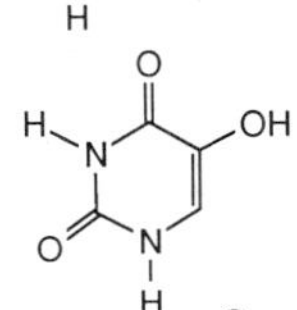	Uracil	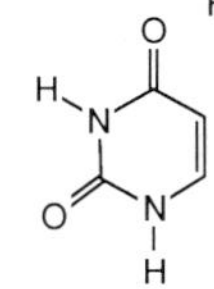
2,4-Diaminopyrimidine	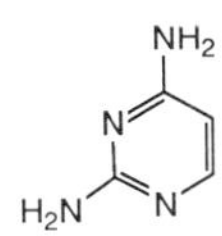	Hypoxanthine	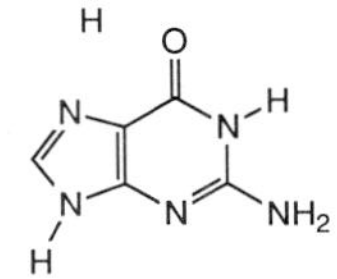	Melamine	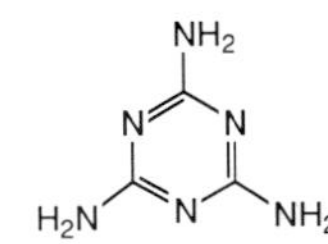
				Orotic Acid	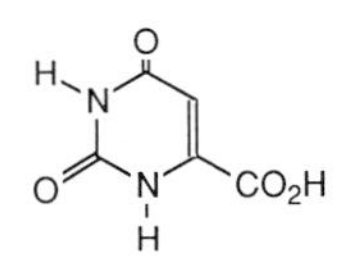

The pyrimidine ring system has also been formed from cyanoacetylene (Section 1.4.2). Reaction of cyanoacetylene with cyanate ($N{=}C{=}O^-$) yields cytosine, a compound which is hydrolyzed to uracil (Ferris *et al.*, 1968). Alternatively, the hydrolysis of cyanoacetylene yields cyanoacetaldehyde which reacts with guanidine to yield diaminopyrimidine which, in turn, is hydrolyzed to cytosine and uracil (equation 13) (Ferris *et al.*, 1974b).

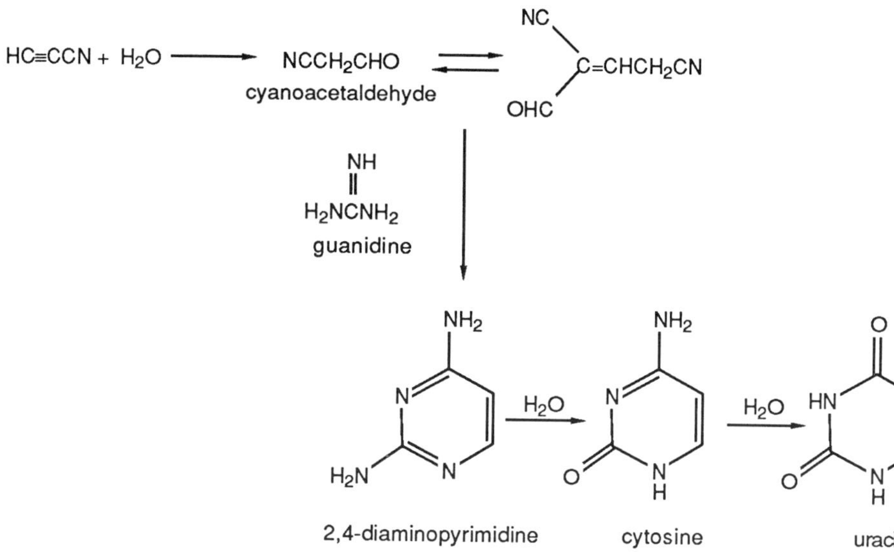

(13)

Adenine, guanine, xanthine, hypoxanthine and uracil are found in small amounts in the Orgueil, Murray and Murchison meteorites (Hayatsu *et al.*, 1975; Van der Velden and Schwartz, 1977; Stoks and Schwartz, 1979, 1981; Cronin *et al.*, 1988).

2.6. SULFUR COMPOUNDS

The presence of an abundance of inorganic sulfides suggests that organic sulfur compounds may have had a central role in the organic chemistry in hydrothermal systems (Shock, Chapter 5). The observation of heterocyclic sulfur compounds in contemporary hydrothermal systems has been discussed by Simoneit in Chapter 4. There have been very few studies on the formation of biomolecules containing sulfur under prebiotic conditions (for a review see Raulin and Toupance, 1977; see Table 2 for the structures of the compounds discussed in this section). When a Miller electric discharge experiment is performed in the presence of H_2S, methionine is observed as a reaction product (van Trump and Miller, 1972). Cysteic acid, taurine and cystamine were detected when CH_4, NH_4OH

and H_2S were irradiated with electrons. It was reported that cysteine and cystine were formed but were oxidized to cysteic acid (Choughley and Lemmon, 1966). Photolysis of mixtures of CH_4, C_2H_6, NH_3, H_2S and H_2O yields cysteine and cystine (Sagan and Khare, 1971; Becker *et al.*, 1974). Thus it appears likely that thiols and sulfur containing amino acids were available on the primitive earth and possibly in hydrothermal systems.

TABLE 2
Sulfur containing amines and amino acids

Compound	Formula	Compound	Formula
methionine	$CH_3SCH_2CH_2CH(NH_2)COOH$	cysteine	$HSCH_2CH(NH_2)COOH$
cystine	$HOOCCH(NH_2)CH_2SSCH_2CH(NH_2)COOH$	cysteic acid	$HO_3SCH_2CH(NH_2)COOH$
taurine	$HO_3SCH_2CH_2NH_2$	cystamine	$H_2NCH_2CH_2SSCH_2CH_2NH_2$

The possible role of hydrogen sulfide, thiols and thioamides on prebiotic reactions in aqueous solution has been studied only briefly. HS^- inhibits the tetramerization of HCN to DAMN (Section II.A.2) (Sanchez *et al.*, 1967). The reaction of DAMN with thioformamide yields 4-aminoimidazole-5-carboxamide and 5-aminothiazole-4-nitrile (equation 14) (Sanchez *et al.*, 1967). The addition of ethanethiol to activated double bonds and nitriles has also been observed (Raulin and Toupence, 1977) (equations 15, 16). Presumably H_2S will undergo similar addition reactions in hydrothermal systems.

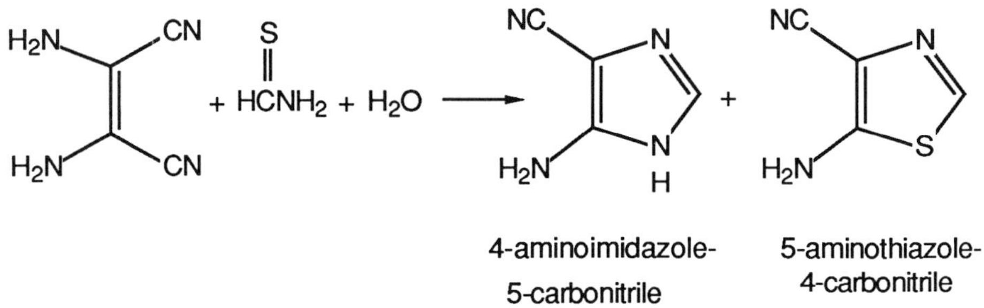

(14)

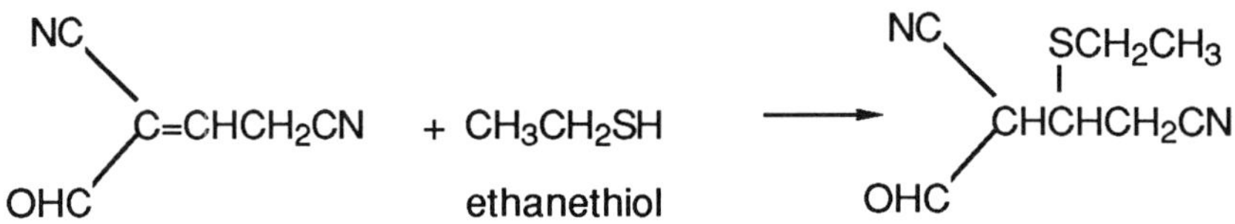

(15)

(16)

The central role of thioesters in contemporary biochemistry prompted a series of studies of the possible role of these compounds in prebiotic processes by A. L. Weber. He observed the photochemical formation of the thioester N,S-diacetylcysteine starting from the corresponding disulfide (equations 17) (Weber, 1981a).

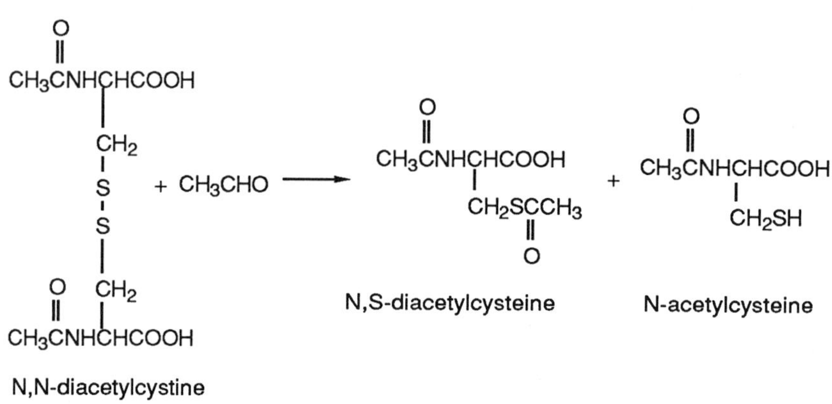

(17)

The direct formation of a thioester in a thermal reaction from a ketoaldehyde and N-acetylcysteine has also been observed (equation 18) (Weber, 1982a). Ketoaldehydes may have been formed from sugars on the primitive earth.

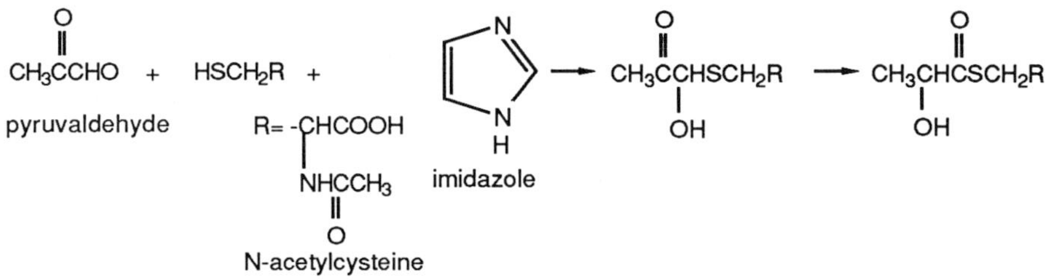

(18)

The pyrophosphate bond, which provides the energy to drive many chemical reactions in contemporary biochemical systems, can be formed from the reaction between phosphate and thioesters (Weber, 1981b). Reaction of the thioester N,S-diacetylcysteamine with phosphate in the presence of imidazole yields pyrophosphate (equation 19). The insoluble phosphate mineral, hydroxyapatite, can be substituted for soluble phosphate in this reaction (Weber, 1982b). Thus, such a reaction of thioesters may generate pyrophosphate in hydrothermal systems.

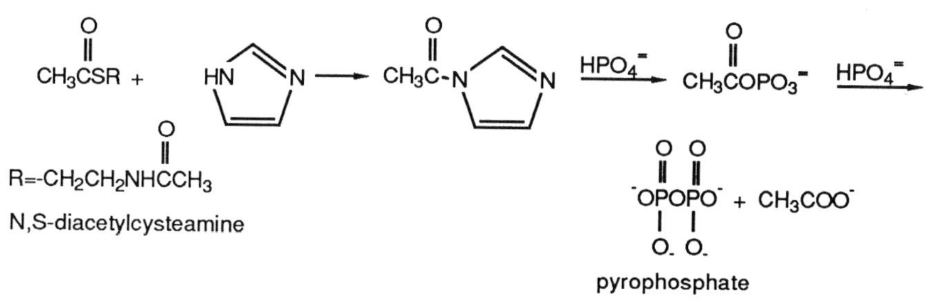

(19)

3. Organic Chemistry in Near- to Supercritical Water

A key point of dispute in the discussion of the role of hydrothermal systems in the origins of life is whether biological molecules survived the high temperatures present there. Miller and Bada (1988) assert that "...the high temperature in vents would not allow synthesis of organic compounds, but would decompose them, unless the exposure time at vent temperatures was short." On the other hand, Shock (1990) reached the conclusion that amino acids attain a metastable equilibrium at high temperatures based on his analysis of the experimental data of Miller and Bada (1988). Furthermore, the properties of supercritical water may favor the formation of complex molecules in hydrothermal systems (Simoneit, Chapter 4). Some insight into the possible prebiotic chemistry in hydrothermal systems can be obtained by inspection of organic chemical reactions investigated in supercritical water (Shaw *et al.*, 1991). It should be noted that the purpose of the studies discussed below was to investigate organic reactions in supercritical water and not to simulate the conditions in hydrothermal systems.

3.1. PROPERTIES OF SUPERCRITICAL WATER

Water is considered to be supercritical if its temperature is above the critical temperature (374°C) and its pressure is above the critical pressure (3200 psia or 22.1 MPa) (Note - 1 psia = 14.7 psi, 2 psia = 15.7 psi). The physical properties of water vary dramatically near the critical pressure and temperature. The variation in the dielectric constant and dissociation constant with the density of water (pressure) at 390°C is shown in Table 3 (Penninger and Kolmschate, 1989).

TABLE 3
Some Physical Properties of Water[a]

Density gm/cm^3	Temperature (°C)	ε	K_w
0.997[b]	25	78.5[b]	1×10^{-14} [b]
0.25	390	3.59	8.91×10^{-18}
0.35	390	5.89	9.91×10^{-16}
0.45	390	8.20	3.34×10^{-14}
-	350[c]	-	$1 \times 10^{-11.8}$

[a]Adapted from Penninger and Kolmschate, (1989)
[b]At 25°C and 1 psia (Weast, 1964)
[c]Marshall and Franck (1981)

At the critical temperature and pressures lower than the critical pressure, water has the properties of steam where it has a low dissociation constant (K_w) and dielectric constant (ε). At pressures above the critical pressure it behaves like liquid water except its dielectric constant and viscosity is much lower because there is less hydrogen bonding between the water molecules (Table 1). The dissociation constant of the water decreases as the density decreases (Table 3).

3.2. REACTIONS IN NEAR- TO SUPERCRITICAL WATER

At water densities greater than 0.4 g cm^{-3} the dielectric constant of water is sufficiently high so that reactions which proceed by ionic or polar transition states are observed. When the density of water is 0.2 g cm^{-3} or less it no longer stabilizes charged species by solvation so free radical reactions are favored over ionic processes. The solubility of organics is greatly enhanced and of inorganics suppressed in supercritical water as compared with ordinary water (Connolly, 1966; Josephson, 1982). It has been suggested that the solvent properties of water at 300°C resemble those of acetone (Siskin and Katritzky, 1991).

The different pathways for the reaction of organics in supercritical water are apparent in the data shown in Table 4. For example, $PhCH_2OH$ is obtained by hydrolysis of the ether grouping in $(PhCH_2)_2O$ when the density is greater than 0.4, while PhCHO and PhH are products of the free radical decomposition when the density is less than 0.4. In general, the greater the liquid-like properties of the water, the more predominant are the hydrolytic reactions.

The ether and amine functional groups in the first seven compounds listed in Table 4 are significantly more stable to hydrolysis than that of the peptide bond linking proteins (equation 20) or the phosphodiester bond linking nucleotides (equation 21). It can be inferred from these data that the reactive functional (e.g., peptides and nucleotide bonds) groups in biomolecules may undergo rapid thermolysis and hydrolysis (equations 20, 21) in hydrothermal systems. This appears to be confirmed by the facile decomposition of benzyl nitrile to benzene and 1,3-butanediol and glycerol to olefins and aldehydes (Table 4). Since nitriles are important products of prebiotic syntheses, it is likely that they would undergo rapid decomposition in supercritical water. The rapid reaction of 1,3-butanediol and glycerol suggests that the alcohol groups in sugars and nucleotides would also decompose rapidly.

$$R\overset{\overset{\displaystyle O}{\|}}{C}NHR' + H_2O \longrightarrow RCOOH + R'NH_2 \qquad (20)$$

peptide bond

$$RO\underset{\underset{\displaystyle O_-}{|}}{\overset{\overset{\displaystyle O}{\|}}{P}}OR + H_2O \longrightarrow RO\underset{\underset{\displaystyle O_-}{|}}{\overset{\overset{\displaystyle O}{\|}}{P}}O^- + ROH \qquad (21)$$

phosphodiester bond

TABLE 4

Some Representative Organic Reactions in Supercritical Water

Reactant	Major Products		Reference
	$\rho > 0.4$	$\rho < 0.4$	
$PhCH_2OCH_2Ph$	$PhCH_2OH$	PhH, PhCHO	a
$(PhCH_2CH_2O)_2$	NR	$PhCH_3$, $PhCH{=}CH_2$	a
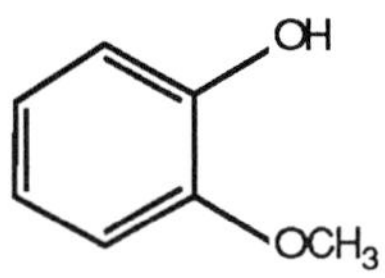	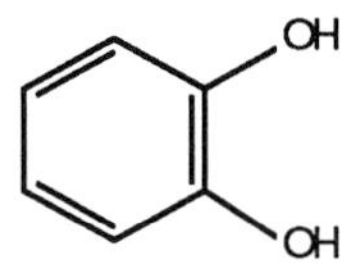 CH_3OH	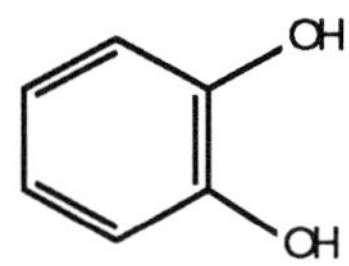 CH_4, Char	
$PhNHCH_2Ph$	$PhNH_2$, $PhCH_2OH$	$PhNH_2$, $PhCH_3$	a
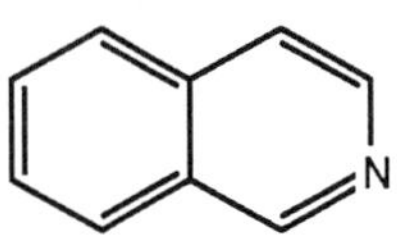	nd	$PhCH_3$, $PhCH_2CH_3$	b
$PhNH_2$	nd	PhOH, NH_3	b
$PhCH_2NH_2$	$PhCH_3$, $Ph_2CH{=}CHPh$, $(PhCH_2)_2$	$PhCH_3$, PhCHO, NH_3	c
PhCN	nd	PhH	b
$HOCH_2CH_2CH_2OH$	CH_2O, CH_3CHO, C_2H_5OH, CH_3COCH_3, polymers	CH_2O, $CH_2{=}CHCH_3$	d
$HOCH_2CH(OH)CH_2OH$ + H^+ (0.005 M)	$CH_2{=}CHCHO + CH_3CHO$	$CH_2{=}CHCHO$	e

nd = Not determined
[a]Townsend *et al.*, 1988
[b]Houser *et al.*, 1986
[c]Houser *et al.*, 1989
[d]West and Gray, 1987
[e]Ramayya *et al.*, 1987

Synthetic processes have been reported in reactions performed in supercritical water which give some hope for the formation of more complex molecules under these conditions. The aldol condensation of 0.5 M acetaldehyde was observed at 385°C in a flow reactor to give a 37% yield of crotonaldehyde (Ramayya *et al.*, 1987) (equation 22). A mixture of acetaldehyde and formaldehyde generates acrolein and crotonaldehyde under the same reaction conditions (equation 23). These studies and other examples listed in Table 1 show that polyhydroxy alcohols are converted to aldehydes and ketones in supercritical water. The aldehydes and ketones may condense with themselves and with HCN and NH_3 to give aminonitriles which hydrolyze to amino acids.

$$2\,CH_3CHO \longrightarrow [CH_3\overset{OH}{\overset{|}{C}}HCH_2CHO] \longrightarrow CH_3CH{=}CHCHO + H_2O \qquad (22)$$

acetaldehyde — crotonaldehyde

$$CH_3CHO + CH_2O \longrightarrow [CH_3\overset{OH}{\overset{|}{C}}HCHO] \longrightarrow CH_2{=}CHCHO + H_2O \qquad (23)$$

acrolein

The second synthetic example is the partial conversion of mixture of 1 M ethanol with 1 M acetic acid to ethyl acetate in a flow reactor where the residence time was less than 1 minute. About half of the ethanol that reacted was converted to ethyl acetate and most of the remainder was converted to ethylene and ethane (Antal *et al.*, 1987). The efficient formation of ethyl acetate in water, a reaction that requires anhydrous conditions and a strong acid catalyst at normal temperatures and pressures is significant (equation 24). This finding suggests that related condensations to form peptide or phosphodiester bonds from the corresponding monomers may occur in supercritical water. It should be emphasized that 1 M concentrations of reactants were used - higher concentrations than would be expected to occur under most prebiotic conditions. Also, the product ethyl acetate has much fewer reactive groups than do polypeptides or polynucleotides so its rate of decomposition in supercritical water may be appreciably slower than would the more complex biomolecules.

$$CH_3COOH + CH_3CH_2OH \rightleftharpoons CH_3\overset{O}{\overset{\|}{C}}OCH_2CH_3 + H_2O \qquad (24)$$

ethyl acetate

The decomposition of lactic acid to an array of other products is an example of the variety of reaction pathways available to a simple multifunctional molecule in supercritical water (Mok *et al.*, 1989). Decomposition to acetaldehyde followed by the further reaction of the acetaldehyde is shown in equation 25. Elimination of water to form acrylic acid followed by its hydrogenation or decarboxylation is shown in equation 26.

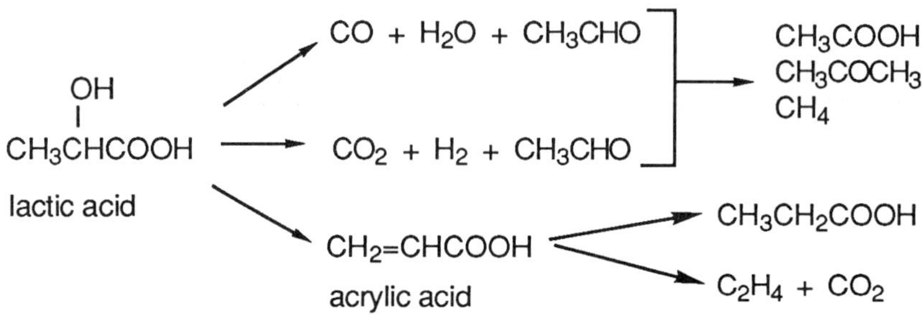

(25)

(26)

4. Fischer-Tropsch Type Reactions (FTT) in Hydrothermal Systems

The reaction conditions in hydrothermal systems precludes the occurrence of many processes discussed in Section 1. For example, it is clear that reactions driven by photons or electric discharges will not be observed. Heat and ionizing radiation were probably the main energy sources driving chemical processes in hydrothermal systems on the primitive Earth. The prebiotic chemistry associated with ionizing radiation has been discussed in Section 2 and these chemical processes should not differ appreciably in hydrothermal systems. The thermally driven FTT reactions merit further discussion because the reaction conditions in hydrothermal systems are different from those described in Section 2 which were designed to simulate chemical events in the solar nebula.

It is possible to prepare amino acids, heterocyclic compounds, and hydrocarbons and fatty acids with limited chain branching by the FTT reaction of H_2, CO and NH_3 (Hayatsu and Anders, 1981). FTT reactions are unique in providing a source of linear fatty acids essential for the formation of the bilayer membranes present in cell walls. It has not been possible to propose a prebiotic synthesis of linear fatty acids because there was no known place on the primitive Earth where both high temperatures and pressures required for FTT processes were present. The requisite conditions are found in hydrothermal systems.

The standard FTT reactions are usually performed using iron, nickel, cobalt and ruthenium catalysts under anhydrous reaction conditions (Anderson, 1958; Falbe, 1980). Platinum metals, Ir, Pd, Pt, Rh and Os, exhibit lower catalytic ability (Falbe, 1980). Basic compounds such as K_2CO_3 and alumina promote some FTT reactions while H_2S and possibly H_2O serve as inhibitors (Asinger, 1968; Storch *et al.*, 1951; Anderson, 1956; Falbe, 1980). Commercial FTT reactions has been optimized for the synthesis of hydrocarbons from CO and H_2. Anders and coworkers (Hayatsu and Anders, 1981) extended the FTT reaction to biomolecules by adding NH_3 to the mixture of gaseous reactants so that nitrogen-containing molecules are produced. Fatty acids are also formed in FTT reactions run under the conditions established by Anders and coworkers but without NH_3 as a reactant (Nooner and Oró, 1979).

The FTT reaction may proceed in hydrothermal systems because the reactants are heated at high temperature and pressure in the presence of minerals. The CO needed may come from degradation of formic acid or its derivatives, formaldehyde (Section 2) or from equilibration processes in magma (Lilley *et al.*, 1982). The H_2 is formed from redox reactions of metals with water and the NH_3 is from the decomposition of nitrogenous organics or by the hydrogenation of nitrogen. Metallic iron, the standard catalyst for FTT reactions, has not been detected in contemporary hydrothermal systems, but it may have been present in such systems on the primitive earth. In addition, a variety of other metals and minerals can also serve as catalysts.

Miller and Bada (1988) concluded that the FTT reaction was not responsible for the synthesis of organic compounds in hydrothermal systems because the process is inhibited by H_2S and H_2O. It is proposed that hydrogen sulfide is an inhibitor because sulfide is chemisorbed to the surface iron atoms responsible for catalysis. The only way to regenerate the catalyst is to dissolve it in HNO_3 and then reprecipitate the metal. The inhibition of iron catalysis by H_2S is permanent, i.e., it cannot be reversed by hydrogen or other FTT reactants (Anderson, 1956). The extent of H_2S poisoning depends on the ratio of H_2S to catalytic sites on the metal. The catalyst will continue to function until most of the catalytic sites have H_2S bound to them. In hydrothermal systems fresh metallic and mineral surfaces are constantly generated deep in the oceanic crust or where fluid mixing occurs (Siskin and Katritzky, 1991). Therefore, it is possible that FTT reactions took place in the hydrothermal systems of the primitive Earth.

There are no data on the role of H_2O as an inhibitor in FTT reactions but it is known to be a reversible inhibitor of the related reduction of nitrogen to ammonia on iron catalysts (Storch, 1951; Anderson, 1956). The reduction of nitrogen is also reversibly inhibited by CO, CO_2, O_2 and SO_2 and is permanently inhibited by H_2S. The observed inhibition by H_2S and CO suggests that the catalysis of nitrogen reduction is comparable to that of CO reduction and oligomerization. If this is correct then it is likely that H_2O is a reversible inhibitor of FTT reactions. What is not clear is whether supercritical water binds to the catalyst strongly enough to inhibit the binding and subsequent conversion of N_2 to NH_3 and CO to hydrocarbons and nitrogen-containing molecules. The reversible nature of water inhibition suggests the temperature may be sufficiently high in the hotter regions of these systems so that water may not bind to the catalyst and block the sites required for CO reduction.

It is of particular interest that the FTT reaction apparently does not require a metal catalyst but can also proceed on mineral surfaces. There are several reports of efficient syntheses proceeding on silica (Asinger, 1968), montmorillonite and montmorillonite-alumina mixtures (Yoshima *et al.*, 1971; Anders *et al.*, 1974). The montmorillonite used in the study by Yoshima *et al.* (1971) contains at most 2% Fe (Kerr *et al.*, 1951). It seems unlikely that this small amount of iron would be released from the octahedral layer of the montmorillonite so that it could serve as a catalyst. Clay minerals and silica are abundant in hydrothermal systems and may provide sites for FTT reactions. The mechanisms of the silica- and clay-catalyzed reactions are not known so it is possible that the extent of inhibition by H_2S and H_2O is quite different from that observed with metallic iron.

In addition, molybdenum sulfide (Storch, 1951), nickel sulfide and tungsten sulfide (Asinger, 1968) have also been shown to be FTT catalysts. This is an especially surprising result since the inhibitory action of H_2S is believed to be due to formation of sulfide derivatives of metal atoms at active sites on the catalyst. Since metal sulfides are formed in great quantities in hydrothermal systems, it is possible that some of these sulfides are FTT catalysts and their formation does not inhibit but rather enhances the prospects for FTT synthesis.

5. Selected Chemical Processes in Hydrothermal Systems of Potential Importance to the Origins of Life

5.1. ASSUMPTIONS

It is assumed that synthetic processes driven by heating in the 100-390°C temperature range and reactions resulting from ionizing radiation were the most likely ones occurring in hydrothermal systems (Section 4). It is difficult to predict the extent of thermal degradation and hydrolysis since there have been only a limited number of experimental studies performed under controlled hydrothermal reaction conditions (Section 3). Consequently, the assumption will be made that only modification of the more reactive functional groups will occur such as hydrolysis of nitrile or amide groups to acids. A related assumption is that the synthetic processes will be rapidly quenched in hydrothermal systems before the reaction products undergo major degradation reactions such as breaking carbon-carbon bonds.

5.2. AMINO ACIDS AND PEPTIDES

Amino acids are formed readily in thermal prebiotic simulation experiments (Ferris *et al.*, 1978; Yanagawa *et al.*, 1984a; Hayatsu *et al.*, 1971; Harada, 1967; Harada and Fox, 1964; Lawless and Boynton, 1973). Glycine, alanine and aspartic acids are among the more abundant amino acid products while smaller amounts of glutamic acid, threonine, serine, valine, leucine, histidine, phenylalanine and tyrosine were also produced in these studies. Glycine, alanine and aspartic acid are also the main amino acids formed by irradiation of a mixture of CO, N_2 and H_2O with protons (Kobayashi *et al.*, 1990). An array of amino acids which is not usually found in biological systems is also produced in the studies cited above. These include β-alanine, α-amino-n-butyric acid, β-amino-n-butyric acid, α-aminoisobutyric acid, N-methyl-β-alanine and diaminosuccinic acid.

There have been few studies of the reactions of amino acids under hydrothermal conditions. In an investigation of amino acid stability a mixture of aspartic acid, serine, alanine and leucine was heated at 250°C and 265 atm. It was observed that aspartic acid and serine decomposed within 1-2 h, alanine increased and then decreased in concentration, leucine decomposed after 5 h and glycine increased to a low steady state concentration (Miller and Bada, 1988). These data suggest the stability of amino acids in hydrothermal systems is different from that observed under standard conditions on the earth's surface (Shock, 1990a). Consequently, a different array of amino acids will probably eminate from hydrothermal systems than are observed in a "conventional" prebiotic syntheses on the earth's surface.

The detection of non-biological amino acids would provide support for the occurrence of prebiotic synthesis in contemporary hydrothermal systems. Such findings would not be conclusive because the non-biological amino acids may be formed from biological precursors. For example, when glycine was heated with $MnCO_3$-$Mn(OH)_2$ and Al_2O_3 at 240°C for 1 h and the product mixture is hydrolyzed with 6 N HCl α-aminobutyric, norvaline, norleucine, sarcosine, N-ethylglycine, N-methylalanine and N-ethylalanine were detected as reaction products (Ivanov and Slacheva, 1977). Extreme care will be required to establish that the biological amino acids formed in simulation studies are not due to terrestrial contamination. The use of ^{13}C- and ^{15}N-labeled starting materials is recommended to prove that the amino acids detected resulted from synthesis and not from decomposition of contemporary biological materials (Briggs *et al.*, 1991).

It is possible that peptides will form in hydrothermal systems if the concentrations of amino acids is sufficiently high. Oligomers (molecular weight-1000-2000), which may contain peptide bonds, were formed by heating glycine, alanine, valine and aspartic acid (see Chapter 7). Glycine oligomers up to the dodecamer were obtained from glycine amide (Yanagawa *et al.*, 1984b). The synthesis of ethyl acetate from ethanol and acetic acid in supercritical water (Antal *et al.*, 1987) is consistent with peptide bond formation at high temperatures and pressures. Oligomer formation may proceed via diketopiperazines (Nagayama *et al.*, 1990; Basiuk *et al.*, 1991) which form spontaneously from dipeptides (equation 27) (Kopple, 1966). Mineral catalysis may be important in these and other condensation reactions discussed in Section 5 (Siskin and Katritzky, 1991).

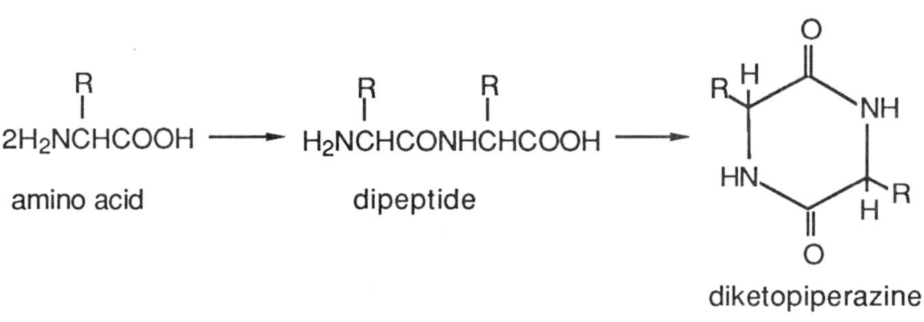

(27)

5.3. HETEROCYCLIC COMPOUNDS

5.3.1. From HCN

5-Aminoimidazole-4-carboxamide, 5-aminoimidazole-4-carboxamidine, adenine, 8-carbamyladenine, orotic acid, 4,5-dihydroxypyrimidine, 5-hydroxyuracil and uracil have all been detected as products of the self condensation of HCN (Section 2.5). Triazine is formed by the acid catalyzed condensation of HCN but it is not likely to be detected in hydrothermal systems because it undergoes rapid hydrolysis in aqueous solution (Smolin and Rappoport, 1959c). Glycolonitrile, the HCN adduct of formaldehyde, dimerizes to an oxazole (Section 2.1.3). No data is available concerning the hydrolysis of this oxazole under conditions prevalent in hydrothermal systems.

The reaction of 5-aminoimidazole-4-carboxamide with cyanogen or cyanate yields guanine and with HCN a small yield of hypoxanthine is obtained (Sanchez *et al.*, 1968).

The main reaction products resulting from the self condensation of 0.1 M HCN *after acid hydrolysis* are adenine and 4,5-dihydroxypyrimidine. An unidentified precursor to 4,5-dihydroxypyrimidine, possibly 6-cyano- or 6-carbamyl-4,5-diaminopyrimidine, was detected before the reaction mixture was subjected to acid hydrolysis (Ferris *et al.*, 1978). 8-Cyano- or 8-carbamyladenine are likely precursors to adenine (Voet and Schwartz, 1983). If these precursors are formed, it is likely that they will be converted to adenine and 4,5-dihydroxypyrimidine when subjected to the reaction conditions in hydrothermal systems. Since 4,5-dihydroxypyrimidine has not been detected in any other prebiotic synthesis its detection in a contemporary hydrothermal system would provide strong evidence that HCN condensation reactions were responsible for it and some of the other hydrothermal products.

5.3.2. FTT Synthesis

The yield of heterocycles formed in the FTT synthesis from CO, H_2 and NH_3 decreases in the order cyanuric acid>melamine>adenine, guanine>ammeline (Hayatsu *et al.*, 1968). Uracil and methylated uracils, thymine, xanthine, 2-methylpyrrole and histidine were also detected (Hayatsu *et al.*, 1972). Heterocycle formation by FTT synthesis requires the presence of NH_3 in the hydrothermal systems while their formation from HCN does not require NH_3. If NH_3 is present then cyanuric acid is likely to be the most abundant heterocycle produced by FTT synthesis since it is also formed by the hydrolysis of the melamine and ammeline. The detection of adenine will not differentiate between its formation by a FTT synthesis or direct formation from HCN (see previous section). Cyanuric acid provides the best evidence for prebiotic synthesis via FTT reactions in contemporary hydrothermal systems but its detection does not provide unambiguous evidence since it is also produced by the thermolysis of urea (Section 2.1.4.5). Its absence could also be due to its thermolysis of the cyanuric acid to cyanate or its hydrolysis to urea (Section 2.1.4.5).

5.3.3. *Thermal Synthesis of Imidazoles From NH_3 and Aldehydes*

Reaction of NH_3 with formaldehyde and glyoxal results in the formation of imidazole (equation 28) (Oró *et al.*, 1984). Substitution of glucose for glyoxal yields 4-methylimidazole (Oró *et al.*, 1984; Bernhauer, 1929). Reaction of formamidine, the NH_3 adduct of HCN, with erythrose yields imidazole-4-acetaldehyde (IAA). Histidine results when the amino nitrile derivative of IAA is hydrolyzed (equation 29) (Shen *et al.*, 1987, 1990).

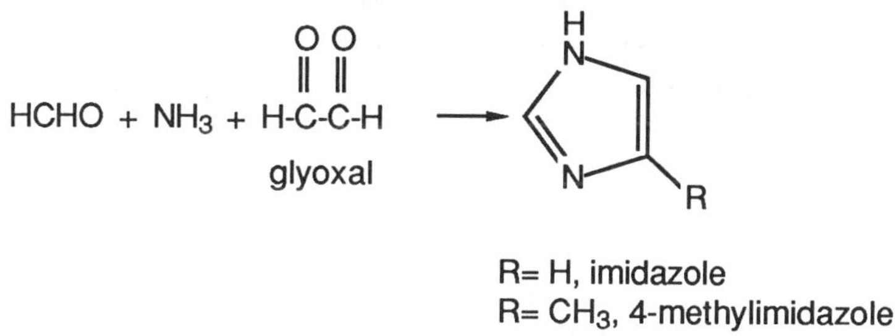

(28)

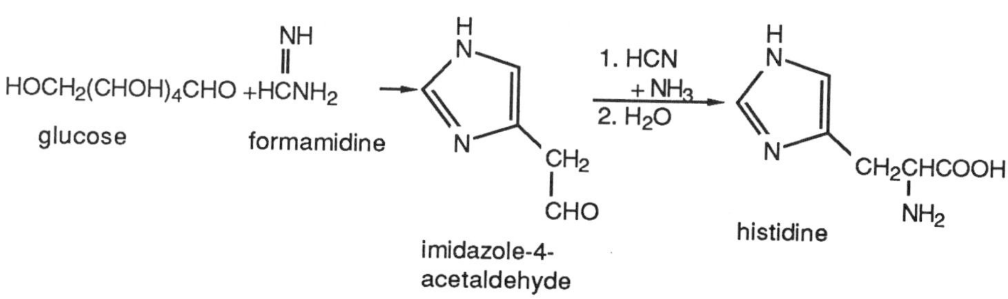

(29)

Imidazoles may result if NH_3 and aldehydes are present in appreciable amounts in hydrothermal systems. Imidazoles are readily detected by simple diazotization after chromatography (Oró *et al.*, 1984).

5.3.4. *Thermolysis of Biological Molecules*

The direct detection of abiotic synthesis in hydrothermal systems will be very difficult because compounds which are not prevalent in biological systems are formed by heating biological molecules (Section 5.2). Heterocycles are produced by the pyrolysis of amino acids and proteins. Heating an equimolar mixture of glycine, alanine, valine, lysine, aspartic acid, and glutamic acid for 4-6 h at 160-200°C gives UV absorbing compounds which appear to be pteridines (Heinz *et al.*, 1979). Thermal degradation of proteins at 840°C generates pyrrole, indoles, pyridines, quinoline, pyrrole and isoquinoline (Higman *et al.*, 1970).

5.4. CARBOXYLIC ACIDS

5.4.1. Ionizing Radiation

Simple acids such as formic and acetic may be formed by the radiolysis of carbonates in hydrothermal systems (Albarrán *et al.*, 1987). More complex (branched) acids may result from the radiolysis of these simple carboxylic acids (Negron-Mendoza and Ponnamperuma, 1982). Dicarboxylic acids are formed by the irradiation of urea (Navarro-Gonzalez *et al.*, 1989).

5.4.2. FTT Synthesis

Linear and branched fatty acids are produced in FTT syntheses (Section 2.3). It will be possible to distinguish those acids produced by FTT processes from those formed from contemporary biological systems by the presence of both an odd and even number of carbon atoms and branched chains in those formed in FTT reactions.

Aliphatic carboxylic acids synthesized in hydrothermal systems will not undergo appreciable decarboxylation (Drummond and Palmer, 1985). The rate of the thermal decarboxylation of acetic acid is dependent on pH and surface catalysis by reaction vessel used (equation 30) (Palmer and Drummond, 1985). The half life for acetic acid decarboxylation is 10-100 years at 300°C. In one study it was established that the rate constant for the decarboxylation of butyric acid was about half that of acetic acid under roughly comparable reaction conditions. These kinetic findings are in agreement with qualitative studies performed in supercritical water (Swanson *et al.*, 1986).

$$CH_3COOH \longrightarrow CH_4 + CO_2 \qquad (30)$$

CHAPTER 7

HYDROTHERMAL ORGANIC SYNTHESIS EXPERIMENTS

EVERETT L. SHOCK

Department of Earth and Planetary Sciences, and McDonnell Center for the Space Sciences, Washington University, St. Louis, MO 63130, U.S.A.

1. Introduction

The serious scientific debate about spontaneous generation which raged for centuries reached a climax in the nineteenth century with the work of Spallanzani, Schwann, Tyndall, and Pasteur. These investigators demonstrated that spontaneous generation from dead organic matter does not occur (see reviews by Conant, 1953 and Kenyon and Steinman, 1969). Although no aspects of these experiments addressed the issue of whether organic compounds could be synthesized abiotically, the impact of the experiments was great enough to cause many investigators to assume that life and its organic compounds were somehow fundamentally different than inorganic compounds.

Meanwhile, other nineteenth-century investigators were showing that organic compounds could indeed be synthesized from inorganic compounds. In 1828 Friedrich Wöhler synthesized urea in an attempt to form ammonium cyanate by heating a solution containing ammonia and cyanic acid (Wöhler, 1828). This experiment is generally recognized to be the first to bridge the artificial gap between organic and inorganic chemistry, but it also showed the usefulness of heat in organic synthesis. Not only does an increase in temperature enhance the rate of urea synthesis, but Walker and Hambly (1895) showed that equilibrium between urea and ammonium cyanate was attainable and reversible at 100°C. Wöhler's synthesis of urea, and subsequent syntheses of organic compounds from inorganic compounds over the next several decades dealt serious blows to the "vital force" concept which held that: 1) organic compounds owe their formation to the action of a special force in living organisms, and 2) forces which determine the behavior of inorganic compounds play no part in living systems. Nevertheless, such progress was overshadowed by Pasteur's refutation of spontaneous generation which nearly extinguished experimental investigations into the origins of life for several decades.

Vitalism was dealt a deadly blow in the 1950's with Miller's famous spark-discharge experiments (Miller, 1953, 1955, 1957a, 1957b; Miller and Urey, 1959) which were undertaken in the framework of the Oparin (1924) and Haldane (1929) hypotheses concerning the origin of life. These hypotheses were constructed on some basic assumptions which included a reduced atmosphere, and a low surface temperature for the early Earth. These ideas meshed well with the prevailing hypothesis of the 1940's and 50's that the Earth had formed through heterogeneous accretion of dust from a condensing solar neb-

Origins of Life and Evolution of the Biosphere **22**: 135–146, 1992.

ula. Miller's experiments were extremely successful, and were followed by numerous other experiments by various investigators who employed a wide variety of energy sources for abiotic synthesis including spark discharges, ultra-violet radiation, heat, shock waves, plasmas, gamma rays, and other forms of energy. The conclusion reached from this body of work is that energy inputs can drive organic synthesis from a variety of inorganic starting materials.

1.1. ENVIRONMENTS FOR ORGANIC SYNTHESIS ON THE EARLY EARTH

A vast majority of the early abiotic synthesis experiments were based on an assumption that the primitive Earth had a reduced atmosphere consisting of gases such as CH_4, NH_3, H_2, CO, and H_2O. This type of primary atmosphere was assumed to exist based on studies of the outer planets which are rich in reduced gases, together with the prevailing hypothesis of the midcentury that the Earth formed through relatively cool, heterogeneous accretion of material which condensed in a cooling solar nebula (Urey, 1952). Major difficulties with the heterogeneous accretion model are outlined by Ringwood (1979) and Stevenson (1983), and reviews of current models of the accretion of the Earth are given by Wetherill (1990), Taylor and Norman (1990) and Boss (1990). In these models, formation of planets through accretion is viewed as a violent process leading to considerable heating, melting and differentiation of plantary materials as planets form (Carmeron, 1986; Benz *et al.*, 1987; Stevenson, 1987; Davies, 1990; Ringwood, 1990).

High velocity impacts of planetesimals onto a growing planet can result in the impact-degassing of volatiles and the formation of an impact-induced atmosphere at high temperature (Fegley *et al.*, 1986; Matsui and Abe, 1986a, 1986b; Holloway, 1988; Kasting, 1990). According to this hypothesis, the resulting impact-induced atmosphere contains carbon monoxide and/or carbon dioxide as the major forms of carbon. In addition, results of photochemical studies of the early Earth's atmosphere show that the presence of hydroxyl radicals from the photodissociation of H_2O, together with the greater flux of UV radiation from the young sun, would have limited the half lives of reduced gases such as methane and ammonia to about 50 years and about 1 week, respectively (Levine, 1982, 1985; Zahnle and Walker, 1982; Canuto *et al.*, 1982, 1983; Kasting *et al.*, 1983). Therefore, for a variety of reasons, the primitive Earth atmosphere may have been only "mildly reduced", which means it may have been similar to the present atmosphere but without high levels of O_2. Some investigators view this as a problem for conventional chemical evolution theories because only a trace of amino acids is obtained in simulated primitive atmosphere experiments when carbon monoxide or carbon dioxide are used as carbon sources (Schlesinger and Miller, 1983).

On the other hand, if carbon in the early atmosphere was present as CO or CO_2, organic molecules required for the origin of life could be formed in local "reducing" environments. Suggestions for such environments include impact sites of comets and/or meteorites, and submarine hydrothermal systems. The production of organic compounds in the cooling gas mixtures formed during impacts has received recent attention (Chyba *et al.*, 1990; Chyba and Sagan, 1992). The oxidation states prevailing during large impacts will be strong functions of the composition of the target, which in the case of the Earth would likely lead to oxidized conditions (Fegley *et al.*, 1986). In contrast, submarine hydrothermal systems may provide the type of reduced environment needed for the abiotic synthesis of various organic compounds. In the present-day oceans, hydrothermal vents are more reduced than their immediate vicinity, and hydrothermal fluids are known to contain reduced gases such as methane, hydrogen, hydrogen sulfide, and ammonia (Lilley *et al.* 1983).

These observations have led several investigators to propose that abiotic organic synthesis can take place in hydrothermal systems (French, 1964, 1971; Ingmanson and Dowler, 1977; Corliss *et al.*, 1981; Baross and Hoffman, 1985; Shock, 1990b, 1992b). These suggestions are strengthened by the wide use of thermal energy for prebiotic synthesis experiments, and it is the purpose of this paper to summarize these experimental results, especially those involving aqueous solutions. However, as emphasized elsewhere (Shock, 1990b, 1992b), there is a great deal more to submarine hydrothermal systems than hot water. Oxidation states, gas fugacities, pH, and activities of aqueous ions are commonly buffered by mineral assemblages and fluid/rock reactions. Differences between the buffered conditions in natural systems and those which may be generated in unbuffered experiments can exceed many orders of magnitude. As a consequence, experiments solely involving heat as a source of energy are not likely to simulate realistic hydrothermal conditions even when an aqueous medium is used. This point is emphasized by reports in the literature of experiments conducted at the same temperature and pressure in which thermal energy has been used to drive aqueous organic synthesis reactions as well as reactions in which aqueous organic compounds are destroyed (Miller and Bada, 1988; Yanagawa *et al.*, 1988). The extent to which heat facilitates the synthesis or destruction of certain organic compounds at realistic hydrothermal conditions is not yet known. However, the body of relevant experimental work summarized below shows that there is great potential for organic synthesis in hydrothermal systems. One conclusion of this review is that this research area should be a prime target of further, well-constrained experiments.

2. Heat as an Energy Source for the Abiotic Synthesis of Organic Compounds

Heat is useful in two major ways in organic synthesis experiments. Reaction rates are enhanced by elevated temperatures, so reactions which may take millennia at room temperature are conducted at higher temperatures in order to conform to human lifespans. This application of thermal energy is often made in prebiotic experiments, and rates obtained at high temperatures are often extrapolated to lower temperatures by application of the Arrhenius equation (Abelson, 1957; Vallentyne, 1964; White, 1984; Miller and Bada,

1988). Additionally, the thermodynamic properties of reactions involving organic compounds are typically strongly temperature dependent, and increased temperatures may favor one set of reaction products over another. Often these two applications of thermal energy are employed simultaneously, as in the use of heat to drive polymerization reactions through dehydration (Schwartz *et al.*, 1975; Schwartz and Chittenden, 1976; Ivanov and Slavcheva, 1977; Epps *et al.*, 1978; Lahav *et al.*, 1978; Lawless and Levi, 1979; Lahav and White, 1980; Hawker and Oró, 1981a,b; Rishpon *et al.*, 1982; Rao *et al.*, 1982; Weber, 1989; among others, see below). Dehydration reactions also proceed in aqueous solution, and it has been shown that those which lead to peptide bonds are energetically more favorable at elevated temperatures than they are at 25°C and 1 bar (Flegmann and Tattersall, 1979; Shock, 1992a). In addition, theoretical calculations show that metastable equilibrium states in which aqueous organic compounds predominate are likely to be attainable at the elevated temperatures and pressures of submarine hydrothermal systems (Shock, 1990b, 1992b). Applications of thermal energy to synthesize organic compounds from inorganic starting materials and to polymerize organic compounds are reviewed in this section. Experiments related specifically to hydrothermal synthesis of organic compounds are described below.

2.1. THERMAL ORGANIC SYNTHESIS FROM INORGANIC COMPOUNDS

Several investigators have shown that thermal energy can be used to synthesize amino acids and other organic compounds from inorganic compounds in prebiotic experiments. For example, Harada and Fox (1964) synthesized amino acids from a mixture of CH_4, NH_3 and H_2O which had been passed through silica sand at 950°C, or through silica gel at 950° and 1050°C. After passing the gases through the hot silica, the products were collected in 3 N aqueous ammonia solutions which were then heated to 75°C for 35 hours. These authors reported the production of fourteen amino acids with glycine and alanine being the most abundant in all cases. Oró (1965) reported that at least a dozen amino acids were produced by reacting CH_4, NH_3 and H_2O at about 1025°C in a "hot tube" of unspecified composition. Glycine and alanine predominated by at least an order of magnitude in concentration over the other amino acids which included β-alanine, threonine, serine, and glutamic acid.

Silica gel was used by Taube *et al.* (1967) who heated a mixture of CH_4, NH_3 and H_2O (at a 5:3:2 ratio) in a streaming apparatus. As in the case of the experiments reported by Oró (1965), the residence time of the gas mixture in the reaction zone was very short (0.125 sec). Taube *et al.* (1967) found that the maximum conversion of NH_3 occurred at 1000°C (95.4%). Glutamic acid, lysine, proline, glycine, β-alanine, alanine, serine, valine, leucine, aminobutyric acid, and phenylalanine were identified, together with nitriles, amines, alcohols, aldehydes, aromatic hydrocarbons, CO, HCN and alkanes. Their yields of amino acids were lower at 900°C, and at 1150°C they observed that methane was converted to graphite. A similar experimental procedure was adopted by

Lawless and Boynton (1973) who used quartz sand and found only glycine and alanine in the samples run at 930°C; glycine, β-alanine, alanine and *N*-methyl-β-alanine at 980°C; and a variety of amino acids at 1060°C among which β-alanine was the most abundant by nearly two orders of magnitude.

These experiments are substantially similar to spark discharge experiments in that the spark temperatures are in the vicinity of 900°C (Fox and Dose, 1977). In both cases, enormous amounts of energy are used to generate high-energy radicals and other species which react on condensation in aqueous solution to yield amino acids and other organic compounds. Only in the vicinity of the magma/hydrothermal fluid interface are conditions in hydrothermal systems likely to be as severe than those imposed in traditional spark discharge or thermal synthesis experiments.

2.2. THERMAL ORGANIC SYNTHESIS FROM OTHER ORGANIC COMPOUNDS

Samochocka *et al.* (1968) synthesized amino acids by heating mixtures of acetylene and NH_3 (1:1 mol ratio) and acetylene, NH_3 and CO_2 (1:1:1 mol ratios) in quartz tubes heated to 600°C. The reacted gas mixtures were collected in H_2O and the resulting solutions were heated at 70°C for 35 hrs, and then evaporated. The residue was dissolved in 5N HCl and heated to 70°C for 5 hrs before analysis. Amino acids were produced from both initial gas mixtures, although a wider variety were found with CO_2 present. However, the CO_2 used was ^{14}C-labeled, and none of the amino or carboxylic acids produced contained radioactive carbon. The lack of radioactive labeling of products indicates that carbon from CO_2 was not incorporated in the products, and the authors state that this demonstrates that a Strecker synthesis did not occur in these experiments.

Other investigators, working at lower temperatures, have synthesized amino acids and other organic compounds from mixtures which include inorganic and simple organic compounds. For example, Oró *et al.* (1959) heated sealed glass vials containing aqueous mixtures of formaldehyde and hydroxylamine to 80°, 90° and 100°C for up to 400 hours. Ammonia was a major product but amino acids, HCN, carboxylic acids, and amides were also produced. The appearance and disappearance of the amides during the course of the experiments suggests that they were reaction intermediates to the formation of amino acids. Among the amino acids, glycine, alanine, and β-alanine were identified and aspartic acid, serine and threonine were tentatively identified. Formic, acetic, glycolic, and lactic acids were also produced in the experiments. Oró and Kamat (1961) report the formation of amino acids by heating solutions of HCN in 3 N ammonium hydroxide to 70°C for 25 days. Most of the HCN was converted to polymers, but the concentration of amino acids in the solution reached 10 mM (as glycine). Glycine, alanine and aspartic acid were tentatively identified by paper chromatography.

Aqueous formaldehyde and hydroxylamine were heated to 100°C for 150 hours in a solution of potassium phosphate and various trace elements by Ventilla and Egami (1977), who found that amino acids and oligomers of amino acids were formed although no specific amino acids were identified. The same reactants were heated to 105°C in an aqueous solution of KCl and many trace constituents including Na^{+}, Ca^{+2}, Mg^{+2}, SO_4^{-2}, HPO_4^{-2}, NO_3^{-}, Zn^{+2}, Fe^{+3}, Cu^{+2}, Co^{+2}, Mn^{+2}, and MoO_4^{-2} by Hatanaka and Egami (1977a, 1977b) who report the formation of many amino acids, and by Kamaluddin *et al.* (1979) who report the production of glycine, serine, aspartic acid and β-alanine, as well as NH_3, urea, glycinenitrile and glycineamide. Amino acids were also produced by Yanagawa and Egami (1981) who heated similar aqueous solutions containing carbon suboxide (C_2O_3) polymers and hydroxylamine or ammonia to 105°C, and from solutions of sugars and ammonia (Yanagawa *et al.*, 1980, 1981), or oxo-acids and ammonia (Yanagawa *et al.*, 1982).

Mixtures of formaldehyde, ammonia and water were heated to 185°C by Fox and Windsor (1970) who report the formation of several amino acids, with glycine again being the predominant amino acid identified. Yields of amino acids were in the range 0.002 to 0.007 percent of the initial formaldehyde, and a major product of these reactions was hexamethylenetetramine. A wider spectrum of products was obtained by Lowe *et al.* (1963) who heated concentrated aqueous solutions of hydrocyanic acid, ammonia and oxalic acid to 90°C for 18 hours. More than ten amino acids were identified and evidence of several others was also found. In addition, ammonia, urea, adenine and formic acid were identified. In other aqueous thermal synthesis experiments, uracil was formed by heating solutions of urea, malic acid and polyphosphoric acid to 100°-140°C for up to 2 hrs (Fox and Harada, 1961), and aspartic acid and alanine were synthesized by heating ammonium malate and ammonium fumarate solutions to 200°C (Fox *et al.*, 1955).

More recent results from Kimoto and Fujinaga (1988, 1990) indicate that elevated temperatures may not be necessary for aqueous organic synthesis given the right initial mixture of compounds. These investigators obtained glycine, serine, and alanine, as well as poorly-defined polymers, at room temperature by mixing H_2S or Na_2S into aqueous solutions of ammonium formate, formaldehyde and magnesium chloride. These results may indicate that organic synthesis is possible in aqueous solution at the low temperatures which characterize seafloor weathering, but the concentrations of the reactants (5M $HCOONH_4$, 0.1M formaldehyde) may greatly exceed the concentrations attainable in natural systems.

None of the experiments described above were designed to evaluate hydrothermal organic synthesis. In fact, most of this work predates the discovery of hot springs on the ocean floor. Although many of these experiments were conducted in aqueous solution, and it might seem attractive to assert that these experiments represent hydrothermal syntheses, many of the features which characterize hydrothermal systems are absent from these experiments. In particular, the buffered fugacities of H_2, CO_2, and activities of aqueous ions which characterize hydrothermal systems are not constrained or approached in these experiments. In addition, the experiments which employ simple organic compounds as starting materials are difficult to mesh with geologic systems unless a source of such starting materials can be established. Nevertheless, these results are tantalizing glimmers of possible hydrothermal synthesis which can only be confirmed once properly constrained experiments have been conducted.

2.3. THERMAL POLYMERIZATION

It has been shown repeatedly that amino acids will polymerize under the influence of heat at elevated temperatures. Many of these experiments have been conducted at anhydrous conditions which strongly favor the dehydration reactions leading to peptide bond formation (Meggy, 1953, 1956; Kovacs *et al.*, 1953; Kovacs and Konyers, 1954; Heyns and Pavel, 1957a, 1957b; Fox *et al.*, 1957, 1959, 1963; Fox and Harada, 1958, 1960; Harada and Fox, 1958, 1960, 1965, 1975; Vegotsky *et al.*, 1958; Fox, 1963, 1964, 1965; Rohlfing, 1967, 1976; Fox and Waehneldt, 1968; Saunders and Rohlfing, 1972, 1974; Phillips and Melius, 1974; Snyder and Fox, 1975; Rohlfing and McAlhaney, 1976; Fouche and Rohlfing, 1976; Temussi *et al.*, 1976; Fox and Dose, 1977; Grote *et al.*, 1978; Kokufuta *et al.*, 1978; Lacey *et al.*, 1979; Harada and Matsuyama, 1979; Hartmann *et al.*, 1981; Fox and Nakashima, 1981; among others). None of these results can be applied directly to hydrothermal systems because of the anhydrous nature of the experiments.

Other investigators have demonstrated polymerization of amino acids at elevated temperatures in aqueous solution. For example, Oró and Guidry (1961) polymerized glycine in aqueous solution at temperatures from 130° to 160°C. Lowe *et al.* (1963) formed polymers of amino acids at 90°C in some of the same aqueous solutions in which they synthesized amino acids. Nagayama *et al.* (1990) have shown that linear glycine peptides can be formed from diketopiperazines in aqueous solutions at 90°C. Rode and Schwendinger (1990) report that glycine will polymerize in 5M NaCl solutions containing Cu^{+2} at 70° to 100°C. The latter experiments demonstrated that the polymerization of glycine was enhanced by increasing the concentration of NaCl, decreasing the pH, and allowing the experiments to run for greater lengths of time. Using aqueous salt solutions including many transition metals, Yanagawa *et al.* (1980, 1988) and Yanagawa and Egami (1980) synthesized polymers of amino acids in solution at 105°C. Yanagawa and Kojima (1985), Yanagawa *et al.* (1988), and Yanagawa and Kobyashi (1992) were successful in synthesizing peptide-like polymers in aqueous solutions at temperatures ≥ 250°C and elevated pressures.

3. Hydrothermal Destruction and Synthesis of Organic Compounds

Hydrothermal systems have been studied for many decades because they are responsible for a majority of the economically mined ore deposits of precious and base metals. In conjunction with field and analytical studies of ore deposits, hundreds of hydrothermal experiments involving minerals have been conducted by petrologists and geochemists. In these experimental efforts it has become crucial, and customary, to obtain reversals of various reactions in order to demonstrate equilibrium. Experiments are commonly conducted with mineral/fluid systems which buffer pH, and/or the oxidation state of the experiment to values obtained from theoretical studies of mineral alteration and ore deposition, as well as analytical studies of present-day hot spring and geothermal systems. Careful attention to detail requires thorough analysis of reactants and products before and after these experiments are conducted. Results can then be interpreted with the aid of thermodynamic and kinetic calculations of speciation, solubility and mass transfer. References to hydrothermal experimental techniques and interpretation of the results can be found in Barnes (1979), Ulmer and Barnes (1987), Holloway and Wood (1988) and Spencer and Chou (1990).

In contrast, the rigor which characterizes the experimental study of hydrothermal minerals is distinctly lacking from most experimental studies of hydrothermal organic compounds. None of the work to date has attempted an experimental reversal of a reaction, and few if any constraints from studies of hydrothermal systems are imposed on these experiments. Typically, the lack of experimental constraint makes it impossible to link experimental conditions with those which exist in nature. Therefore, it comes as little surprise that there are reports in the literature of destruction and production of aqueous amino acids at the same temperatures and pressures. Descriptive variables other than temperature control the results of these experiments. Likely candidates include the fugacities of H_2, CO_2, and other gases, and the activities of aqueous species. The experiments leading to the destruction of amino acids, which represent one direction of reaction progress, are summarized in the next section, and those representing the opposite direction of reaction progress, amino acid production, are described below. It may be that well-constrained experiments will be undertaken in the near future, as many investigators appear to be realizing that their recent experiments do not come close to representing natural conditions or processes.

3.1. HYDROTHERMAL DESTRUCTION EXPERIMENTS

Many investigators have conducted experiments in which amino acids are altered into other organic or inorganic compounds at elevated temperatures (Abelson, 1957; Jones and Vallentyne, 1960; Vallentyne, 1964, 1968; Povoledo and Vallentyne, 1964; White, 1984; Bernhardt *et al.*, 1984, Lavrent'yev *et al.*, 1985; Miller and Bada, 1988; Bada *et al.*, 1991), but none of the conditions imposed in these experiments can be considered to be close analogues of hydrothermal systems (Shock, 1990a, 1990b). The typical experimental procedure consists of preparing an aqueous solution of amino acids, enclosing the solution in a vessel, and heating the vessel--usually under pressure--to temperatures of 200°C or higher. No efforts are made to buffer fugacities of H_2, CO_2, NH_3, N_2, or any other component of the system. Occasionally efforts are made to buffer pH. However, this has been done with organic buffers, about which the temperature dependence of pKa is often unknown. Therefore, it is unclear to what extent the pH at elevated temperatures resembles those evaluated for hydrothermal systems. In addition, the buffers apparently fail as demonstrated by experiments conducted by Miller and Bada (1988) who used HEPES (N2-hydroxyethylpiperazine N-2 ethane sulphonic acid) to buffer the 25°C pH to 7.0. After the experiment the pH was reported to be 7.9.

Reaction progress in these experiments is typically monitored by the disappearance of the amino acid(s) of interest. Only rarely are reaction products searched for and identified and none of the investigations referred to above made any effort to demonstrate mass balance. It is often established that heating solutions of amino acids causes the amino acid concentrations to decrease. These decreases are due to unknown reactions which may include decarboxylation, deamination, oxidation, reduction, polymerization or other processes. Lack of knowledge of the reactions involved has not inhibited several investigators from extracting rate data from these experiments and using such data to make inferences about the geochemistry of amino acids (Abelson, 1957; White, 1984; Miller and Bada, 1988). It is not apparent how useful a rate for an unknown, and possibly unnatural, reaction can be in understanding hydrothermal organic geochemistry. It should be noted that similar mysteries shroud the vast majority of hydrous pyrolysis experiments conducted in an effort to understand the generation of petroleum from organic-rich source rocks.

3.2. HYDROTHERMAL SYNTHESIS EXPERIMENTS

In contrast to the thermal synthesis and polymerization experiments discussed above, several researchers have synthesized organic compounds abiotically, and/or polymerized monomers at conditions much more reminiscent of hydrothermal systems. Probably the first study to document hydrothermal abiotic synthesis was strictly serendipitous. During his Ph.D. thesis research on the synthesis of siderite ($FeCO_3$) at metamorphic and hydrothermal conditions, French (1964) noted distinctive odors on opening reaction vessels. Gas chromatography indicated the presence of numerous organic compounds including ethanol; acetic, propanoic, and butanoic acids; and ketones. The purpose of the investigation was to synthesize siderite at hydrothermal conditions, and not organic compounds. As a consequence, these results receive only a brief mention in a later paper (French, 1971). Nevertheless, it would appear that the first proposal of organic synthesis in hydrothermal systems in the context of the origin of life was made by French who wrote: "The syntheses described here, which occur at elevated temperatures and pressures, may indicate that organic compounds could be produced by hydrothermal activity at shallow depths within the crust. If this were the case, the presence of such compounds on the surface would not reflect the composition of the atmosphere into which they were introduced." (French, 1964, p. 68). French (1971) expanded on this point and suggested that organic compounds synthesized in hydrothermal systems could reach the surface and provide "material for biological development even in cases where such synthetic reactions were not possible in the planetary atmosphere or oceans themselves." The perspective of this statement is extraterrestrial, but at present it has terrestrial implications in light of the mounting evidence that a reduced atmosphere was unlikely on the early Earth.

In an effort to simulate the action of an "under water volcano," Mukhin (1976) exposed mixtures of H_2O (90%), NH_3 (5%) and CH_4 or CO (5%) to 1100-1200°C for ~0.1 sec in "quartz tubes with volcanic lava." Analysis of these solutions detected aldehydes and amino acids as well as many aromatic hydrocarbons and polycyclic aromatic hydrocarbons (PAH). Synthesis of PAH from CH_4, and mixtures of CH_4, NH_3 and H_2O at 1000°C is also reported by Oró and Han (1966). The appearance of PAH in these experiments bears a striking similarity to submarine hydrothermal systems on sedimented ridge systems (Kawka and Simoneit, 1990; Simoneit, 1990; Kvenvolden and Simoneit, 1990), although it is generally believed that PAH in hydrothermal systems are not synthesized from methane, but represent pyrolysis products of detrital organic matter. The total concentration of the amino acids in Mukhin's experiments was estimated to be 10^{-4} g l^{-1}. Mukhin (1976) suggests that amino acids formed through the hydrolysis of nitriles generated in the reaction zone. He supports this suggestion with the observation that the concentration of HCN in the solutions is nearly 10^{-2} g l^{-1}, and indicates that a Strecker mechanism may be involved in amino acid production.

Synthesis of amino acids at elevated temperatures (200-300°C) is apparently also possible by reacting ammonium carbonate solutions with carbides. This process is related to those involving carbides suggested by Oparin (1924), but which have never been adequately studied owing to arguments against the presence of carbides in the geologic record and the early success of spark-discharge and other synthesis methods. Marshall (1987) synthesized amines and hydrocarbons using Fe_3C as a reactant, and found that amino acids could be generated from CaC_2 at more oxidized conditions obtained by the addition of H_2O_2 (Marshall, 1992). Several amino acids were produced with this method, among which glycine, alanine and lysine typically predominated over aspartic acid, isoleucine, phenylalanine and other products (Marshall, 1992). Carbides are found in some meteorites, and it may be that these minerals are exposed to hydrothermal alteration. It is doubtful whether such reactions occur in present-day, terrestrial hydrothermal systems, but these results may have some bearing on organic synthesis on meteorite parent bodies. To the extent that processes which occurred on the early Earth can be understood from the study of meteorites, these experiments may be applicable to the study of the origin of life.

Recently, several investigators have begun efforts to synthesize amino acids and other organic compounds at conditions more representative of hydrothermal systems. Some of these results are reviewed here, and a summary of recent results by Yanagawa and Kobayaski is given in another paper in this volume. Preliminary results obtained by Hennet *et al.* (1992) indicate that several amino acids including glycine, alanine, aspartic acid and isoleucine can be produced in aqueous solution at 150°C through a Strecker-like synthesis involving formaldehyde and HCN. The main feature which distinguishes these results from other hydrothermal synthesis experiments is that an attempt was made to buffer the oxidation state with pyrite, pyrrhotite and magnetite, a mineral assemblage common to hydrothermal systems. Equilibration of an aqueous fluid with this buffer assemblage is likely to be sluggish at 150°C (Barnes, 1987), but experiments at higher temperatures may circumvent this kinetic barrier.

Despite the focus on amino acids in the preceding discussion, which stems from the emphasis placed on searching for these compounds in abiotic synthesis experiments, hydrothermal synthesis is not limited to amino acids. This was demonstrated recently in a study by Yamagata *et al.* (1991) who showed that polyphosphates can be generated by the action of H_2O vapor on a basaltic magma spiked with calcium phosphate. The same authors have found evidence of polyphosphate in volcanic gases exiting fumaroles at temperatures of 540-690°C.

4. Concluding Remarks

Thermal energy has a long and distinguished history in organic synthesis experiments, extending back at least to the early 1800's. The frenzy of experimental work which followed Miller's spark discharge experiments showed that heat could be used as an energy source in abiotic synthesis, and many of those experiments were done in aqueous solution. The success of these experiments is the basis of the many suggestions and scenarios in the literature for organic synthesis in hydrothermal systems (French, 1964, 1971; Ingmanson and Dowler, 1977; Corliss *et al.*, 1981; Baross and Hoffman, 1985; Shock, 1990b, 1992b). There is ample heat in submarine hydrothermal systems for driving abiotic synthesis reactions. Macdonald *et al.* (1980) calculated that the heat flow from a single black-smoker chimney was equivalent to the total heat loss for a 3- to 6-km segment of spreading center out to 10 km on either side. Hot springs on the East Pacific Rise at 21°N operate at temperatures up to and above 350°C with flow rates from 1 to 5 m sec^{-1}. Therefore, if heat can overcome kinetic barriers to the formation of metastable states from reduced or oxidized starting materials, abiotic synthesis at hydrothermal conditions is a distinct possibility (Shock, 1990b, 1992b). However, carefully controlled experiments which replicate the descriptive variables of natural hydrothermal systems (temperature, pressure, oxidation state, pH, inorganic fluid composition, etc.) have not yet been conducted with the aim of testing the hypothesis of hydrothermal organic synthesis. There are tantalizing results, especially in the experiments conducted by French (1964, 1971), which are consistent with metastable synthesis of organic compounds as postulated by Shock (1990b, 1992b). Well-constrained experimental tests of these hypotheses should be the next step.

Acknowledgements

This review benefitted from discussions with Dale Ingmanson, Bevan French, Mike Russell, Hiroshi Yanagawa, Bill Marshall, Nils Holm and Remy Hennet. Thanks are due to Carla Koretsky and Patty DuBois for technical assistance, and Dale Ingmanson for his review of the manuscript. This work was funded in part by NSF grant EAR-9018468 and NASA grant NAGW-2818.

CHAPTER 8

AN EXPERIMENTAL APPROACH TO CHEMICAL EVOLUTION IN SUBMARINE HYDROTHERMAL SYSTEMS

HIROSHI YANAGAWA

Mitsubishi Kasei Institute of Life Sciences, 11 Minamiooya, Machida-shi, Tokyo 194, Japan

KENSEI KOBAYASHI

Department of Physical Chemistry, Yokohama National University, Hodogaya-ku, Yokohama 240, Japan

1. Introduction

The discovery of submarine hydrothermal vents in the late 1970's (Corliss *et al.*, 1979) stimulated geophysical, geochemical, microbiological, ecological and ore deposit research on submarine hydrothermal systems. Since these systems represent reducing, energy-rich, and metal ion-rich conditions in the present terrestrial environment, they are considered to be ideal sites for present-day abiogenic synthesis of organic compounds and have been suggested as a possible environment for chemical evolution toward the origin of life (Corliss *et al.*, 1981; Baross and Hoffman,1985). However, it has also been suggested before the discovery of submarine hydrothermal vents that an important relationship may exist between the evolution of the Earth's crust such as a hydrogeothermal zone associated with axes of plate spreading and the process of chemical evolution (Ingmanson and Dowler, 1977; Degens, 1979).

From the point of view of chemical evolution, reducing environments are attractive because amino acids can be synthesized abiotically from reduced gas mixtures with spark discharges (Miller, 1953), heat (Harada and Fox, 1964), ultraviolet light (Sagan and Khare, 1971), and shock waves (Bar-Nun *et al.*, 1970). These experiments demonstrate that if the primitive earth atmosphere was reduced, consisting of a mixture of methane, ammonia and water, amino acids and other organic compounds could have been easily obtained with the available energy sources on the early Earth.

Recent studies on the formation of planets have suggested, however, that high velocity impacts of small bodies into a growing planet can result in impact-degassing of volatiles and formation of an impact-induced, high temperature atmosphere (Abe, 1986; Holloway, 1988; Kasting, 1990; Matsui and Abe, 1986a,b). According to this hypothesis, such an atmosphere contains carbon monoxide or carbon dioxide as a major carbon source. In addition, results of photochemical studies with an early Earth atmosphere show that the presence of hydroxyl radicals from the photodissociation of H_2O, together with the greater flux of UV radiation from the young sun, would have limited the half

Origins of Life and Evolution of the Biosphere **22**: 147–159, 1992.

lives of reduced gases such as methane and ammonia to about 50 years and about 1 week, respectively (Canuto *et al.*, 1982, 1983; Kasting *et al.*, 1983; Levine, 1982, 1985; Zahnle and Walker, 1982). Therefore, the primitive Earth atmosphere may have been only "mildly reduced". Some researchers view this as a problem for conventional chemical evolution theories because only traces of amino acids are obtained in simulated primitive atmosphere experiments when carbon monoxide or carbon dioxide are used as the carbon source (Schlesinger and Miller, 1983).

In a mildly reduced atmosphere, organic molecules required for the origin of life could be formed in local "reducing" environments like submarine hydrothermal systems, high latitude areas irradiated with cosmic rays and solar flare particles, and impact sites of comets and meteorites. Recently, Kobayashi *et al.* (1989, 1990) found that proton irradiation, simulating the action of cosmic rays and solar flare particles, formed imidazole and proteinaceous and nonproteinaceous amino acids such as glycine, alanine, aspartic acid, and β-alanine, from a mixture of carbon monoxide, carbon dioxide, nitrogen, and water. The production of organic compounds in the cooling gas mixtures formed during impacts has also been considered (Chyba *et al.*, 1990). The oxidation states prevailing during major impacts will be determined by the composition of the target, which in the case of the Earth would likely lead to oxidized conditions (Fegley *et al.*, 1986). Submarine hydrothermal systems may provide the type of reduced environment needed for the abiotic synthesis of various bioorganic compounds. In the present-day oceans, hydrothermal vents are more reduced than their immediate vicinity, since hydrothermal fluids contain reduced gases such as methane, hydrogen, hydrogen sulfide, and ammonia (Lilley *et al.*, 1983).

Although methane in the Earth's crust is formed biologically and abiologically from organic matter, it can also form at magmatic and metamorphic conditions from carbon dioxide and hydrogen (Wakita and Sano, 1983; Holloway, 1984; Schoell, 1988; Wakita *et al.*, 1990). There is also ample heat in submarine hydrothermal systems for driving abiotic synthesis reactions. MacDonald *et al.* (1980) calculated that the heat flow from a single chimney was equivalent to the total heat loss for a 3- to 6-km segment of spreading center out to 10 km on either side. Black smokers on the East Pacific Rise at 21°N operate at temperatures up to and above 350°C with high flow rates. Therefore, if heat can overcome kinetic barriers to the formation of metastable states, abiotic synthesis under hydrothermal conditions is a possibility (Shock, 1990b).

Submarine hydrothermal fluids contain high concentrations of metal ions such as iron, manganese, copper, and zinc (Edmond *et al.*, 1982; Von Damm *et al.*, 1983), which may be able to catalyze organic synthesis reactions, and which are essential to many present-day biochemical processes. The role of trace elements in chemical evolution has been emphasized by Egami (1974) who found that a close correlation exists between the concentrations of transition elements in contemporary sea water and their biological importance (Table 1). Egami (1974) suspected that transition elements relatively abundant

TABLE 1

Concentrations of transition elements in seawater and their biological importance

Concentration in seawater	Transition element	Biological importance
About 100-40 nM	Mo	Probably essential to all organisms including strict anaerobes
	Zn	Probably essential to all organisms including strict anaerobes
	V	Essential to a wide variety of organisms
	Fe	Essential to all organisms including strict anaerobes
About 40-10 nM	Ni	Probably essential to higher land animals and plants
	Ti	No evidence that it is a bioelement
	U	No evidence that it is a bioelement
	Cu	Essential to all organisms, probably with the exception of anaerobes
About 10-1 nM	Cr	Probably essential to higher land animals
	Mn	Essential to a wide variety of organisms including certain bacteria
	Cd	No evidence that it is a bioelement
	Co	Essential to a wide variety of organisms including certain bacteria
About 1-0.1 nM	W	W can replace Mo in several enzymes
	Ag	No evidence that it is a bioelement
	Zr	No evidence that it is a bioelement
	Hg	No evidence that it is a bioelement
	Nb	No evidence that it is a bioelement
	Ce	No evidence that it is a bioelement

in sea water such as molybdenum, iron, and zinc must have played important roles in the course of chemical evolution in the primeval sea, and proposed that the inorganic composition of the primeval sea was essentially similar to that of the present sea.

Based upon this idea, Egami, Yanagawa and coworkers have been trying an experimental approach to chemical evolution in the primeval sea. A "modified sea medium" was designed to accelerate possible chemical evolution reactions in the laboratory. It was enriched with 1,000-100,000 times higher concentrations of six transition elements: iron, molybdenum, zinc, copper, cobalt, and manganese. As mentioned above, it was found that amino acids and related compounds could be formed from formaldehyde and hydroxylamine (Hatanaka and Egami, 1977; Kamaluddin *et al.*, 1979), sugars and ammonia (Yanagawa *et al.*, 1980a, 1981), and carbon suboxide (C_3O_2, O=C=C=C=O) (Yanagawa and Egami, 1981) in the modified sea medium. Furthermore, it was found that marigranules, highly organized particles, were produced from a reaction of glycine and acidic, basic, and aromatic amino acids at 105°C in the modified sea medium (Yanagawa and

Egami, 1980; Yanagawa *et al.*, 1980b, 1988). We believed that metal ions in the modified sea medium played important roles in the formation of amino acids, peptides, and peptide-like polymers. When Egami and co-workers designed the modified sea medium enriched with high concentrations of six metal ions in 1973, they were unaware of a natural, near surface environment containing such high concentrations of metal ions. Submarine hydrothermal vents were discovered later and their metal ion concentrations are similar to those of the modified sea medium. Results of recent experiments using similar aqueous solutions are described in the next two sections.

2. Formation of Amino Acids under Simulated Submarine Hydrothermal System Conditions

A medium that approximated hydrothermal vent sea water, designated the modified hydrothermal vent medium (MHVM), contained 2mM $Fe(NH_4)_2(SO_4)_2$, 0.6 mM $MnCl_2$, 0.1 mM $ZnCl_2$, 0.1 mM $CuCl_2$, 20 mM $CaCl_2$, 0.1 mM $BaCl_2$, and 50 mM NH_4Cl. Its pH was adjusted to 3.6 with 1 N HCl. The metal ion concentrations and pH were set close to the reported value of hydrothermal vent sea water (Edmond *et al.*, 1982; Von Damm *et al.*, 1983). All water used was double distilled in glassware. All glassware was acid (7 M nitric acid) washed and rinsed repeatedly with double distilled water. All chemicals were analytical grade. The MHVM (30-50 ml) was put into a pyrex glass tube (22 mm x 190 mm), which was placed in a stainless steel autoclave (inner volume: 100ml). The autoclave was purged with nitrogen five times. A gas mixture of methane (40 kg/cm^2) and nitrogen (40 kg/cm^2) was then put into the autoclave at room temperature, which was heated at 325°C for 1.5-12 hours. After cooling to room temperature, the resulting aqueous solution was filtered through a membrane filter of 0.2 μm pore size, freeze-dried and hydrolyzed with 6 N hydrochloric acids for 24 hours. Amino acids were extracted from the acid-hydrolyzate with an AG-50WX8(Bio-Rad) resin column, and analyzed by ion-exchange chromatography (Irica Automated Amino Acid Analyzed Model A-5500), by gas chromatography (GC: Shimadzu Model GC-9A with a FID detector, column: Chirasil-Val capillary column) and by gas chromatography combined with mass spectrometry (GC/MS; Shimadzu Model GCMS-QP1000 column: Chirasil-Val capillary column).

Fig. 1 shows an ion exchange chromatrogram of (a) the amino acid fraction of the product compared to (b) a procedural blank. Amino acids such as glycine, alanine, aspartic acid, serine, and glutamic acid, together with non-proteinaceous amino acids (e.g., 2-aminobutyric acid and sarcosine) were detected in the chromoatogram. In other experiments, the yield of aspartic acid and glutamic acid increased when carbon dioxide was added to the starting reaction mixture. The major amino acid products: glycine, alanine, and sarcosine, were also identified by GC (Fig. 2) and/or GC/MS (Fig. 3). Evidence that these amino acids were synthesized abiotically from methane, includes: (1) the gas chromatographic analysis with an optically-active column indicated that the resulting

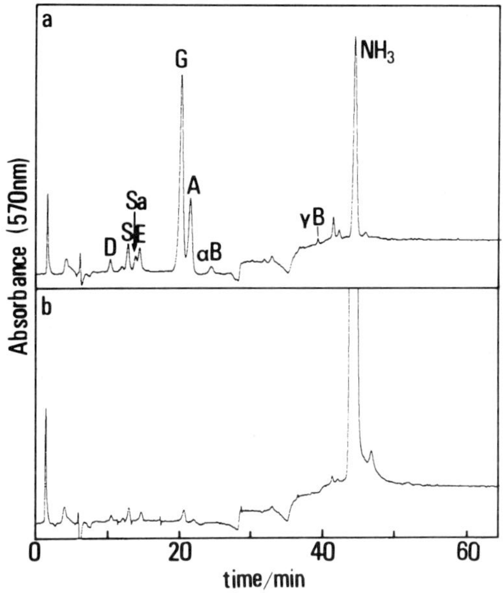

Fig. 1. Amino acid analysis of the products formed from methane and nitrogen in a modified hydrothermal vent medium by heating at 325°C and 200 kg/cm^2 for 1.5 hours (a) and a procedural blank (b) Abbreviations: D, aspartic acid; S, serine; Sa, sarcosine; E, glutamic acid; G, glycine; A alanine; αB, 2-aminobutyric acid; γB, 4-aminobutyric acid.

amino acids were racemic mixtures (Fig. 2); (2) non-proteinaceous amino acids such as sarcosine and 2-aminobutyric acid were detected (Fig. 1a); and (3) the procedural blank gave much lower amounts of amino acids (Fig. 1b). The yield of minor amino acids such as aspartic acid, serine, and glutamic acid were of the order of those in a procedural blank (Fig. 1a). Table 2 shows the yield of major amino acids synthesized in the present experiments. Larger amounts of glycine, alanine, and sarcosine were obtained when a longer heating period was applied. After the reaction, silica was found with Fe(III) and Mn(IV) in the precipitate. The silica was apparently derived from the pyrex glass wall. Effects of metal ions on the formation of amino acids was examined by removing metal ions from the MHVM, in which case the yield of amino acids was remarkably decreased. Some trace metals are now known to be essential to present-day life, and they may act as catalysts for prebiotic reactions in the course of chemical evolution (Kobayashi and Ponnamperuma, 1985).

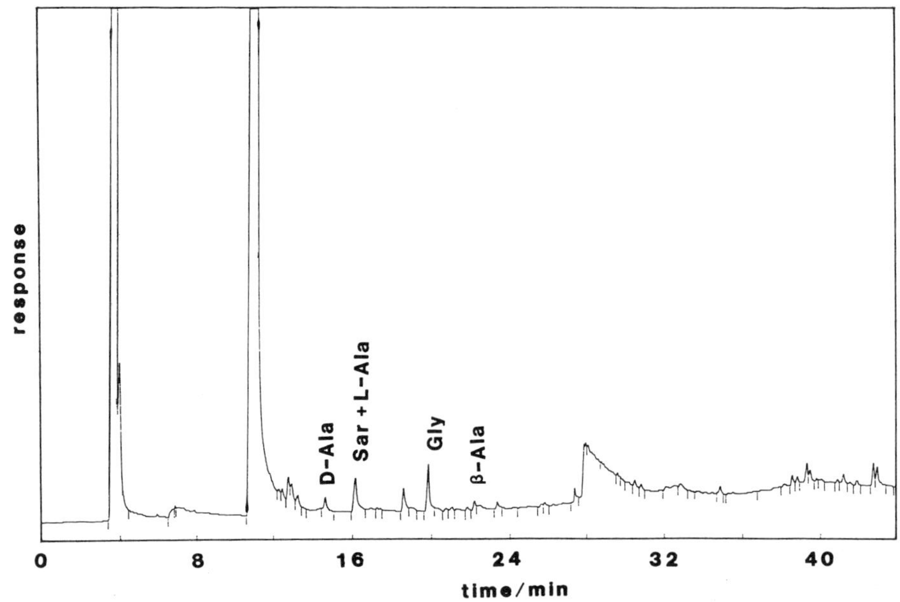

Fig. 2. Gas chromatogram of amino acids synthesized under hydrothermal vent condition. Reaction condition was the same as the run No. 3 in Table 2. Amino acids were derivatized with 2-propanol and N-trifluoroacetyl anhydride before injection.

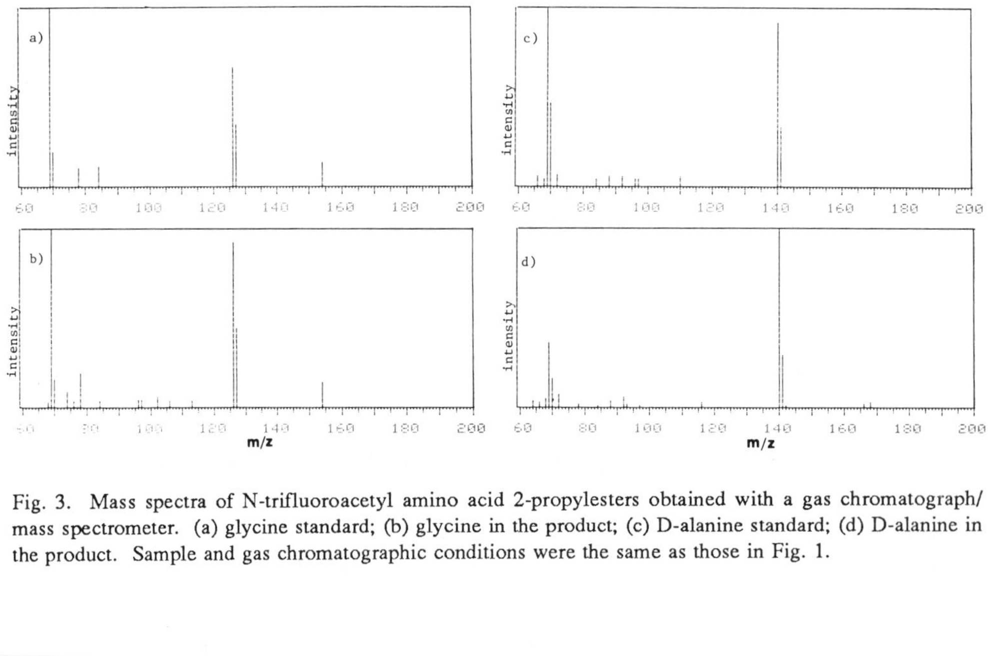

Fig. 3. Mass spectra of N-trifluoroacetyl amino acid 2-propylesters obtained with a gas chromatograph/ mass spectrometer. (a) glycine standard; (b) glycine in the product; (c) D-alanine standard; (d) D-alanine in the product. Sample and gas chromatographic conditions were the same as those in Fig. 1.

TABLE 2

Formation of amino acids in superheated hydrothermal environments

			Reaction System			
Reaction condition	Procedural blank		1	2	3	4
Medium	MHVM	MHVM	H_2O	MHVM	MHVM	MHVM
Volume (ml)	40	40	50	50	50	30
CH_4 (kg/cm^2)	40	0	40	40	40	40
N_2 (kg/cm^2)	40	80	40	40	40	20
CO_2 (kg/cm^2)	0	0	0	0	0	12
Temperature (°C)	25	325	325	325	325	260
Pressure (kg/cm^2)	80	200	200	200	200	140
Time (h)	12	3	12	1.5	12	6
Major amino acids produced (pmol/ml)						
Glycine	54	47	57	1190	1990 ($2.4x10^{-4}$)*	2400
Alanine	29	21	trace	600	750 ($1.4x10^{-4}$)	630
Sarcosine	0	0	trace	640	1480 ($2.7x10^{-4}$)	**

* The yield(%) on methane.

**Not separated from a large peak of glutamic acid.

The results summarized here support the possibility that amino acids such as glycine and alanine are produced abiotically in present-day submarine hydrothermal systems (Ingmanson and Dowler, 1980). The yield of amino acids was not high, because the experimental set up is a closed system, and amino acids synthesized in the thermal reaction would be thermally decomposed. If reactions were carried out in flow reactors like actual submarine hydrothermal systems (Corliss, 1986), the yield would be expected to increase since the products would be moved away from the threat of thermal decomposition.

3. Formation of Thermophilic Microspheres and Peptide-Like Polymers under Simulated Submarine Hydrothermal System Conditions

Experimental results summarized above indicate that abiotic synthesis of amino acids and other organic compounds from inorganic starting materials may be possible in submarine hydrothermal systems. Questions as yet unanswered concern the role of hydrothermal environments in the abiotic synthesis of polypeptides. Results summarized below show that extremely thermophilic microspheres and peptide-like polymers can be formed in aqueous solution at 250-350°C.

The four amino acids glycine, alanine, valine, and aspartic acid have been used as starting materials for prebiotic synthesis of polypeptides (Yanagawa *et al.*, 1990; Sakurai and Yanagawa, 1984; Yanagawa, 1984). These amino acids were chosen because they are the more abundant of the amino acids formed in prebiotic synthesis experiments (Miller and Orgel, 1973), and are found in the Murchison meteorite (Kvenvolden *et al.*, 1970). Thus glycine, alanine, valine, and aspartic acid are likely to have been present in the early stages of chemical evolution (Eigen and Shuster, 1978; Shimizu, 1981; Egami, 1981).

When an aqueous solution (15 ml) containing 0.3 M glycine, 0.1 M L-alanine, 0.3 M L-valine, and 0.1 M L-aspartic acid was put in a glass tube and heated at 250°C and 134 atm for 6 hours, numerous microspheres were formed (Yanagawa and Kojima, 1985). These are organized particles of 1.5 to 2.5 μm in diameter. Fig. 4a shows the microspheres observed on scanning electron microscopy. The microspheres had a boundary layer structure whose thickness was about 30 nm (Fig. 4b). They were also obtained in reactions at 300°C, though their spherical structures were deformed (Fig. 4c and d).

Microspheres were not obtained in the reaction at 200°C. However, IR and NMR studies on the heated mixture revealed the presence of amide bonds and the near absence of imide bonds. After hydrolysis of products, glycine, alanine, valine, and aspartic acid were found in nearly equal amounts. When the reaction mixture was heated at 250°C and 134 atm for 6 hours in a stainless steel vessel, no microspheres were observed. Microspheres were not obtained from mixtures of glycine alone, of glycine-alanine or of glycine-alanine-valine heated at 250°C for 6 hours, suggesting that the presence of aspartic acid is crucial for microsphere production. Other polar amino acids such as glutamic acid, lysine, arginine, histidine, serine, threonine, and 4-hydroxyproline could be used instead of aspartic acids to form microspheres (data not shown). Especially, basic amino acids (lysine, arginine, and histidine) resulted in the formation of larger microspheres (4-8 μm in diameter). These results indicate that a polar amino acid, a glass tube, and a reaction temperature of 250°C or above are necessary for the formation of the microspheres.

Elemental analysis of the microspheres showed a percentage composition of C, 41.7; H, 4.3; N, 5.0; Si 12.0%; and others, e.g. oxygen and ash. The IR spectrum of the microspheres showed strong adsorption bands at 3400, 2970, 2940, 2870, 1670-1630, 1090, and 470 cm^{-1}. When the microspheres were treated with hydrofluoric acid, two clear bands appeared at 1670 and 1630 cm^{-1} that can be attributed to an imide bond and an amino group (NH_3^+ deformation vibration of an amino acid). The HF treatment suggested the possibility that the amino group was linked with silicates. Two bands at 1090 and 470 cm^{-1} could be attributed to the Si-O-Si bond. The characteristic band (amide II band) of a peptide bond at 1550 cm^{-1} was absent.

The microspheres were dissolved with 1% SDS-8 M urea or 0.5% hydrofluoric acids. The microspheres completely disappeared on treatment with 1% SDS-8 M urea at 100°C

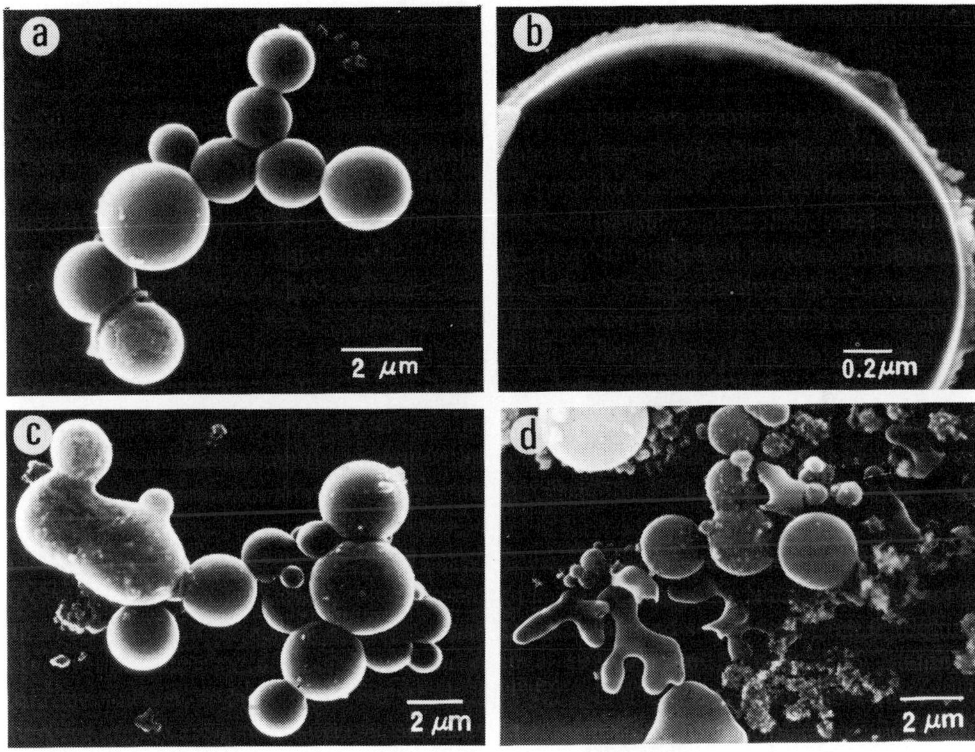

Fig. 4. Formation of microspheres in a reaction of mixture containing glycine, alanine, valine, and aspartic acid at 250°C(a), 300°C(c), and 350°C(d) under a hydrostatic pressure of 134 atm, as well as the interior structure of the microspheres produced at 250°C(b). A reaction mixture (15 ml) containing 0.3 M glycine, 0.1 M L-alanine, 0.3 M L-valine, 0.1 M L-aspartic acid, and 0.1 M $KHCO_3$-NaH_2PO_4 buffer (pH 7.2) was put into a pyrex tube (20 x 105 mm), which was capped and placed in a stainless steel autoclave (70 ml) which was encased in an aluminum heating block. The autoclave was heated at 200°C, 250°C, 300°C, or 350°C for 6 hours. The resulting solution was centrifuged and the precipitate was placed on a clean glass coverslip. The dried specimens were examined under a scanning electron microscope (JEM 100 CX-ASID). In order to observe the interior of the microsphere, freeze dried specimens were cracked using two pieces of adhesive vinyl tape, and then examined under the microscope.

for 10 min or 0.5% hydrofluoric acid at room temperature for 10 min. Polyacrylamide gel electrophoresis of the 1% SDS-8 M urea-solubilized components of the microspheres gave, after silver staining, a broad band with molecular weights of 1,000-2,000 daltons (Fig. 5a). The microspheres were partially hydrolyzed with 6 N HCl at 110°C for 72 hours. After the hydrolysis, 10% of the total amide bonds were cleaved and silica was found in the hydrolyzate. These results suggest that the microspheres are made of peptide-like polymers with the following characteristics: 1) resistance to acid hydrolysis, 2) silicon bonds are included, and 3) molecular weights of 1,000-2,000.

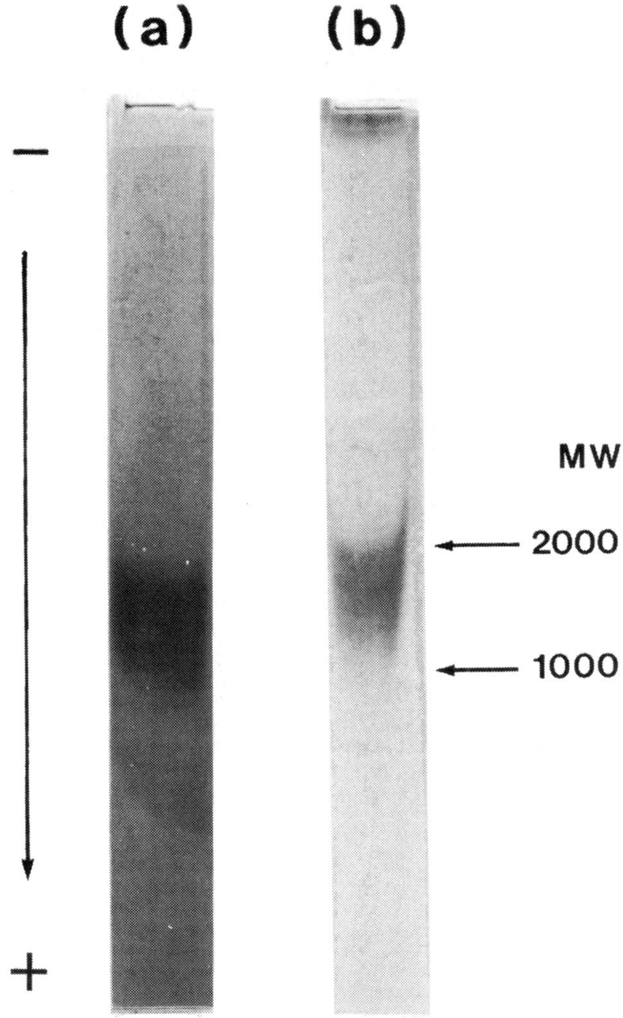

Fig. 5. Polyacrylamide gel electrophoresis of thermostable microspheres and supernatants of reaction mixture. (a) 1% SDS-8 M urea-solubilized components of microspheres produced at 250°C; (b) supernatants of reaction mixture heated at 250°C.

Peptide-like polymers were also present in the supernatant of the reaction mixture, and were found to remain on an Amicon YM-2 membrane after ultrafiltration. Polyacrylamide gel electrophoresis of peptide-like polymers in the supernatant gave, after silver staining, a broad band with molecular weights of 1,000-2,000 daltons (Fig. 5b). The electrophoretic pattern was very similar to that of the 1% SDS-8 M urea-solubilized component of the microspheres. The IR spectrum of the peptide-like polymers obtained from the supernatant showed characteristic bands at 1670 (broad) and 1570 cm^{-1} which can be attributed to the imide and peptide bonds, respectively. They were partially hydrolyzed with 6 N HCl at 110°C for 72 hours. After such acid hydrolysis, 50% of the total amide bonds were cleaved and silica detected.

^1H NMR in D_2O of the peptide-like polymers found in the supernatant showed a signal at 0.82-1.10 ppm that suggested the presence of valine residues. Similarly, β-CH_3 of alanine residues, α-CH_2 of aspartic acids residues, and β-CH_2 of glycine residues appeared at 1.44-1.52 ppm, 2.42-2.98 ppm , and 3.58-2.62 ppm, respectively. The intensity of signals from valine residues was 4-fold that for glycine and that for alanine. ^1H NMR in DMSO-d_6 showed a signal at 8.07 ppm that suggested the presence of peptide bonds. Signals on ^{13}C NMR in D_2O at 18.5-19.3 ppm, 19.4-22.2 ppm., 44.7-46.1 ppm, and 52.7-56.9 ppm were assigned to β-CH_3 of alanine residues, γ-CH_3 of valine residues, α-CH_2 of glycine residues, and α-CH of aspartic acid residues, respectively. Similarly, the resonances at 173.7-179.2 ppm were assigned to the carbon of amide bonds, and the resonances at 183.7-188.3 ppm to the carboxyl group. These NMR studies showed that the peptide-like polymers contain valine, alanine, glycine, and aspartic acid residues which are linked by imide bonds and peptide bonds. Judging from the intensity of the signals, the content of valine was 4-fold that of any other amino acid.

Peptide-like polymers were also present in the supernatant of the reaction mixture. Fig. 6 shows gel chromatograms of the products of several reaction mixtures heated at different temperatures. The reaction mixture at 200°C gave a single peak with a molecular weight of 1,100 daltons. This peak can barely be detected at 100°C, increases at 150°C and is the sole product at 200°C (Fig. 6a). At 250°C several peaks with molecular weights up to 4,000 daltons appear in the chromatogram (Fig. 6b). A gel chromatogram of the product of the reaction mixture heated at 250°C in a stainless steel vessel was much the same as that of the reaction mixture heated at 250°C in a glass tube. At 300°C these peaks disappeared and a peak with a large molecular weight and a peak with a small molecular weight survived. (Fig. 6c). Judging from the fact that polyacrylamide gel electrophoresis of the large molecular weight peak gave a broad band with a molecular weight of 1000-2000 daltons (Fig. 5), this large molecular weight peak seems to be an aggregated form of peptide-like polymers. Apparently at 250°C and 300°C aggregation and degradation of the peptide-like polymers occurs simultaneously. At 350°C these peaks decreased considerably (Fig. 6d). As in the results discussed above, the NMR spectrum of the peptide-like polymers showed a high content of valine. The resulting peptide-like polymers were resistant to acid hydrolysis. The fraction of the polymers susceptible to hydrolysis decreased with a rise in the temperature of hydrolysis.

Based upon the physical data described above, a possible chemical structure of the peptide-like polymers of microspheres is considered. They consist of imide bonds and amino acid residues having an abundance of valine. The presence of imide bonds results in the branched structure of the polymer. Some amino groups near the branching point are linked by an aminosilyl (N-Si) bond to silica. The silica may contribute to the hy-

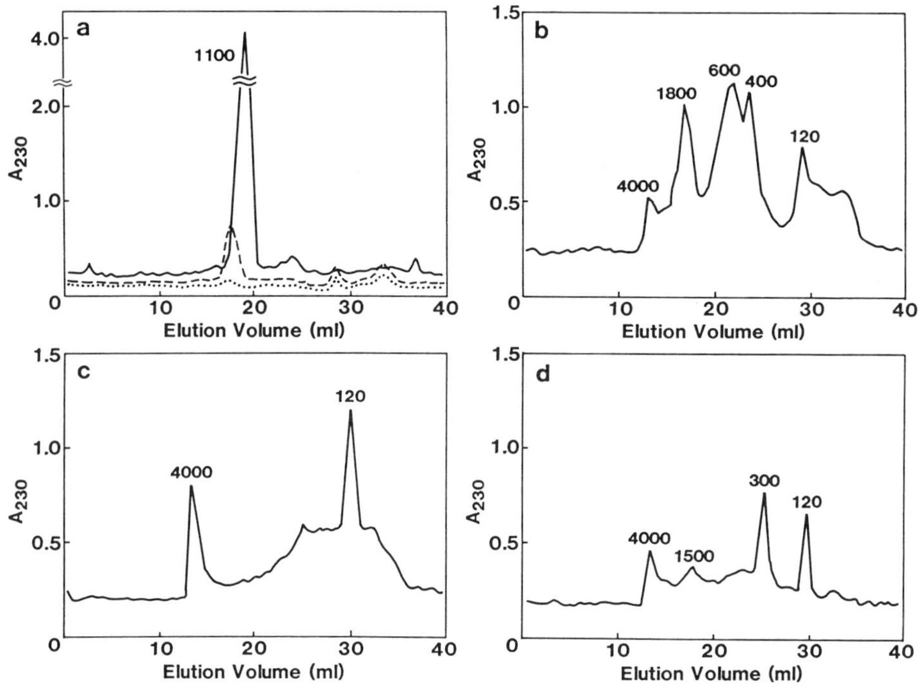

Fig. 6. Bio-Gel P-4 chromatography of supernatants of a reaction mixtures heated at different temperatures. (a) Supernatants of reaction mixtures heated at 100°C (...), 150°C(----), and 200°C(——) for 6 hours; (b), (c), and (d), supernatants of reaction mixtures heated at 250°C, 300°C and 350°C for 6 hours, respectively. Numerical values over peaks represent molecular weight of the peaks.

drophilic moiety of the polymers. The peptide-like polymers in the supernatant may be precursors of the peptide-like polymers found in the microspheres.

Concerning the stability of the aminosilyl bond, there must be some protective mechanisms. A peptide bond between two valines was very resistant to acid hydrolysis. For example, only 5% of the total peptide bonds of polyvaline were hydrolyzed with 6 N HCl at 110°C for 72 hours: only with trifluoroacetic acid-12 N HCl (1:1) at 175°C for 24 hours was polyvaline completely hydrolyzed. Thus, the aminosilyl bond may be protected by the hydrophobic moiety of valine residues from attack by water.

It may be that at lower temperatures of the reaction, polypeptides are produced that are rich in glycine, alanine, valine, and aspartic acid. (Yanagawa et al., 1990; Ito *et al.*, 1990). Peptide bonds composed of glycine, alanine, and aspartic acid may be preferentially cleaved because of their relative instability at 250°C as compared to peptide bonds between valine residues which can withstand hydrolysis at high temperature. Thus, the content of valine increases and peptide-like polymers having a preponderance of valine survive.

4. Implications for Hydrothermal Synthesis of Organic Compounds and Thermophilic Life

The experimental results summarized above indicate that abiotic synthesis of organic compounds may be possible in submarine hydrothermal systems. In addition, these results show that polymerization reactions may occur among organic compounds in hydrothermal fluids. These results are consistent with theoretical calculations which show enhanced stability of peptide bonds with increasing temperature in aqueous solution (Shock, 1992a). Further experiments on systems in which the pH, oxidation state and activities of inorganic ions are buffered to values consistent with observations of submarine hydrothermal systems will help clarify the extent to which the present results reflect what can happen in natural systems. One reason why these experiments demonstrate amino acid production rather than high temperature destruction documented by others (White, 1984; Miller and Bada, 1988; see above) may be the capacity of the high concentrations of transition metals in the solutions to constrain the oxidation state to fairly reduced levels.

These experiments may also have implications for life at high temperatures. Numerous studies have shown that extremely thermophilic organisms are found in submarine springs at temperatures of at least 110°C (Stetter, 1982; Jones *et al.*, 1983, 1989; Fiala and Stetter, 1986; Stetter *et al.*, 1983; Jannasch *et al.*, 1988; Zhao *et al.*, 1988; Huber *et al.*, 1989; Zillig *et al.*, 1990). Although some investigators argue against the possibility of life at temperature in excess of 200°C (Trent *et al.*, 1984; White, 1984; Miller and Bada, 1988, 1991a,b) recent results from submarine hydrothermal vents along the Endeavor Ridge show that particulate DNA is recovered from the fluids and survives incubation for several hours at vent temperature (Deming, 1991).

These results may indicate that life exists in present-day submarine hydrothermal systems at temperatures exceeding those for which bacteria have so far been cultured. Thermophilic bacteria capable of growing in hydrothermal environments are expected to have special mechanisms for adaptation to these environments (Baross, 1991). The results summarized above indicate that extremely thermophilic cellular structures may be possible if they are somehow analogous to the microspheres produced from amino acids at 250°C.

Acknowledgements

We thank Dr. T. Onoda, Dr. K. Wada, Dr. M. Dokiya, and Y. Kasori for performing reactions at high temperatures, Dr. T. Hayase and H. Kosono for obtaining the NMR spectra, S. Kondo for obtaining the electron micrographs. We express our gratitude to Prof. E. L. Shock for helpful discussion and providing some text. This work was partly supported by a Grant-in-Aid (No. 63540445) from the Ministry of Education, Science and Culture, Japan.

CHAPTER 9

MINERAL THEORIES OF THE ORIGIN OF LIFE AND AN IRON SULFIDE EXAMPLE

A. GRAHAM CAIRNS-SMITH

Department of Chemistry, University of Glasgow, Glasgow G12 8QQ, Scotland

ALAN J. HALL AND MICHAEL J. RUSSELL

Department of Geology and Applied Geology, Glasgow G12 8QQ, Scotland

"Oh! let us never, never doubt
What nobody is sure about"

Hilaire Belloc.

1. Introduction

If we really believe that a process of Darwinian evolution started on the Earth without the help of miracles, spacemen or whatever, then the first organisms (by which we mean the first systems able to evolve indefinitely through natural selection) must have been made of materials that were, broadly speaking, geochemicals.

The usual view has been to suppose that the critical geochemical materials were a special class of organic compounds generated by atmospheric, oceanic and other geochemical processes that were special to the early Earth. The classical picture is of life arising from an oceanic primordial soup, a vast stockpile of the right sorts of organic molecules that had been built up as a result of various energy inputs to a reducing atmosphere. Almost without discussion "the right sorts of organic molecules" for the most primitive organisms were taken to be the same as those molecules which are universal to life now. Experiments have confirmed that amino acids at least can be produced from mixtures of reducing gases subjected to artificial lightning (Miller's famous sparking experiment, 1953). Helped by the simplicity and dramatic qualities of Miller's beautiful experiment the "primordial soup" doctrine became established to the point at which most scientists seem to believe it. But there have always been doubts, particularly more recently (Hull, 1960; Hulett, 1969; Cairns-Smith, 1982; Shapiro, 1986). For one thing the idea that there was ever a strongly reducing atmosphere has little to support it, beyond a belief that such an atmosphere would have been essential for life to have started (Chang *et al.*, 1983). Even if there had been such an atmosphere it would probably not have lasted long since it would have been photochemically unstable (Wayne, 1991). Although

Origins of Life and Evolution of the Biosphere **22**: 161–180, 1992.

the matter is still being hotly debated, most current models of planet formation give the early Earth a more or less redox neutral atmosphere, an atmosphere more like that of Mars or Venus than of Jupiter, with CO_2, H_2O and perhaps N_2 rather than CH_4 and NH_3 as dominant constituents (e.g. Kerr, 1980). Large scale organic synthesis becomes progressively more difficult as conditions become less reducing. As Miller has shown more recently yields from sparking experiments with mixtures of H_2 and CO_2 become negligible as H_2:CO_2 ratios fall below unity (Chang *et al.*, 1983; Schlesinger and Miller, 1983). In any case the Isua rocks with their carbonates and banded iron formations speak of an Earth whose atmospheric and hydrospheric chemistry was not strongly reducing (Allaart, 1976).

Then again ultraviolet light, in spite of its potential for prebiotic photosynthetic activity in surface waters (Getoff, 1962; Akermark *et al.*, 1980) would actually have tended to *clear* the oceans of organic molecules through oxidation by photochemically generated ferric iron - crudely:

$$Fe^{2+} + H^{+} + h\nu \rightarrow Fe^{3+} + \tfrac{1}{2}H_2 . \qquad (i)$$

Under near neutral conditions of pH this kind of reaction is particularly efficient and leads to ferric precipitates (Braterman *et al.*, 1983; Borowska and Mauzerall, 1987). These sinking precipitates would have tended to adsorb and sediment organic materials, sooner or later oxidizing them back to CO_2 (Braterman and Cairns-Smith, 1987). So a tendency for surface waters to become slightly oxidizing under the Archaean sunlight should be set against the optimistic assessments of Mauzerall (1990) and de Duve (1991) that the atmosphere would have been more reducing than it might otherwise have been due to hydrogen production through the above kind of process. The present surface of Mars is a bleak reminder that although sunlight can produce intricate and interesting local effects - such as life now on Earth - an overall effect of, especially, its ultraviolet component on the surface of a planet is to destroy organic molecules (Armstrong *et al.*, 1966). The harder u.v. photons that can get through most ozone-free atmospheres can break many kinds of covalent bonds. In particular hydrogen atoms are broken off, and these have a net tendency to escape to space. The result is a surface oxidizing effect. If we want reducing conditions to favour the production and survival of organic molecules we should be thinking about places away from the surface, more in touch, perhaps, with the vast reservoir of reducing power in the form of, especially, iron (II), and sulfide within and beneath the crust. Hence the interest in hydrothermal systems and our emphasis later in this chapter on iron sulfide minerals.

2. Mineral Theories - Conservative and Radical

The geochemist V. M. Goldschmidt wrote a short speculative piece in 1945, two years before his death, on "Geochemical aspects of the origin of complex molecules on the

Earth, as precursors to organic life". This was published in 1952, but has received comparatively little attention. Goldschmidt suggested that carbon dioxide was probably "the main material available for the formation of primitive organic molecules, apart from subordinate hydrides of carbon, and very subordinate amounts of CO". He went on to suggest that concentrations of organic molecules would have been rather low, and that adsorption of simple organic molecules at phase boundaries, and especially on the surfaces of mineral crystals, would thus have been critical in giving rise to organized higher order aggregates. He wrote: "One most important factor for selection and arrangement of adsorbed molecules is their size and shape as related to the atomic arrangement of the inorganic substratum. Besides such two-dimensional layers of more or less loosely fixed molecules there certainly must exist some analogous linear phenomena along the edges of crystals, and this would make monodimensional strings or chains of adsorbed particles". He then went on to consider potential catalytic effects of "local fluctuations and concentrations of vibrational energy, which are possible in crystal lattices...." and pointed to the potential for quartz crystals to confer enantiomeric specificity.

Bernal had independently speculated along similar lines and in 1951 stressed the potential importance of clay minerals. Montmorillonite especially was known to have particular affinities for organic molecules, adsorbing them on and between their unit silicate layers. Bernal suggested that these minerals might have acted as heterogeneous catalysts in the building up of larger molecules. This idea has been used in many subsequent experiments and speculative theories.

Most theories of this sort have been *conservative*. By this we mean they hang on to the idea that the first forms of life, the first organisms able to evolve through natural selection, were made, broadly speaking, of the same materials as are organisms today. Minerals were environmental agents of a purely "chemical evolution" which provided an initial feedstock of, it is usually supposed, such familiar "biochemicals" as amino acids, sugars and so on.

Pirie (1959) had a more radical view. He suggested that to begin with there might have been many different biochemistries based on different elements, which settled selectively on our kind of biochemistry at a later stage. He saw the unity of biochemistry now as attesting to the effectiveness of natural selection rather than to the idea that life is only possible, or only ever happened, in one way, with one set of substances. He was particularly critical of the idea that early biological catalysts would have been made of protein: proteins are used now because the means to produce them has been evolved and because they work better, but are less plausible under primitive conditions.

It seems indeed entirely reasonable to suppose that the first organisms would have been made in a way that was particularly accessible, expecially in keeping with a current local geochemistry. Then later on initial subsystems would have been replaced by more efficient ones. This is how evolutionary and other forms of technological innovation commonly proceed - starting with what is easiest, ending with what is most efficient.

The extraordinary fixation of all life on Earth on a particular set of biochemical molecules is now generally agreed to be the result of the descent of all life now on Earth from a last common ancestor - sometimes called the universal ancestor. According to this view we all now share the same central biochemistry because the universal ancestor had this system already irreversibly built-in: DNA, RNA, proteins made from 20 amino acids, a code to translate from the nucleic acids to protein, and complex machine tools - ribosomes - to do it. The sophistication of that common ancestor has been becoming more and more apparent (Cairns-Smith, 1985). It is evidently a substantial evolutionary distance from the first organisms: the universal ancestor was surely the result of a major earlier phase of the (fully Darwinian) evolution. It seems likely that this was the phase of evolution *during which* the biochemical system of the universal ancestor was being invented - and that part of this process would have been the choice of appropriate molecules. It is out of keeping with the well understood lack of foresight in evolutionary processes (e.g. Dawkins, 1986) to insist that this choice would have been made at the outset.

So we can describe *radical* mineral theories of the origin of life as being those that would put minerals into the first organisms as part of their structure. The most radical suggestion has been that the major *control structures*, catalysts, membranes and above all genes, were all of them inorganic in nature in the first organisms, and that these were gradually replaced during the early phases of evolution as competence in the control of organic chemical reactions improved (Cairns-Smith, 1982). We will not pursue this "take over" idea in detail here except to note that given such means of updating early biochemical technologies we must be interested in the entire geochemistry of the hydrothermal systems. It is not just a matter of searching for "the right molecules". Although knowing which if any of the classes of current biochemicals can be made easily or survive well under hydrothermal conditions remains of interest, if our primary concern is with the very earliest stages of evolution we need to know without prejudice which molecules, minerals and higher order structures, are easily formed and/or reasonably stable under hydrothermal conditions. Then we can begin to assess the most likely materials for the earliest membranes, catalysts, genes and so on. So which minerals forming in hydrothermal systems look promising?

3. Layer Silicates

"Conservatives" and "radicals" can agree that clay minerals are of interest in that they interact with organic molecules, as already discussed. In general, however, simple clays have not proved to be particularly effective adsorbers of "biochemicals" from dilute aqueous solutions (Lahav and Chang, 1976) or for that matter as catalysts for making basic "biochemicals" under aqueous conditions, as commented on by Cairns-Smith (1982) and Miller and Bada (1988). They have, however, proved useful in providing surfaces to help bring about limited amino acid condensations in the laboratory through forced cycles of wetting and drying (Lahav *et al.*, 1978) and they can adsorb nucleotides very effectively in the presence of transition metal cations (Odom *et al.*, 1979; Lawless *et al.*, 1985; Lawless, 1986).

At more sophisticated levels montmorillonite clays have been shown to exert control on the polymerization of activated amino acids - aminoacyl adenylates (e.g. Paecht-Horowitz *et al.*, 1970; Paecht -Horowitz and Eirich, 1988) and to assist in the oligomerization of various activated nucleotides (Ferris *et al.*, 1989, 1990; Ferris and Kamaluddin 1989). Then again Mortland (1984) has demonstrated selective deamination of glutamic acid with a "pesudo-enzyme" consisting of a smectite clay with bound Cu^{2+} and pyridoxal phosphate (see Pinnavaia and Mortland, 1986 for a review).

"Conservatives" tend to view such experimental results with caution: aminoacyl adenylates, pyridoxal phosphate, activated nucleotides; such molecules are now widely and properly seen as unlikely prebiotic geochemicals, whether in an ocean or a hydrothermal system. So maybe these are just model experiments, helping us to see the kinds of things that clays and other such minerals can do in the way of controlling organic reactions. (This is the attitude taken by Ferris and Kamaluddin (1989) for example.) No doubt such caution is wise, but radical theorists have reason to be more positive. If minerals were for a long time components of those early evolving systems, components that would only have been displaced once something like the modern system of genetic control had become established, then we would expect a synergy between mineral components of the old system and molecular components of the new. For a considerable time they must have had to work together.

For organisms elaborating pre-protein techniques of organic synthesis, barriers to diffusion - inert "glassware" - might have been even more important than in organisms now, and it is perhaps not too fanciful to think of montmorillonite and similar clays as standing in for lipid bilayer membranes which predominantly serve this function now. Individual silicate layers of clays are tough and flexible and, incidently, of similar thickness to lipid bilayers. Imogolite, halloysite and allophane naturally form tubes and vesicles, and clay minerals generally create complex compartmentalized microenvironments (Cairns-Smith, 1982).

Conditions in hydrothermal systems are conducive to the formation of relatively well crystallized clays and other related layer silicates such as micas. Phlogopite, for example, forms rapidly under hydrothermal conditions (*ca.* 500°C) in the laboratory as Baronnet (1972, 1980) has demonstrated. Special interest attaches here to the phenomenon of long range polytypism in connection with a possible primitive genetic system. We will now discuss this briefly.

The nanometer thick unit layers of micas have a directional structure such that each successive layer in a mica crystal can be stacked on top of the layer below it in one of three orientations. Often the stacking sequence is simple - e.g. always with the same orientation or always alternating between two orientations - but complicated more random-like sequences are common and, interestingly, some quite complicated sequences are often repeated again and again within a single stack. Apparently this arises through the operation of a screw dislocation growth mechanism (Baronnet, 1980). A random-like stacking sequence can hold "information", rather as the sequence of base pairs stacked inside a DNA molecule can, and the repetition of the same such sequence in a crystal is strongly suggestive of a replication of this "information" during crystal growth. This raises the possibility that a mica-like mineral could have been a primitive genetic material operating perhaps on surrounding organic molecules in the kind of way that Goldschmidt (1952) suggested, but through sequence dependent groove patterns that generally develop on the edge surfaces of polytype stacks (Cairns-Smith, 1988).

4. Zeolites

Zeolites can form easily under mild hydrothermal conditions. In addition to their well known size selectivity for organic molecules these minerals have the potential for information storage - especially the ABC-6 type zeolites where slabs of subtly different crystal structure can be stacked on top of each other to form an infinity of possible mixed intergrowths (Milward *et al.*, 1985) some perhaps with distinctive adsorptive and catalytic properties. This is similar to polytypism, so the possibility of a screw replication mechanism exists here too (Cairns-Smith, 1988). Their catalytic properties and concentration in the vicinity of "black smokers" have been remarked on with respect to their possible implication in the origin of life by Nisbet (1986).

5. Oxides/Hydroxides

Of considerable interest is a large class of layer structures with adsorptive, catalytic and membrane-forming properties which have recently been studied in detail by Arrhenius and his group (Kuma *et al.*, 1989) and discussed by them in relation to their possible significance for the origin of life. These are mixed valence hydroxides of the pyroaurite type. They can be formed in the laboratory (at pH 8-9 or above, and at temperatures ranging from 25-200°C according to type) and can be made much more quickly and easily than most clays. They can be thought of as modifications of the basic brucite structure (which consists of a stack of uncharged $M(OH)_2$ layers where M is a divalent ion such as Mg, Ca, Fe, Mn...). Substitution of some of the M^{2+} ions by trivalent ions such as Fe^{3+} or Al^{3+} creates positively charged layers between which are located anions such as Cl^- or CO_3^{2-} to balance the charges. The interlayer anions can often be exchanged - for example with CN^- or organic anions such as acetate, benzoate, dodecylsulfate, or the nucleotide adenosine monophosphate.

Oxide-hydroxides of iron and manganese are another group of minerals with anion exchange capacities, often membranous structure, and similar conditions of formation to the pyroaurite group. Arrhenius (1986, 1990) has suggested key probiotic roles for both of these general classes of minerals - especially the iron ones - in building organic molecules from cyanide and formaldehyde.

Granick (1957) suggested a role for magnetite as part of a primitive photosynthetic device. This was never realized experimentally, but the potential of defect magnetite for reduction of carbon dioxide in the dark has recently been demonstrated (Tamaura and Tabata, 1990).

Using Inelastic Electron Tunnelling microscopy, Field and Spencer (1990) have demonstrated the production of monolayer amounts of formic acid on alumina surfaces from carbon dioxide and water; glycine from carbon dioxide, water and ammonia; and sugar-like products from carbon monoxide and water.

6. Phosphates

We should add phosphates to our list of especially interesting minerals in view of the key position of phosphate in our biochemistry. Phosphorous is often the limiting nutrient for life now on Earth (the concentrations of phosphate in bulk natural waters is of the order of 10^{-7} molar; Li, 1982). One might speculate that this now inconvenient dependence on phosphate arose because in the environment within which the first organisms evolved phosphate was relatively abundant. Certainly phosphate approaches a concentration of about 5 wt% in the iron and zinc rich layers of the most ancient of the giant exhalative-sedimentary base-metal orebodies, Broken Hill in Australia (Stanton, 1976). Stanton *et al.* (1978) speculate that the phosphorous here may have been introduced as the pentasulfide and upon hydrolysis and/or double decomposition it deposited as apatite or an iron phosphate. The requirement may have been, as now, for soluble phosphate, although there have also been speculations about the possible roles of hydroxyapatite and other phosphate mineral surfaces (see Orgel, 1986). Yamagata *et al.* (1991) have shown both from experiments at around 1300°C simulating magmatic conditions, and from volcanic gas analysis, that water-soluble polyphosphates may be produced by the partial hydrolysis of P_4O_{10}.

7. Sulfides

Sulfides too, although not as prominent as phosphates in our present biochemistry, have central roles (e.g. in ferredoxins and the thioester linkage of coenzyme A). These appear also to be ancient and sulfides may have been more important in the earliest phases of evolution (Hartman, 1975; Wächtershäuster, 1988a,b; de Duve, 1991). In relation to theories of the origin of life in hydrothermal systems they have an added interest since sulfides, especially iron sulfides, are dominant minerals here. As with other minerals the interest of these materials is not *only* that they might have been catalysts and reagents, that is to say early versions of the sulfide components of present day biochemistry, but again for possible wider roles connected with those altogether more primitive biochemistries which surely preceeded ours. For example we have speculated (Russell *et al.*, 1988), that the demonstrable tendencies for iron sulfides to form tubes and botryoids (hollow pyrite hemispheres <1 cm diameter) might have provided complex reaction vessels on scales from microns to millimeters to help in the control of organic reactions. Such control is part and parcel of normal organic chemical manipulations and is in addition to catalytic control. It operates by separating solutions of different chemical composition, different pH, Eh and so on, as well as helping to arrange for solutions to be appropriately mixed - and for products to be purified through chromatographic effects. In present day organisms, maintained chemical potentials across membranes, e.g. proton potentials in mitochondria, serve as immediate power sources for redox and other reactions. Such potentials will inevitably exist within the complex plumbing of hydrothermal springs where, for example, solutions flowing within sulfide tubes have a different pH and/or Eh from the surrounding seawater - suggesting the possibility of reactions being locally driven against thermodynamic gradients.

Effects like this can be imagined from the more "conservative" point of view as having helped to provide dynamic sources of feedstock and/or niches for early organisms. On the more "radical" tack touched on earlier we should keep in mind the possibility that at some stage a geochemical evolution became a biological evolution too, that materials were being made and higher order structures were being formed under some kind(s) of genetic control - and that (for example) sulfide tubes and more especially bubbles (protobotryoids) might have become actual structural components of early organisms themselves, standing in, so to speak, for the protein-lipid membranes that constitute most of the containment, energy transducing, and other "apparatus" in organisms today.

Let us move now to a more detailed - "radical" - model of the origin of life in hydrothermal systems based on one of the most abundant kinds of minerals that are being made in them - iron sulfides.

8. A Mineral Model Based on Iron Sulfides

Corliss and his co-workers (1981), spurred on by the discovery of present day hydrothermal systems along ocean ridge spreading centers (Corliss, 1979), especially "black smokers" (Spiess *et al.*, 1980) suggested that such hot springs, which discharge ~150 liters of recycled, modified seawater a second at up to 360°C at about 250 bars, would have been ideal places to focus the chemical constituents and thermal energy required for life's origins. More recently Corliss (1990) has argued that in the hot spring environment thermal energy may be converted to high energy chemical bonds during rapid quenching and that the steep thermal and chemical gradients in this environment would encourage the formation of complex molecules. As a generality this is an attractive idea but it assumes a similar hydrothermal chemistry at the time when life was initiated, about four billion years ago. An example of evidence against this assumption is that the formation of sulfide chimneys at black smokers depends on replacement of calcium sulfate which precipitates directly from cold ocean water as it is heated through entrainment in the hot water (Haymon and Kastner, 1981). Calcium sulfate has an inverse solubility with respect to temperature and therefore is physically dumped as a sleeve or chimney around the hot plume. The chimney supplies a protective environment and is replaced by sulfides which are precipitated as the acid hydrothermal solution mixes with alkaline seawater (Fig. 1a).

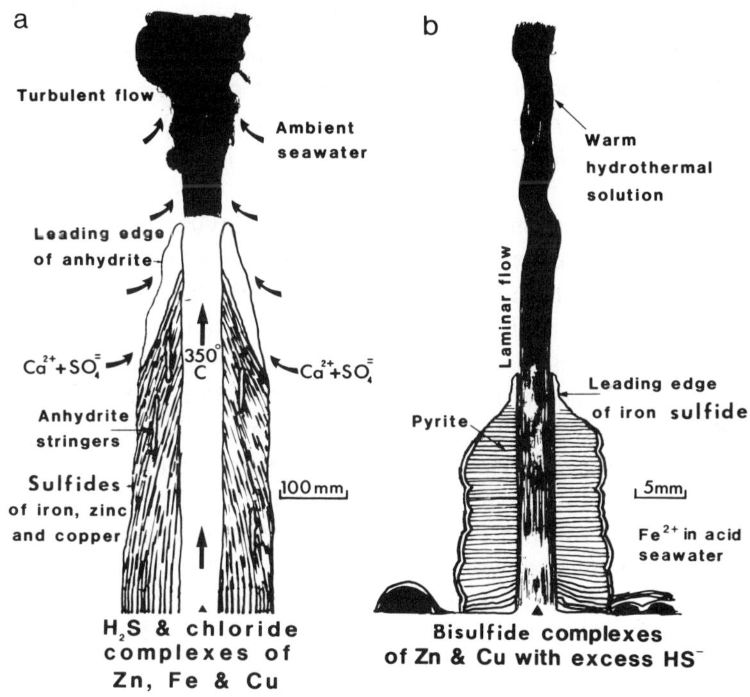

Fig. 1. Contrasting processes producing the two types of hydrothermal chimneys mentioned in the text. a) Growth of a black smoker by preliminary precipitation of anhydrite ($CaSO_4$) from ambient cold seawater on heating by the 350°C hydrothermal solution. Once this sleeve has been produced then sulfides are precipitated in this protected environment as the hydrothermal solution meets cold alkaline seawater (Haymon and Kastner, 1981). b) Model for the growth of fine chemical gardens in the Hadean ocean away from ridges and volcanoes. The fine chimney spires grow in the absence of calcium sulfate by a reaction between alkaline hydrothermal bisulfide bearing solutions entering through fine cracks in the ocean floor into rather acid iron bearing seawater. Such chimneys could grow in the absence of sulfate but because of the kinetics of this reaction a chimney is only produced when the injection of the hydrothermal solution into the seawater is slow and well focused. The fine tubes act as nucleation points for further precipitation of iron monosulfide. This monosulfide then ozxidizes to the disulfide. Botryoids or bubbles of iron sulfide may have developed in place of chimneys or preceded chimney growth where fluid pressures were relatively low, as sketched to the left of the diagram (cf. Fig. 4) (Russell *et al.*, 1989).

Although one may assume a myriad of hot springs in the earliest oceans, this particular process is unlikely in that the oceans probably contained only minor concentrations of sulfate (Walker and Brimblecombe, 1985). Also the high iron to sulfur atomic ratio in lavas would have been reflected in black smokers generated at submarine spreading centers or volcanoes associated with mantle plumes. So black smokers would have supplied Fe^{2+} and Fe^{3+} to the early oceans governed by the dissolution of magnetite. In 1979 Degens suggested that prebiotic C-H-N-O-P-S compounds would have been discharged to the sea along the same channels as vesiculating magma and he envisaged a series of organic fronts fractionating in the same way as that observed in ion exchange or gas chromatographic patterns as a function of temperature gradients within the crust. Thus he imagined a wide spectrum of new 'organic' compounds being generated and promoted by mineral catalysis to be transported to the hydrosphere. Indeed, Mg-smectites, zeolites and sepiolite (e.g. Evarts and Schiffman, 1983; Bonatti *et al.*, 1983) in the crust might well have filtered tars and caramels from the rising organics just as they do in oil refining (Barrer, 1978).

The nature of key components of the terrestrial environment during the early Archaean when life originated remains a problem: the atmosphere, seawater, crust and mantle have all evolved and been recycled to such an extent that we have few constraints on their compositions. We can only proceed using "best estimates" but the versatility of hydrothermal processes gives us the flexibility to adapt the hydrothermal model as new evidence and constraints emerge. For example, Miller and Bada (1988) seized on another problem, arguing that far from being encouraged to form in black smokers, amino acids would, if in existence, be destroyed by the high temperature solutions.

In response to this Russell *et al.* (1988) pointed out that there were other submarine hot spring scenarios and recalled the discovery of 350 million year old pyrite (FeS_2) chimneys, along with botryoidal pyrite, at Silvermines in Ireland which had formed in an entirely different chemical environment and at lower temperatures to those of black smokers (Larter *et al.*, 1981; Boyce *et al.*, 1983). In Ireland pyritiferous lead and zinc sulfide deposits were generated at sea floor springs as a result of free convection of seawater, chemically modified as it circulated in the top twelve kilometers of the crust (Russell, 1978). Similar chimneys and botryoids have since been discovered 45 km to the north of Silvermines, at the Tynagh orebody (Banks, 1985). The chimneys at Silvermines and Tynagh had a maximum internal diameter of a millimeter or so, ranging down to micrometers, two orders of magnitude narrower than those of black smokers (Figs. 2 and 3). Millions of them formed like natural "chemical gardens" in the hydrothermal systems feeding the Silvermines and Tynagh orebodies in Ireland. This kind of system could have lasted much longer than black smokers, probably of the order of a hundred thousand years or more (Russell, 1988).

Such chemical gardens, essentially composed of iron monosulfide (Fig. 3) are formed easily in the laboratory at STP (Fig. 4) (Russell *et al.*, 1984) much like silica gardens (Coatman *et al.*, 1980). We made them by introducing a solution of ferrous chlo-

Fig. 2. Pyrite chimney formed by the egress of hydrothermal solutions at around 150°C into a brine pool 350 million years ago. Sample is from the Ballynoe Open Pit Barite Deposit, Silvermines, Ireland.

ride (made acid by hydrolysis) through an aperture 0.5 mm in diameter into a vessel containing a solution of sodium sulfide (rendered alkaline for the same reason) notionally:

$$FeCl_2 + Na_2S \rightarrow FeS + 2\,NaCl\ . \qquad \text{(ii)}$$

Bubbles and fine chimneys composed of iron sulfide formed first with both the iron and the sulfur in reduced form probably as $Fe(SH)_2$ or mackinawite (~FeS) (Russell, 1986, 1988; and see Rickard, 1989). These oxidized to greigite (Fe_3S_4) and to pyrite (FeS_2) within a few months (cf. Kribek, 1975). Rigorous anoxic conditions which would have been necessary to prevent iron sulfide oxidation (Berner, 1967; Taylor *et al.*, 1979) were impractical in these experiments. In this analogy the iron chloride deputizes for the hydrothermal solution responsible for the Silvermines orebody which, according to our fluid inclusion analyses, ranged in temperature between ~100-220°C (Samson and Russell, 1987); the sodium sulfide solution is the equivalent of a highly evaporated seawater enriched in (in the Silvermines case bacteriogenic) H_2S occupying a brine pool on the sea bottom (Boyce *et al.*, 1983). The sulfide gardens grown thus in the laboratory are remarkably similar to the fossil pyrite botryoids (hardened bubbles) and chimneys found at Silvermines and Tynagh (Russell *et al.*, 1989 and Figs. 2 and 3).

Fig. 3. View of 350 Ma-old chimneys and botryoids (hemispherical pyritic structures) from the Tynagh, ore-body, Ireland.

Conversely, exactly the same structures are produced if the alkaline sodium sulfide solution is passed through a hole 0.5mm across at the bottom of a visi jar, into a solution of acid ferrous chloride; again chimneys and bubbles of an iron sulfide gel are formed (Fig. 4). These conditions are more similar to those that we imagine pertaining four billion years ago because at that time, although much of the crust may have been basaltic (Bickle, 1986; Sleep and Windley, 1982; but see Nisbet, 1987) igneous sills would, in places, have intruded the upper oceanic crust, composed chiefly of forsterite olivine [$(Mg>Fe)_2\ SiO_4$], pyroxenes [$(Mg,Fe,Ca)_2\ Si_2O_6$] and minor plagioclase [$(Ca>>Na)(Si,Al)_4O_8$] or their hydrated (serpentinized) equivalents, as well as minor chrome spinels [$(Mg,Fe)(Cr,Fe,Al)_2O_4$]. Several percent of pyrrhotite ($Fe_{1-x}S$) and pentlandite $(Fe, Ni)_9S_8$ would also have been carried within the magma (Burns, 1988) along with other base metal monosulfides, and probably accessory sulfides such as rasvumite (KFe_2S_3) (Sokolova *et al.*, 1970; Czamanske *et al.*, 1979), djerfisherite (K_3[Na,Cu]

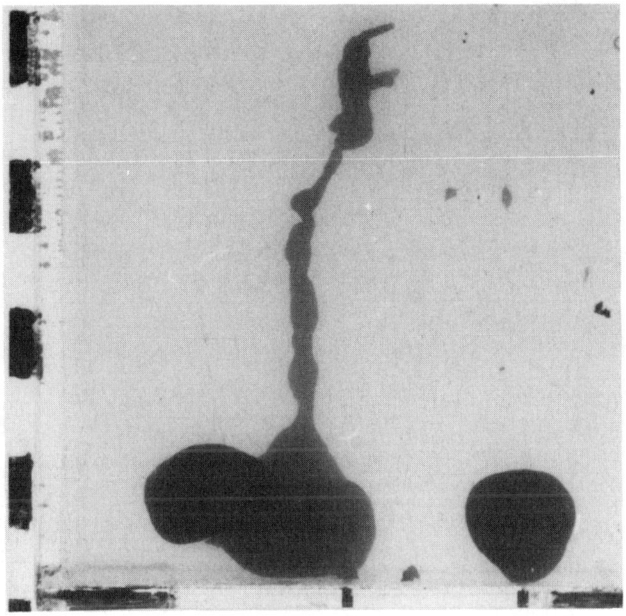

Fig. 4. Chimney-like structure of iron monosulfide formed by passing sodium sulfide solution upwards through three holes 0.5 mm in diameter in a visi jar containing ferrous chloride. Note the filled globules at the base which are compared to the pyrite botryoidals of Fig. 3. The central "proto" botryoidal bubble has "spawned" another such protobotryoid to the left before rupturing to produce the fine chimney. Further buoyant globules formed at the top of the tube, become severed, float upwards and disperse. (Vertical scale in centimeters.)

$[Fe,Ni]_{12}S_{14}$) (Genkin *et al.*, 1970), erdite ($NaFeS_2{\cdot}2H_2O$) (Konnert and Evans, 1980; Czamanske *et al.*, 1980), chalcopyrite ($CuFeS_2$) and possibly molybdenite (MoS_2) (Fig. 5). Sulfides are especially concentrated at the base of the komatiitic lava piles where they were introduced as immiscible melts (Gresham and Loftus-Hills, 1981). We know from analysis of ground water and warm springs from modern mafic and ultramafic complexes of similar chemistry that their pH is likely to have been high (Neal and Stanger, 1984) because of the partial dissolution of magnesium and calcium oxides on hydrolysis of silicates:

$$(MO)_{\text{silicate}} + H_2O \rightarrow M^{2+} + 2OH^- \quad . \tag{iii}$$

This type of hydrothermal solution would have comprised seawater convecting thermally in cracks in the oceanic crust much as today similar mass transfer of heat takes place throughout the ocean floors away from the ridges (Anderson *et al.*, 1977). The mechanism of circulation we envisage involves natural or free thermal convection driven by a

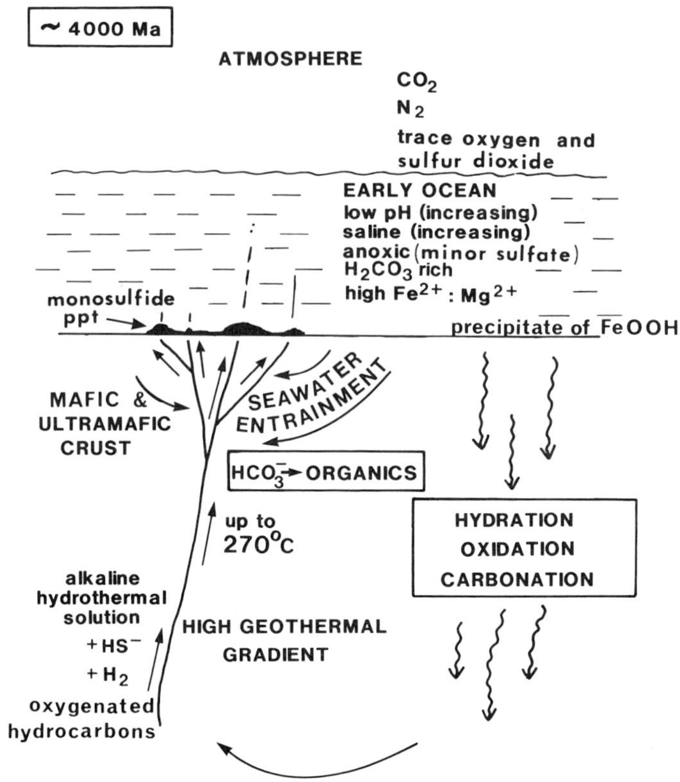

Fig. 5. Speculative sketch of the natural convection system envisaged for the Hadean Sea unrelated to hot spot volcanism oceanic ridges. The hydrothermal solution is seawater modified by interaction with the mafic and ultramafic crust and rendered alkaline by dissolution of calcium and some sodium and potassium. The early ocean itself has a relatively low pH because of the solution of atmospheric carbon dioxide and hydrogen chloride released from the mantle during massive volcanic activity.

very high regional geothermal gradient (i.e. unrelated to a magma chamber). Permeability through the top few kilometers of the crust would have been allowed by fracturing generated by differential stresses, with fracturing being further encouraged by the internal hydrostatic pressures which increased the effective stress allowing migration of joints down to depths where the temperature approached 270°C. Above this temperature porosity would have been very low indeed as a result of crack sealing. The cells would have operated to a greater depth (ultimately about 12 km) within the oceanic crust than the black smoker's convective systems, the hot water interacting with sulfide-bearing source rocks. They could have continued to operate for a million years or so in any one place at a temperature of around 150°C.

Today, seawater circulating in the crust at the oceanic ridges becomes acid (Seyfried and Dibble, 1980; Lecuyer *et al.*, 1990):

$$Mg^{2+}(\text{seawater}) + 2H_2O \rightarrow Mg(OH)_2\,(\text{silicate}) + 2H^+ \quad \cdot \qquad \text{(iv)}$$

but in the early Archaean such a reaction would have been muted *off-ridge*, because of lower operating temperatures and the presence of $Ca(OH)_2$ in solution. Calculations modelling anoxic sulfate-free "Archaean seawater" with mafic to ultramafic rock using the EQ3/6 codes of Wolery *et al.* (1984) suggest a hydrothermal fluid pH about one to two units alkaline (McLeod *et al.*, 1992). Crerar *et al.* (1978) demonstrate that iron is insoluble in these conditions, though we calculate a release of ~0.001m of HS^- as pyrrhotite in the crust converts to magnetite (McLeod *et al.*, 1992).

In contrast to the ~150°C springs, the early ocean itself was probably neutral or slightly acid because of the CO_2 rich atmosphere and the precipitation of ferric hydroxide (equation i). Buffering by magnetite would have prevented large scale excursions to lower pH. Iron (II) derived from the high enthalpy springs (black smokers) would have been present to the extent of at least a few parts per million (Holland, 1973; Walker and Brimblecombe, 1985). Moreover there were no large continental masses at that time (McLennan and Taylor, 1982; Moorbath, 1985) and land would have been restricted to ephemeral volcanoes and mantle plume bulges. The reaction of the seawater with the oceanic crust would have been through further hydration, carbonation (from a portion of the dissolved carbon dioxide) and oxidation (using combined oxygen coming from the water and carbon dioxide). Hydrogen would have been released during these reactions at high temperature (Coveney *et al.*, 1987), and Fischer-Tropsch type syntheses (Storch *et al.*, 1951) could have produced "organic" molecules. French (1964) recognized this possibility following hydrothermal synthesis of siderite from ferrous oxalate dihydrate at low oxygen fugacity. His experiments resulted in the formation of complex oxygenated hydrocarbons. As the alkaline water returned to the sea floor in a convective updraft it would have cooled and mixed with acid seawater occupying the uppermost part of the crust where sulfides would have been precipitated below and at the site of exhalation (de Wit *et al.*, 1980) (Figs. 1b and 5).

An aspect of the chemistry of iron and other sulfides is made use of in mineral separation technology. Sulfides have a general affinity for dithiocarbonates and dithiophosphates and even amines (Wark and Wark, 1935; Leja, 1982). Pyrite shows a particular affinity for these surfactants which makes this mineral a nuisance during separation of zinc, lead and copper sulfide ores (Ball and Rickard, 1976).

Another aspect of mineral processing, biohydrometallurgy, exploits the propensity of *Thiobacillus ferrooxidans* to catalyze the oxidation of pyrite to ferrosulfate and sulfuric acid. Use of the energy therefrom is a further indication of the significance of redox reactions for the most primitive extant forms of life (Temple and Colmer, 1951):

$$FeS_2 + 3\tfrac{1}{2}\,O_2 + H_2O \xrightarrow[\text{ferrooxidans}]{\text{Thiobacillus}} FeSO_4 + H_2SO_4 \,. \qquad \text{(v)}$$

That sulfide mounds and chimneys are an attractive proposition for the site for life's early evolution gains support from the biochemical work of Woese (1987) and Lake (1988). Woese particularly has argued that the earliest known ancestor of life is an archaebacterium which is both thermophilic and sulfur metabolizing. Woese *et al.* (1990) have named such bacteria the Crenarchaeota and note that *Pyrodictium* may be related to a particularly ancient example. Intriguingly Stetter *et al.* (1983) have reported that *Pyrodictium occultum* tends to precipitate pyrite or sphalerite (ZnS) as waste around its cell walls during growth.

In thinking about the significance of iron sulfides for the origin of life we can see three intersecting lines. First the geological arguments emphasize the role of iron monosulfides and disulfides at the focal point of a medium temperature hydrothermal convection system on the primitive earth in the first oceans. This is the view forward, as it were, from conditions on the early Earth: here is a place generally suitable for the origin of life and it is a place where sulfides abound.

Then, second, there are some general properties of sulfides to consider. For example industrial experience shows that at 200°C or more, sulfide catalysts are effective in organic processing where they resist poisoning, for example in the treatment of high sulfur petroleum and oil shale (Weisser and Landa, 1973). They are semiconductors and can drive both oxidations and reductions, as well as catalyze dehydration, isomerization and cleavage reactions (Weisser and Landa, 1973), so one can imagine iron sulfides as primitive reagents ("coenzymes") and catalysts. Then again the covalent character of iron sulfide crystals and their defect structures (Ball and Rickard, 1976; Murrowchick and Barnes, 1986) make them potential information carriers as remarked earlier.

Finally there is the view backwards from current biochemistry. As we have seen earlier, this gives several indications that sulfides were important early components (Hartman, 1975; de Duve, 1991). Indeed Hall and his co-workers had suggested as early as 1971 that ferredoxins were the first enzymes and that a precursor could have been an iron monosulfide/disulfide redox catalyst. As we have seen, recent phylogenetic analyses have implicated a hot metabolism involving iron sulfides at the lowest junction of the evolutionary tree (Woese, 1987; Lake, 1988).

Taken together these three lines of exploration suggest that life may have originated in a sulfide rich environment - a hydrothermal system - and stayed in such an environment right up to the emergence of, essentially, life as we know it. Yet the gap between anything that sulfides and sulfide surfaces could do and the working of, say, the most basic protein synthetic machinery of modern organisms remains enormous. Can it really be bridged?

Wächtershäuser believes it can. In 1988 he proposed that the first energy source for reducing carbon dioxide to organic molecules was the formation of pyrite from hydrogen sulfide and ferrous ions:

$$Fe^{2+} + 2H_2S \rightarrow FeS_2 + 4H^+ + 2e^- \quad . \tag{vi}$$

Wächtershäuser (1988a, 1990b) has developed his ideas, speculating (after Hartmann, 1975) that the first autocatalytic cycle fixing carbon was the extant reductive citric acid cycle. He has further suggested (1990b) that reduction of a carboxylic acid takes place by pyrite formation via a thioacid (Fig. 6).

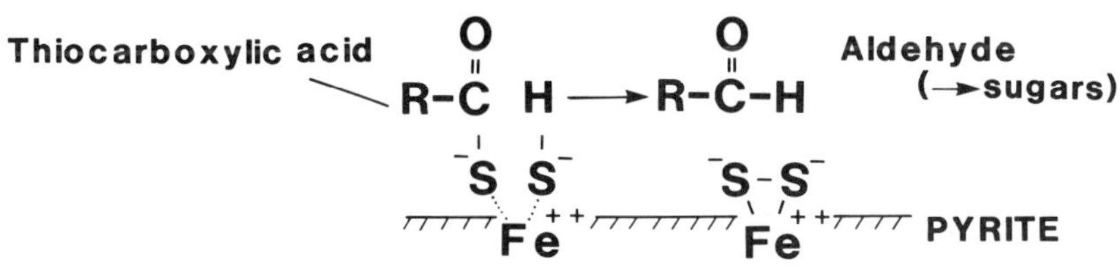

Fig. 6. Wächtershäuser's suggestion for pyrite formation. A redox reaction involving the concomitant reduction (hydrogenation) of a thiocarboxylic acid to produce an aldehyde: a route to sugars.

The thermodynamic feasibility of the reduction of CO_2 by pyrite formation is of great significance for mineral theories of the origin of life in a hydrothermal setting. It must be stressed that as yet there has been no laboratory demonstration of this process. Yet if this process is part of the metabolism of an organism we should not be discouraged by the evident existence of kinetic barriers. As Wächtershäuser (1990a) points out it is a positive *requirement* of such a chemoautotrophic redox energy source that it should not be tapped too easily or it will have been released within the general environment before the organisms can make use of it.

The reduction of carbon dioxide to organic material concomitant upon the oxidation of iron monosulfide to disulfide would only take place below about 220°C. Above that temperature carbon dioxide is stable in the iron monosulfide field. This is because of a cross-over of field boundaries at this temperature as illustrated in Fig. 7. We know from "burning the toast" that carbon is relatively stable even in oxidizing conditions.

More generally, Shock (1990b) has explained how abiotic synthesis of organic compounds of the C+H+O+N system could occur through the reduction of both CO_2 and N_2 within a seafloor hydrothermal system; the optimum temperature for high activities of "metastable" aqueous organic species is 200°C, between the oxygen fugacities of the

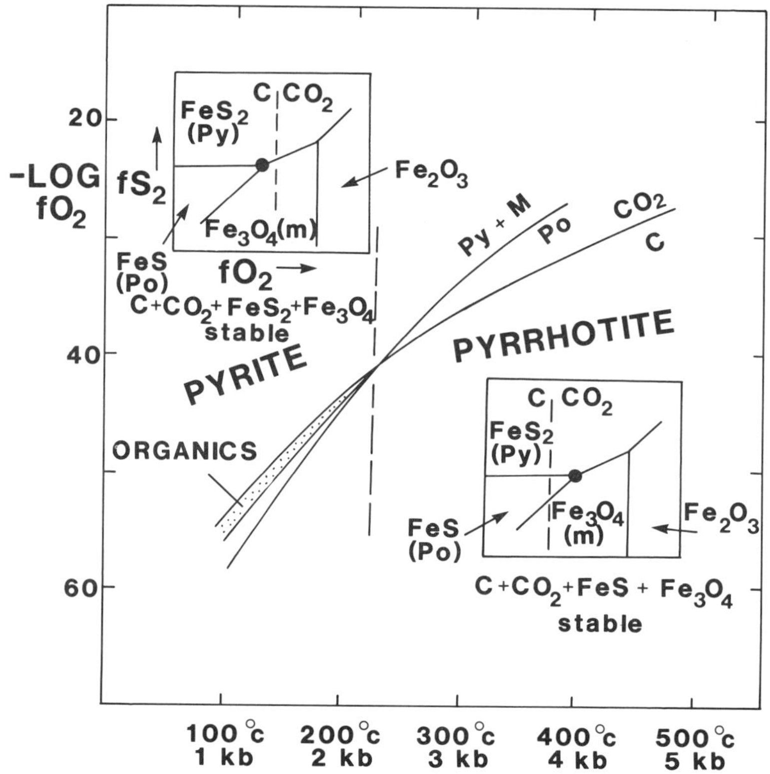

Fig. 7. With C/CO_2 as the control buffer, pyrrhotite is stable at high temperature and pyrite at low temperature. The insets illustrate how the C/CO_2 buffer migrates to lower oxygen fugacity relative to the pyrrhotite/magnetite + pyrite reaction, with increase in temperature. Curves calculated using SUPCRIT (Helgeson *et al.*, 1978) for a geothermal gradient of 100°C/Kbar. The "organics" field represents tentative phase relations at low temperature where reduction of CO_2 by FeS could produce organics (partly based on Hall, 1986 and Shock, 1990b).

pyrite + pyrrhotite + magnetite and the fayalite + magnetite + quartz buffers. The low temperature "field of organics" is illustrated notionally on Fig. 7.

Although no laboratory reduction of CO_2 by pyrite formation has been demonstrated, our geological studies of retrogressive metamorphic reactions (Hall *et al.*, 1987) have provided circumstantial evidence for such a reduction. In the Ballachulish Slates pyrrhotite clearly disappears around growing pyrite porphyroblasts (Fig. 8) and the reaction is likely to have been 'pyrite-pulled'. The C/CO_2 buffer becomes more reduced relative to other buffers with increase in temperature (French and Eugster, 1965) and this leads to progressive reduction on progressive regional metamorphism (Hall, 1986). On

Fig. 8. Slab of Ballachulish slate cut parallel to cleavage and across a late pyrite veinlet (left). Pyrite in the veinlet and as porphyroblasts has grown at the expense of metamorphic pyrrhotite blebs. In the depletion zones around pyrite, pyrrhotite is replaced by quartz. (From Hall *et al.*, 1987, published with permission.)

cooling therefore, a C-O-H fluid buffered by C/CO_2 would become relatively oxidizing (Fig. 7). This is the process envisaged for the Ballachulish slates where near-pervasive post-metamorphic fluid has resulted in the oxidation and dissolution of pyrrhotite and the formation of pyrite. CO_2 in the fluid is the possible oxidizing agent while CH_4 is a product:

$$8FeS + 6H_2O + 9CO_2 \rightarrow 4FeS_2 + 4Fe^{2+} + 8HCO_3^- + CH_4 \quad . \qquad \text{(vii)}$$

Difficulty in pyrite nucleation is evident from the relative scarcity of pyrite porphyroblasts and their similar large size. This is not surprising given the complexity of the FeS_2 structure, the stereochemistry of which is illustrated in Fig. 9 (Finklea *et al.*, 1976).

This apparent high pressure and temperature reduction of CO_2 by pyrrhotite has prompted us to speculate (Russell and Daniel, 1992) that, somewhat contrary to Wächtershäuser's view, the production of simple organic molecules took place in quite

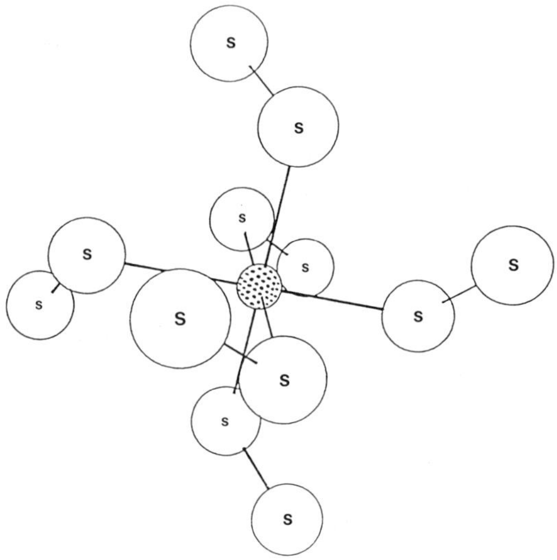

Fig. 9. Perspective drawing illustrating the stereochemistry of the $Fe(S_2^{2-})_6$ site in pyrite. Fe is in 6-fold octahedral coordination bonded to 6 S atoms. Each S is in 4-fold tetrahedral coordination bonded to 3 Fe and 1S.

extensive hydrothermal systems deeper within the oceanic crust, and that this provided feedstock for the cooler iron sulfide botryoidal "culture chambers". Within these further elaborations and ultimately some kind of genetically controlled metabolism became established.

Our use of an iron sulfide example for more particular speculations arose, as we discussed, from the recent interest in iron and sulfur chemistry in relation both to hydrothermal systems and to the origins of life. It is perhaps worth noting that Herrera, after more than 40 years of study into the origin of life, concluded that sulfur merited special attention because of its propensity for "cellular" development (Herrera, 1942). Little attention has been paid to Herrera's work, no doubt because of the still prevailing opinion that the first "cells" would have been made of something like lipids - or anyway that they should have been organic. But to reiterate a point made earlier, we cannot assume that the first evolving microsystems on Earth were simply simpler versions of the only evolving systems that we now know. There would, most likely, have been critical organic and inorganic components that are no longer important, or even present in our biochemistry. But perhaps iron and sulfur are among the survivors from earliest times.

Acknowledgements

We thank Roy Daniel for discussions and the SCOR and the NERC (grant number GR3/7779) for financial support.

CHAPTER 10

FUTURE RESEARCH

NILS G. HOLM

Department of Geology and Geochemistry, Stockholm University, S-106 91 Stockholm, Sweden

A. GRAHAM CAIRNS-SMITH

Department of Chemistry, University of Glasgow, Glasgow G12 8QQ, Scotland

ROY M. DANIEL

Thermophile Research Group, University of Waikato, Private Bag 3105, Hamilton, New Zealand

JAMES P. FERRIS

Department of Chemistry, Rensselaer Polytechnic Institute, Troy, NY 12180 U.S.A.

REMY J.-C. HENNET

S.S. Papadopulos and Associates, Inc., 7944 Wisconsin Avenue, Bethesda, MD 20814, U.S.A.

EVERETT L. SHOCK

Department of Earth and Planetary Sciences and McDonnell Center for the Space Sciences, Washington University, St. Louis, MO 63130, U.S.A.

BERND R. T. SIMONEIT

Petroleum Research Group, College of Oceanography, Oregon State University, Corvallis, OR 97331, U.S.A.

HIROSHI YANAGAWA

Mitsubishi Kasei Institute of Life Sciences, 11 Minamiooya, Machida-shi, Tokyo 194, Japan

1.A. Overall Definition of Natural Systems

What field studies need to be carried out in order to test the hydrothermal model for the origin of life? This question was already raised in 1989 by the workshop in Prague on 'Chemical Evolution and Neo-Abiogenesis in Marine Hydrothermal Systems' (Holm, 1990). The full range and diversity of hydrothermal environments can only be revealed if more exploration work is carried out on a world-wide basis. Much progress has been made in understanding the controls on the hydrothermal fluids of individual areas, but extension to a global context remains difficult (cf. Von Damm, 1990). In Chapter 2 off-axis hydrothermal systems are discussed, pointing out that only minor surveys have been directed towards this type of hydrothermal activity, even though the amount of water which circulates through off-axis systems is estimated to exceed that of on-axis systems by a factor of between 20 and 25.

Origins of Life and Evolution of the Biosphere **22**: 181–190, 1992.

A list of possible studies was compiled at the Prague meeting in order to identify future research opportunities (field studies in italics):

1. *Chemical monitoring of deep-sea hydrothermal systems.*
2. *Comparative studies of terrestrial systems.*
3. *Comparative studies of shallow marine systems.*
4. *On-station experiments.*
5. *Injection-capture experiments.*
6. *On-line pumping of deep-sea hydrothermal water to the sea surface for analysis and experiments.*
7. *Drilling activities on the seafloor.*
8. *Survey of areas other than ridge axes - primarily ridge flanks* (see above).
9. Modelling of hydrothermal systems with O_2 excluded.
10. Modelling of hydrothermal conditions in laboratory experiments.
11. Adsorption/desorption and catalysis experiments at high pressures and temperatures.
12. Construction and operation of quenching systems.

Edmond (1991), in his recent review article about research on hydrothermal chemistry, has remarked that on a general level information is lacking from the super-fast spreading centers which occur mainly in the southern oceans. It is possible that hydrothermal systems present at fast-spreading centers are more representative of systems on the early Earth than other, better studied settings on the present Earth. Information is also scarce from most of the numerous types of environments encountered in back-arc basins. Diversity in substrate composition and in the rate and extent of reactions may broaden the constraints on chemical and thermodynamic interpretations. For example, the report by Fouquet and coworkers (1991) on high temperatures (400°C) and high proton activities (pH 2) in hydrothermal fluids of the Lau back-arc spreading center indicates that these systems may differ considerably in chemical composition and hydrologic variability from many of the best-studied ridge-crest environments.

The report of the first meeting in 1990 of InterRIDGE, the international framework of research related to the U.S. Ridge Inter-Disciplinary Global Experiments (RIDGE) science program, concluded that 'biological investigations of hydrothermal systems thus directly concern an understanding of the origin and evolution of life on Earth'. Some of the central program components for future research of InterRIDGE concern the development of an inventory of hydrothermal activity on a global scale and the characterization of environments in which life forms can be sustained. The second InterRIDGE meeting in 1992 identified a set of current scientific problems that are essential components of a comprehensive understanding of temporal patterns of ridge processes. One such problem is whether or not the chemical environments of hydrothermal systems yield viable environments for the origin of life. This is, of course, in line with the spirit of the work of SCOR Working Group 91. Within SCOR there is a continuous discussion of future participation in large-scale science programs, and one such program that has been mentioned lately is InterRIDGE. It is the opinion of Working Group 91 that such an engagement should be strongly encouraged.

One of the first steps of the spatial characterization of mid-ocean ridges on a global scale would be to carry out off-axis bathymetric and across-axis geophysical surveys. The bathymetry and geophysics should eventually be followed by investigations of individual hydrothermal (off-axis) sites for the evaluation of the ranges of, for instance, temperature, pH, fH_2, and flow rates. On the basis of mineral alteration, maximum values for the longevity of hydrothermal systems at off-axis positions have been estimated to reach up to and beyond 10^7 years (Fehn and Cathles, 1986). Persistent upwelling flow in the temperature range 14-40°C has been demonstrated in 11 x 10^7 years old oceanic crust of the Bermuda Rise in the Atlantic (Muehlenbachs, 1979; Lawrence, 1979; Alt and Honnorez, 1984). According to some hydrologic models, the temperatures reached by convecting fluids in off-axis systems are not expected to exceed 200°C (Fehn and Cathles, 1986). A vast distribution system in that temperature region has, however, yet to be confirmed. The East Pacific Rise Detailed Planning Group (ERDPG, 1991) of the Ocean Drilling Program (ODP) has recommended a strategy of cross-axis array drill sites at approximately 9°30'N on the East Pacific Rise. The strategy includes a submersible fluid sampling program and hydrological sealing of the drilled holes for re-entry and multiple sample recovery. The drill-ship JOIDES Resolution will temporarily leave the Pacific by January 1993 and work in the Atlantic Ocean for perhaps about three years. Plans for participation in further East Pacific Rise drilling legs should thus be developed within the next few years.

1.B. Physical and Chemical Characterization of Natural Systems

To study the possibility that complex organic syntheses and perhaps life itself could have originated within the Earth's crust, it is necessary to acquire data from laboratory and in situ experiments to advance present knowledge. What follows is a non-comprehensive set of suggestions for further experimental research activities which are considered to be 'reasonable, feasible and timely' enterprises to better characterize the physics and chemistry of natural systems within the Earth's crust.

The detailed study of hydrologic regimes (temperature, pressure, permeability, porosity, preferential pathways, fracture systems, fluid velocities, etc.) operating within small segments (m to 100 m) of the Earth's crust is warranted. This is necessary because an understanding of fluid dynamics at small scales is required for the study of abiotic organic synthesis and the origin of life. It is important to understand whether and where fluids can mix, where they can be trapped, and where boiling or degassing can occur. In particular, the migration velocity of fluids in the crust is of importance because it determines, among other things, how far a fluid can travel before approaching chemical and/or physical equilibrium. It is also important to realize that hydrology, based on a fluid reference frame, in the Earth's crust is strongly scale dependent. For example, a large (km scale) hydrothermal field may be quite stable for long periods of time and one could assume that within such a system most physical and chemical equilibria could easily be approached. However, at a smaller scale (m) within some portions of the same hydrothermal system, it becomes difficult to defend the idea that any physical or chemical equilibrium could be approached, due to the ever changing dynamic conditions. In addition, little is known of the longevity of hydrologic regimes, particularly in proximity to active hydrothermal systems. Hydrologic information should be acquired preferentially from in situ measurements. The experience acquired by hydrologists and petroleum reservoir scientists should be extensively applied. Drilling log tests (gamma-ray, X-ray, resistivity, etc.), tracer tests and long term pumping tests should be conducted at selected sites. Long term observation (years) of small hydrothermal features such as hydrothermal seeps or vents should also be conducted.

Characterization of the types of fluids circulating in the Earth's crust would be of great interest. Aqueous and non-aqueous phases are circulating in the crust. For example, the magnitude of circulation of supercritical CO_2 or other supercritical non-aqueous fluids is virtually unknown. Sampling technology designed for collecting non-aqueous fluids should be developed. Laboratory studies should characterize non-aqueous fluid migration in material representative of the Earth's crust. The detailed chemical composition of non-aqueous phases should be determined, and chemical reaction mechanisms within such phases should be investigated.

Laboratory studies should be conducted to investigate abiotic organic synthesis in all the phases known to exist in the Earth's crust, as well as at interfaces of immiscible phases. The effects of phase transition (appearance or disappearance of a phase) on organic syntheses deserves to be studied. In addition, analytical techniques should be developed to identify and quantify organic compounds within naturally occurring fluid inclusions, since fluid inclusions may represent encapsulated environments for the study of natural abiotic organic chemistry.

1.C. Supercritical Fluids

Supercritical fluid research is a fascinating field. Observed and/or postulated supercritical fluids in hydrothermal systems are CO_2, H_2O, CH_4 (with C_2H_6, C_3H_8, etc.), H_2S, and perhaps NH_3, as well as mixtures of these components. The divergence between the characteristics of supercritical fluids and liquids at room temperature and atmospheric pressure has been discussed in several of the previous chapters (see Chapters 2, 4, 5 and 6). Supercritical fluids can be excellent solvents and extractors of organic compounds (Hawthorne, 1990); they have much better mass transfer characteristcs than liquid solvents (Hawthorne, 1990); the heat capacity of any substance approaches infinity at its critical point; both oxidative and reductive reactions can easily be carried out in modified supercritical fluids (see Chapter 4); etc. Much fundamental research needs to be done, however, in order to achieve an improved understanding of the thermodynamic and molecular properties in the supercritical region. Shaw and coworkers (1991) remarked that we must study in supercritical water reactions that are well understood in other media. We also have to elucidate the effects of pressure on the kinetics of well-characterized reactions of simple molecules, and separate such effects from the interaction with solvents and solvent properties. In addition, it is necessary to understand the extent to which diffusion controls the rate of reactions and the associated problem of cages of solvent molecules that surround solute molecules. Such solvation cages may hinder molecular breakup by promoting the recombination of solute fragments (Shaw *et al.*, 1991).

1.D. Mineral Catalysis of Organic Transformations in Hydrothermal Systems

It is most likely that organic compounds percolating through hydrothermal systems will be adsorbed and desorbed many times on and off mineral surfaces. Some organic compounds may adhere to the mineral until they undergo a chemical transformation to another compound which does not bind. The mineral surfaces may catalyze the transformation of some of these organic compounds to higher molecular weight compounds, and surfaces may affect decomposition to simpler derivatives. An example of the formation of higher molecular weight compounds is the conversion of activated mononucleotides to oligonucleotides in aqueous solution in the presence of montmorillonite clay (Ferris *et al.*, 1990). This reaction has, however, not been investigated under hydrothermal conditions.

Experimental studies of the mineral catalyzed transformations of organic compounds at the higher temperatures and pressures present in hydrothermal systems should be explored in the laboratory. Minerals which are present in hydrothermal systems such as zeolites, smectite clays and metal sulfides (cf. Chapter 9) should be investigated as possible catalysts. Care should be taken to use a chemically inert reaction vessel to avoid catalysis by the container material at high temperatures and pressures (Shaw *et al.*, 1991). Examples of chemical transformations to be studied include:

1. Ester formation between an alcohol and carboxylic acid, since this process has been reported to occur in supercritical water (Chapter 6, Section 3.2). If ester formation is successful then one should proceed to investigate amide formation between simple acids and amines as the first step towards a more complicated and detailed study of polypeptide formation from amino acids.
2. An approach similar to that outlined for polypeptide formation could also be devised for the investigation of oligonucleotide synthesis.
3. The potential for Fischer-Tropsch type (FTT) syntheses on montmorillonite, metal sulfides and other minerals should be high on the list for investigation, especially if such reactions are studied at realistic hydrothermal conditions. FTT synthesis may provide a route to linear fatty acids, compounds which have not been detected in the potential prebiotic reactions that have been studied to date. The FTT synthesis of linear hydrocarbons and acids has been reported to take place using montmorillonite

and metal sulfides as catalysts (Chapter 6, Sections 4 and 5.4.2). The formation of linear fatty acids is necessary for the construction of lipid membranes that may have been required to encapsulate the first life. FTT reactions using a mixture of CO and NH_3 as reactants may provide the heterocyclic bases and amino acids necessary for the formation of polypeptides and polynucleotides (Chapter 6, Section 5.3.2).

2.A. Organic Synthesis-Simulations

Abiotic synthesis experiments are notorious for their lack of geochemical or geologic relevance. Synthesis experiments at properly controlled hydrothermal conditions would represent a significant break with this historical trend. However, it must be kept in mind that there is considerably more to hydrothermal systems than hot water. Compositional constraints, including oxidation states, are imposed by fluid/rock reactions. Therefore a constructive effort to test the possibility of hydrothermal synthesis of organic compounds should emerge from the long history of hydrothermal experimentation developed by geochemists and petrologists who have studied these systems for many decades.

Design of hydrothermal synthesis experiments can be guided by theoretical calculations which identify the ranges of the many descriptive variables of hydrothermal systems (temperature, pressure, salinity, metal-ion composition, oxidation state, pH, etc.) which are conducive to organic synthesis. Calculations of this type are available and numerous further calculations are possible with data already available (see Chapter 5). Coupling theoretical predictions with experimental tests will facilitate enormously the general understanding of the geochemical processes which control whether or not organic synthesis is possible in hydrothermal systems.

The primary experiments which should be done would be designed to answer the fundamental yes-or-no questions concerning hydrothermal organic synthesis. Promising indications from work by French (1964) and others (see Chapters 7 and 8) show that hydrothermal organic synthesis can occur. Maybe confirmation of some of these early results should be sought in similarly designed experiments, where the analytical emphasis would be shifted more strongly to the organic compounds.

The key first step to test the hypothesis of hydrothermal organic synthesis is to demonstrate the conversion of CO_2 to organic compounds. This should be followed with experiments on the fixation of N_2 and CO_2 to organic products. If these are successful then all the other reactions described above should also be possible.

2.B. Experimental/Analytical Issues

Initial experimentation will probably be carried out in batch reactors such as various types of autoclaves/bombs. However, flow reactors need to be adapted and developed in order to better simulate the natural systems. This permits single pass interaction of reagents and immediate removal of the products from the reaction zone. Such systems could also be cycled in terms of thermal regime. Reaction and contact times need to be maximized for the highest product yield by matrix experimentation. The temperature range under typical marine hydrothermal system pressures (100-400 bars) for optimal organic synthesis must also be established by the same experimental design.

As mentioned above, container effects such as catalysis or redox reactions should be minimized by choosing inert materials or appropriately buffering the reaction medium. Artifact formation should be prevented and rigorously proven that it does not occur. The minimization of organic contamination in the experimental and analytical procedures is absolutely critical. For example, the report by French (1964) of organic matter formation in the hydrothermal experimentation with ferrous oxalate could also be interpreted as alteration of traces of organic contaminants such as lubricating hydrocarbons if they were in the system. Hydrocarbons are ubiquitous contaminants and are converted in part to oxygenated derivatives such as ketones, aldehydes, and acids by water in the 250-350°C range (Leif *et al.*, unpublished data). Such compounds were detected in the organic products reported by French (1964). Thus control of potential organic contamination is a key aspect in the future experimental design to test this hypothesis. Organic contamination in the analytical procedures can be minimized by applying the established microanalytical techniques of organic geochemistry.

Once organic products have been demonstrated to be synthesized in the laboratory simulations, their confirmation must be extended to the natural systems. This would involve field checks and field simulations with isotope labeled tracers.

3. Macromolecules and Membranes

3.1. SYNTHESIS

It is axiomatic that in order to generate enzymes non-enzymatic synthesis must have occurred. At the moment there is some evidence demonstrating the synthesis of polypeptide chains from amino acids linked by peptide bonds. What is needed is a demonstration not only that such a process can occur but also careful definition of the conditions that are favorable to such polymerization. An indication of the temperature, and of which amino acids are most likely to form peptide bonds would be significant factors for the design of experiments focussed on examining the stability and catalytic ability of minimal enzymes (see below).

Similarly, the generation of nucleotides from simpler chemicals, and the aggregation of phospholipids and their synthesis will also be significant. Another area which is relatively unexplored at elevated temperature is the way in which phospholipids, or indeed any amphipathic compounds, are capable of forming membranes or vesicles at temperatures around 100°C. Given the ubiquity of hydrocarbon-containing compounds (lipids) in modern membranes this seems to be a sensible starting compound class in this area. The difficulties of carrying out such experiments should not be underestimated. It may be necessary to explore an appreciable temperature and pressure range (probably from 40°C to 300°C and from 1 to 300 atmospheres), as well as trying to control the redox and pH of the system throughout this range.

3.2. THE MINIMAL REQUIREMENT OF A FUNCITONAL SYSTEM

The minimal requirement for the origin of life can be said to be the appearance of systems able to evolve under natural selection, suggesting that genetic, catalytic and membrane materials of some sort were provided by the hydrothermal environment at the outset. The usual assumption that any materials must have been similar to nucleic acids, proteins and lipids is by no means clear. As discussed in Chapter 9 one of the exciting prospects for hydrothermal research is in looking carefully at the components of mineral assemblages for possible materials from which components of the first evolvers might have been made. One could then see, through laboratory simulations, if such materials have genetic, catalytic or membrane forming potential. For example, it is known that thin membranes of colloidal iron sulfide vesicles are semipermeable. The significance of conditions under which present day biochemical materials can survive and function remains: the idea is that such primitive mineral components would subsequently have been updated when the means for making polynucleotides or peptides had evolved.

In this context it is reasonable to expect 'protoenzymes' to be much simpler in composition and structure than modern day enzymes, the ubiquity of enzymes suggests that the development of effective protein catalysts is likely to have been an early stage in the development of life. While the discovery of ribozymes is exciting, and has apparently offered the possibility of an evolutionary track that started with nucleic acids rather than with enzymes, the relative inefficiency of ribozymes and their very limited functions still point to the early development of catalytic proteins. Although a considerable amount of work is now being carried out in this field we still do not have a good understanding of, for example, how a few different amino acids can be used to make an effective catalytic protein, and how small a polypeptide chain is likely to be functional. Research on catalytic antibodies will certainly help to define the minimal requirements for catalytic function. It may be that tracing back the evolutionary tree of enzymes may give us some clues about which enzymes or enzyme types were most dominant or important in very early life forms. Certainly within hydrothermal systems we may expect hydrolysis/condensation reactions to be extremely important. It is difficult to see what other constraints hydrothermal systems may have placed on the functions of protoenzymes. However, the ready availability of metals and the general involvement of these in modern enzymes participating in redox reactions may also be suggestive and of importance.

3.3. STABILITY

Much more information is needed on the hydrothermal stability of the small biomolecules (amino acids, sugars, nucleotides) as well as the larger biomolecules (polypeptide chains). For example, it is currently an open question whether, among all the protein structures available, there exist proteins of a nature which will resist the temperatures to be found in some hydrothermal systems (e.g., 150°C upwards) and which will also retain a structure capable of functioning catalytically. The evidence we have on protein stability is mostly based on proteins which have evolved to function well below 105°C. These are presumably highly adapted to these conditions and to their particular functions, so the results may not be applicable to much more stable proteins. In general even high temperature stability experiments have not been carried out with much pH or redox buffering. A variety of artificial peptides need to be examined for resistance to degradative reactions under conditions approximating those in thermal systems. The next stage would be to determine the upper temperature limit for the maintenance of some degree of tertiary structure.

Experiments involving the stability of membranes, both natural and artificial, and of polymeric information storage systems such as DNA and RNA are also needed to fill gaps in our knowledge. Again the conditions under which the stabilities are examined will be crucial. It is likely to be necessary to consider the effects of high pressure as well as temperature although to date the evidence concerning the stabilizing effect of pressure is unconvincing. pH and redox buffering, however, will be crucial and to date it is not possible to measure pH directly much above 120°C. We know, however, that pH is likely to have a large influence on stability to hydrolytic degradation.

4. The Ultimate Limits for Life

The first pure culture of an organism from a hot spring was grown in a laboratory in the late 1960s. Since then we have seen a steady rise in the upper temperature at which we can culture micro-organisms, from 75°C or so in those early days to 110°C at present. We have only recently had access to high temperature (over 100°C) aqueous environments stable over substantial periods, and our techniques for handling organisms at these high temperatures are relatively undeveloped. There is every prospect that the upper growth temperature for life can be raised from its present 110°C to a figure significantly higher. Efforts in this field will influence our views regarding the original temperature at which life evolved, not least because current results indicate that it is these high temperature organisms that are the most primitive found so far.

Not many laboratories are equipped to grow organisms above 80°C, and even fewer to do so at significant pressures. Nevertheless, since many of these high temperature systems are at substantial pressures, it may be that growth experiments will have to be carried out at significant pressures if it is hoped to culture successfully the most extreme thermophiles. To date it has been possible to culture all of the known extreme thermophiles at atmospheric pressures, but it is difficult to be sure whether this is a convenient feature of these organisms or whether we have simply failed to grow those organisms that demand high pressures.

It is imperative that more field and laboratory experiments are carried out not only to seek organisms from high temperature hydrothermal systems but also to find a wider variety of such systems including those 'off-axis' and in sedimentary basins. Non-marine hydrothermal systems will continue to be an extremely important area for exploration, because of the relatively low cost, ready accessibility, and wide variety of such systems.

Many key aspects of thermotolerance and/or thermophily remain to be identified. So far, no significant biochemical differences between extreme thermophiles and mesophiles have been found that can convincingly be attributed to temperature rather than taxonomy. Nevertheless, differences between modes of life at 20°C and 110°C, if they can be found, will certainly shed an important light on the origin of life.

Although the evidence to date seems relatively convincing, further confirmation that thermophily is a primitive characteristic should be sought. Not all of the RNA sequences available have been flawless, many others are not generally available, and taxonomic trees for the archaebacteria are not only variable year to year but are also the subject of some contention. Given the recent development of these trees and the relatively small number or organisms which have been worked on from what is certainly a large population, this is hardly surprising.

LITERATURE CITED (CHAPTERS 1-10)

Abe, Y.: 1986, Early evolution of the terrestrial planets: Accretion, atmosphere formation, and thermal history, Ph.D. Dissertation, University of Tokyo, Tokyo.

Abelson, P. H.: 1957, Organic constituents of fossils, *Geol. Soc. America Memoir* **67**, 87-92.

Achenbach-Richter, L., Gupta, R., Stetter, K. O. and Woese, C. R.: 1987, Were the original eubacteria thermophiles?, *Syst. Appl. Microbiol.* **9**, 34-39.

Agarwal, V. K., Schutte, W., Greenberg, J. M., Ferris, J. P., Briggs, R., Connor, S., Van De Bult, C. P. E. M. and Baas, F.: 1985, Photochemical reactions in interstellar grains, photolysis of CO, NH_3 and H_2O, *Origins Life Evol. Biosphere* **16**, 21-40.

Ahern, J. J. and Klibanov, A. M.: 1985, The mechanism of irreversible enzyme inactivation at 100°C, *Science* **228**, 1280-1284.

Ahern, J. J. and Klibanov, A. M.: 1986, Why do enzymes irreversibly inactivate at high temperatures?, in *Protein Structure, Folding and Design: GENEX-UCLA Symposium Vol. 39* (ed. D. L. Oxender), Allan R. Liss, New York, pp. 283-289.

Airey, P. L. and Dainton, F. S.: 1966, Photochemistry of aqueous solutions of Fe(II), I. Photoelectron detachment from ferrous and ferrocyanide ions, *Proc. Roy. Soc.* **A291**, 340.

Akermark, B., Ecklund-Westlin, U., Baeckström, P. and Löf, R.: 1980, Photochemical, metal-promoted reduction of carbon dioxide and formaldehyde in aqueous solution, *Acta Chemica Scandinavica,* ser B. **34**, 27-30.

Albaiges, J., Borbón, J. and Walker, W., II: 1985, Petroleum isoprenoid hydrocarbons derived from catagenetic degradation of Archaebacterial lipids, *Org. Geochem.* **8**, 293-297.

Albarrán, G., Collins, K. E. and Collins, C. H.: 1987, Formation of organic products in self-radiolyzed calcium carbonate, *J. Molec. Evol.* **25**, 12-14.

Albers, T.: 1989, Mutational effects on protein stability, *Ann. Rev. Biochem.* **58**, 765-798.

Allaart, J. H.: 1976, The pre-3760 m.y. old supracrustal rocks of the Isua area, Central West Greenland, and associated occurrence of quartz-banded iron-stone, in *The Early History of the Earth* (ed. B. F. Windley), John Wiley & Sons, London, 177-189.

Allamandola, L. J., Sandford, S. A. and Valero, G. J.: 1988, Photochemical and thermal evolution of interstellar/precometary ice analogs, *Icarus* **76**, 225-252.

Allen, W. V. and Ponnamperuma, C.: 1967, A possible prebiotic synthesis of monocarboxylic acids, *Curr. Mod. Biol.* **1**, 24-28.

Alt, J. C. and Anderson, T. F.: 1991, Mineralogy and isotopic composition of sulfur in layer 3 gabbros from the Indian Ocean, Hole 735B, *Proc. Ocean Drilling Program, Scientific Results* **118**, 113-125.

Alt, J. C. and Honnorez, J.: 1984, Alteration of the upper oceanic crust, DSDP site 417: mineralogy and chemistry, *Contrib. Mineral. Petrol.* **87**, 149-169.

Alt, J. C., Anderson, T. F. and Bonnell, L.: 1989, The geochemistry of sulfur in a 1.3 km section of hydrothermally altered oceanic crust, DSDP Hole 504B, *Geochim. Cosmochim. Acta* **53**, 1011-1023.

Origins of Life and Evolution of the Biosphere **22**: 191–242, 1992.

Alt, J. C., Honnorez, J., Laverne, C. and Emmermann, R.: 1986a, Hydrothermal alteration of a 1 km section through the upper oceanic crust, Deep Sea Drilling Project Hole 504B: mineralogy, chemistry, and evolution of seawater-basalt interactions, *J. Geophys. Res.* **91,** 10,309-10,335.

Alt, J. C., Muehlenbachs, K. and Honnorez, J.: 1986b, An oxygen isotopic profile through the upper kilometer of oceanic crust, DSDP Hole 504B, *Earth Planet. Sci. Lett.* **80,** 217-229.

Alvarez, J., Crovetto, R. and Fernandez-Prini, R.: 1988, The dissolution of N_2 and H_2 in water from room temperature to 640K, *Ber. Bunsenges. Phys. Chem.* **92,** 935-940.

Anders, E.: 1989, Pre-biotic organic matter from comets and asteroids, *Nature* **342**, 255-257.

Anders, E., Hayatsu, R. and Studier, M. H.: 1974, Interstellar molecules: Origin by catalytic reactions on grain surfaces?, *Astrophys. J.* **192**, L101-L105.

Anderson, R. B.: 1956, The thermodynamics of the hydrogenation of carbon monoxide and related reactions, in *Catalysis, Vol. 4, Hydrocarbon Synthesis Hydrogenation and Cyclization* (ed. P. H. Emmett), Reinhold, New York, pp. 1-27; Catalysts for the Fischer-Tropsch synthesis, pp. 29-255; Kinetics and reaction mechanism of the Fischer-Tropsch synthesis, pp. 257-371.

Anderson, R. N., Langseth, M. G. and Sclater, J. G.: 1977, The mechanisms of heat transfer through the floor of the Indian Ocean, *J. Geophys. Res.* **82**, 3391-3409.

Antal, M. J., Jr., Brittain, A., DeAlmerda, C., Ramayya, S. and Roy, J. C.: 1987, Heterolysis and homolysis in supercritical water, in *Supercritical Fluids - ACS Symposium Series 329* (ed. M. J. Comstock), American Chemical Society, Washington, p. 77-86.

Antal, M. J., Jr., Mok, W. S. L., Roy, J. C. and T-Raissi, A.: 1985, Pyrolytic sources of hydrocarbons from biomass, *J. Appl. Anal. Pyrol.* **8**, 291-303.

Appel, P. W. U.: 1979, Stratabound copper sulfides in a banded iron-formation and in basaltic tuffs in the early Precambrian Isua supracrustal belt, West Greenland, *Econ. Geol.* **74**, 45-52.

Armstrong, F. A. J., Williams, P. M. and Strickland, J. D. M.: 1966, Photooxidation of organic matter in seawater by ultra-violet radiation, analytical and other applications, *Nature* **211,** 481-483.

Arrhenius, G.: 1986, Dysoxic environments as models for primordial mineralisation, in *Clay Minerals and the Origin of Life* (eds. A. G. Cairns-Smith and H. Hartman), Cambridge University Press, Cambridge, pp. 97-104.

Arrhenius, G.: 1990, Sources and geochemical evolution of cyanide and formaldehyde, Fourth Symp. on Chemical Evolution and the Origin and Evolution of Life, *NASA Ames Research Center, Moffett Field, CA. Reprint.*

Arrhenius, G., Bachman, J., Gedulin, B., Hui, S. and Paplawsky, W.: 1989, Anion selective minerals as concentrators and catalysts for RNA precursor components, *Origins Life Evol. Biosphere* **19,** 235.

Ashworth, M. R. F.: 1971, *Analytical Methods for Organic Cyano Groups*, Pergamon Press, Oxford.

Asinger, F.: 1968, *Paraffins Chemistry and Technology* (translated B. J. Hazzard; ed. H. M. E. Steiner), Pergamon Press, Oxford.

Aswad, D.: 1990, Formation of isoaspartate in human growth hormone and bovine brain calmodulin, *Ann. New York Acad. Sci.* **613**, 26-36.

Bada, J. L., Zhao, M. and Miller, S. L.: 1991, Alanine stability in aqueous solutions at 250°C (abs.), *Geol. Soc. Amer. Abstr. & Prog.* **23**, A25.

Ball, B. and Rickard, R. S.: 1976, The chemistry of pyrite flotation and depression, in *Flotation*, A.M. Gaudin Memorial Volume (ed. M. C. Fuerstenau), Vol. 1, American Inst. Mining, Metall. Petrol. Engrs., pp. 458-484.

Ballard, R. D.: 1983, Exploring Our Living Planet, National Geographic Society, Washington, D.C.

Banks, D. A.: 1985, A fossil hydrothermal worm assemblage from the Tynagh lead-zinc deposit in Ireland, *Nature* **313**, 128-131.

Bar-Nun, A. and Shaviv, A.: 1975, Dynamics of the chemical evolution of Earth's primitive atmosphere, *Icarus* **24**, 197-210.

Bar-Nun, A., Bar-Nun, N., Bauer, S. H. and Sagan, C.: 1970, Shock synthesis of amino acids in simulated primitive environments, *Science* **168**, 470-473.

Barnes, H. L. (Ed.): 1979, *Geochemistry of Hydrothermal Ore Deposits,* Second Ed., John Wiley & Sons, New York, 798 pp.

Barnes, H. L.: 1987, in *Hydrothermal Experimental Techniques* (eds. G. C. Ulmer and H. L. Barnes), John Wiley & Sons, New York, pp. 507-514.

Baronnet, A.: 1972, Growth mechanisms and polytypism in synthetic hydroxyl-bearing phlogopite, *Amer. Mineral.* **57**, 1272-1293.

Baronnet, A.: 1980, Polytypism in micas, A survey with emphasis on the crystal growth aspect, *Current Topics in Materials Science* **5**, 447-549.

Baross, J. A.: 1991, The origin and early evolution of life in the subsurface crustal environments of submarine hydrothermal vents: An assessment (abs.), *Geol. Soc. Amer. Abstr. & Prog.* **23**, A19.

Baross, J. A. and Deming, J. W.: 1983, Growth of 'black smoker' bacteria at temperatures of at least 250°C, *Nature* **303**, 423-426.

Baross, J. A. and Hoffman, S. E.: 1985, Submarine hydrothermal vents and associated gradient environments as sites for the origin and evolution of life, *Origins Life Evol. Biosphere* **15**, 327-345.

Barrer, R. M.: 1978, *Zeolites and Clay Minerals as Sorbents and Molecular Sieves*, Academic Press, London, New York, San Francisco, 497 pp.

Barrett, T. J. and Jambor, J. L. (Eds.): 1988, Seafloor hydrothermal mineralization, *Can. Mineral.* **26**, 429-888.

Basiuk, V. A., Gromovoy, T. Y., Glukhoy, A. M. and Golovaty, V. G.: 1991, Chemical transformations of proteinogenic amino acids during their sublimation in the presence of silica, *Origins Life Evol. Biosphere* **21**, 129-144.

Becker, K., Sakai, H., Adamson, A. C., Alexandrovich, J., Alt, J. C., Anderson, R. N., Bideau, D., Gable, R., Herzig, P. M., Houghton, S., Ishizuka, H., Kawahata, H., Kinoshita, H., Langset, M. G., Lovell, M. A., Malpas, J., Masuda, H., Merrill, R. B., Morin, R. H., Mottl, M. J., Pariso, J.

E., Pezard, P., Phillips, J., Sparks, J. and Uhlig, S.: 1989, Drilling deep into young oceanic crust, Hole 504B, Costa Rica Rift, *Rev. Geophys.* **27,** 79-102.

Becker, R. S., Hong, K. and Hong, J. H.: 1974, Hot hydrogen atom reactions of interest in molecular evolution and interstellar chemistry, *J. Molec. Evol.* **4**, 157-172.

Belkin, S., Wirsen, C. O. and Jannasch, H. W.: 1986, A new sulfur-reducing, extremely thermophilic eubacterium from a submarine thermal vent, *Appl. Environ. Microbiol.* **51**, 1180-1185.

Belloc, H.: 1970, The microbe. *In Complete verse,* Gerald Duckworth, London, 246 pp.

Benz, W., Slattery, W. and Cameron, A. G. W.: 1987, The origin of the moon and the single impact hypothesis II, *Icarus* **66**, 515-535.

Berkowitz, N. and Calderon, J.: 1990, Extraction of oil sand bitumens with supercritical water, *Fuel Proc. Techn.* **25**, 33-44.

Bernal, J. D.: 1951, *The Physical Basis of Life*, Routledge and Kegan Paul, London.

Berndt, M. E. and Seyfried, W. E.: 1990, Boron, bromine, and other trace elements as clues to the fate of chlorine in mid-ocean ridge vent systems, *Geochim. Cosmochim. Acta* **54,** 2235-2246.

Berndt, M. E., Seyfried, W. E. and Beck, J. W.: 1988, Hydrothermal alteration processes at mid-ocean ridges: Experimental and theoretical constraints from Ca and Sr exchange reactions and Sr isotopic ratios, *J. Geophys. Res.* **93,** 4573-4583.

Berndt, M. E., Seyfried, W. E. and Janecky, D. R.: 1989, Plagioclase and epidote buffering of cation ratios in mid-ocean ridge hydrothermal fluids: Experimental results in and near the supercritical region, *Geochim. Cosmochim. Acta* **53,** 2283-2300.

Berner, R. A.: 1967, Thermodynamic stability of sedimentary iron sulfides, *Amer. J. Sci.* **265,** 773-785.

Bernhardt, G., Ludemann, H.-D., Jaenicke, R., König, H. and Stetter, K. O.: 1984, Biomolecules are unstable under "black smoker" conditions, *Naturwissenschaften* **71**, 583-586.

Bernhauer, K.: 1929, Reaktionen zwischen Zuckerarten und deren Abbau Produkten mit stickstoffhaltigen Substanzen, *Z. Physiol. Chem.* **183**, 67-73.

Bickle, M. J.: 1986, Implications of melting for stabilization of the lithosphere and heat loss in the Archaean, *Earth Planet. Sci. Lett.* **80,** 314-324.

Biggerstaff, D. R.: 1986, The thermodynamic properties of aqueous solution of argon, ethylene, and xenon up to 720K and 34 MPa, Ph.D. Thesis, University of Delaware.

Bird, D. K., Manning, C. E. and Rose, N. M.: 1988, Hydrothermal alteration of Tertiary layered gabbros, East Greenland, *Amer. J. Sci.* **288,** 405-457.

Bird, D. K., Rogers, R. D. and Manning, C. E.: 1986, Mineralized fracture systems of the Skaergaard intrusion, East Greenland, *Meddelelser om Grønland, Geoscience* **16,** 68p.

Bischoff, J. L. and Pitzer, K. S.: 1989, Liquid-vapor relations for the system NaCl-H_2O: Summary of the P-T-x surface from 300° to 500°C, *Amer. J. Sci.* **289**, 217-248.

Bischoff, J. L. and Rosenbauer, R. J.: 1983, A note on the chemistry of seawater in the range 350°-500°C, *Geochim. Cosmochim. Acta* **47,** 139-144.

Bischoff, J. L. and Rosenbauer, R. J.: 1984, The critical point and two-phase boundary of sea water, 200-500°C, *Earth Planet. Sci. Lett.* **68**, 172-180.

Bischoff, J. L. and Rosenbauer, R. J.: 1987, Phase separation in seafloor geothermal systems: an experimental study of the effects on metal transport, *Amer. J. Sci.* **287**, 953-978.

Bischoff, J. L. and Rosenbauer, R. J.: 1988, Liquid-vapor relations in the critical region of the system $NaCl-H_2O$ from 380 to 415°C: A refined determination of the critical point and two-phase boundary of seawater, *Geochim. Cosmochim. Acta* **52**, 2121-2126.

Bischoff, J. L. and Seyfried, W. E.: 1978, Hydrothermal chemistry of seawater from 25° to 350°C, *Amer. J. Sci.* **278**, 838-860.

Bither, T. A., Bouchard, R. J., Cloud, W. H., Donahue, P. C. and Siemons, W. J.: 1968, Transition metal pyrite dichalcogenides, high-pressure synthesis and correlation of properties, *Inorg. Chem.* **87**, 2208-2220.

Blair, J. S. and Braham, J. M.: 1924, Preparation of guanidinium salts from calcium cyanamide, *Ind. Eng. Chem.* **16**, 848-852.

Blumer, M.: 1975, Curtisite, idrialite and pendletonite, polycyclic aromatic hydrocarbon minerals: Their composition and origin, *Chem. Geol.* **16**, 245-256.

Blumer, M.: 1976, Polycyclic aromatic compounds in nature, *Sci. Amer.* **234**, 34-45.

Boak, J. L. and Dymek, R. F.: 1982, Metamorphism of the ca. 3800 Ma supracrustal rocks at Isua, West Greenland: implications for early Archean crustal evolution, *Earth Planet. Sci. Lett.* **59**, 155-176.

Böhlke, J. K., Honnorez, J., Honnorez-Guerstein, B.-M., Muehlenbachs, K. and Petersen, N.: 1981, Heterogeneous alteration of the upper oceanic crust: Correlation of rock chemistry, magnetic properties, and O isotope ratios with alteration patterns in basalts from site 396B, DSDP, *J. Geophys. Res.* **86**, 7935-7950.

Bohon, R. L. and Claussen, W. F.: 1957, The solubility of aromatic hydrocarbons in water, *J. Amer. Chem. Soc.* **73**, 1571-1578.

Bonatti, E., Simmons, E. C., Bregar, D., Hamlyn, P. R. and Lawrence, J.: 1983, Ultramafic rock/sea water interaction in the oceanic crust: Mg-silicate (sepiolite), deposit from the Indian Ocean floor, *Earth Planet. Sci. Lett.* **62**, 229-238.

Borowska, Z. and Mauzerall, D.: 1987, Efficient near ultraviolet induced formation of hydrogen by ferrous hydroxide, *Origins Life Evol. Biosphere* **17**, 251-259.

Boss, A. P.: 1990, 3D solar nebula models: implications for Earth origin, in *Origin of the Earth* (eds. H. E. Newson and J. H. Jones), Oxford University Press, New York, pp. 3-15.

Bowers, T. S.: 1989, Stable isotope signatures of water-rock interaction in mid-ocean ridge hydrothermal systems: sulfur, oxygen, and hydrogen, *J. Geophys. Res.* **94**, 5775-5786.

Bowers, T. S. and Taylor, H. P., Jr.: 1985, An integrated chemical and stable-isotope model of the origin of mid-ocean ridge hot springs, *J. Geophys. Res.* **90**, 12,583-12,606.

Bowers, T. S., Campbell, A. C., Measures, C. I., Spivak, A. J., Khadem, M. and Edmond, J. M.: 1988, Chemical controls on the composition of vent fluids at 13 - 11°N and 21°N, East Pacific Rise, *J. Geophys. Res.* **93**, 4522-4536.

Bowers, T. S., Von Damm, K. L. and Edmond, J. M.: 1985, Chemical evolution of mid-ocean ridge hot springs, *Geochim. Cosmochim. Acta* **49**, 2239-2252.

Boyce, A. J., Coleman, M. L. and Russell, M. J.: 1983, Formation of fossil hydrothermal chimneys and mounds from Silvermines, Ireland, *Nature* **306**, 545-550.

Bragger, J. M., Daniel, R. M., Coolbear, T. and Morgan, H. W.: 1989, Very stable enzymes from extremely thermophilic archaebacteria and eubacteria, *Appl. Microbiol. Biotechnol.* **31**, 556-561.

Brandts, J. F.: 1967, Heat effects on proteins and enzymes, in *Thermobiology* (ed. A. H. Rose), Academic Press, New York, pp. 25-72.

Braterman, P. S. and Cairns-Smith, A. G.: 1987, Iron photo-precipitation and the genesis of the banded iron-formations, in *Precambrian Banded Iron-Formations* (eds. P. W. Uitterdijk Appel and G. L. LaBerge), Theophrastas Publications, S.A., pp. 215-245.

Braterman, P. S., Cairns-Smith, A. G. and Sloper, R. W.: 1983, Photo-oxidation of hydrated Fe^{2+} - significance for banded iron formations, *Nature* **303**, 163-164.

Brault, M. and Simoneit, B. R. T.: 1988, Steroid and triterpenoid distributions in Bransfield Strait sediments: Hydrothermally-enchaned diagenetic transformations, in *Advances in Organic Geochemistry 1987, Org. Geochem.* **13**, 697-705.

Brault, M. and Simoneit, B. R. T.: 1989, Trace petroliferous organic matter associated with hydrothermal minerals from the Mid-Atlantic Ridge at the Trans-Atlantic Geotraverse 26°N site, *J. Geophys. Res.* **94**, 9791-9798.

Brault, M., Simoneit, B. R. T., Marty, J. C. and Saliot, A.: 1985, Les hydrocarbures dans le système hydrothermal de la ride Est-Pacifique, à 13°N, *C.R. Acad. Sci. Paris* **301**, II, 807-812.

Brault, M., Simoneit, B. R. T., Marty, J. C. and Saliot, A.: 1988, Hydrocarbons in waters and particulate material from hydrothermal environments at the East Pacific Rise, 13°N, *Org. Geochem.* **12**, 209-219.

Brault, M., Simoneit, B. R. T. and Saliot, A.: 1989, Trace petroliferous organic matter associated with massive hydrothermal sulfides from the East Pacific Rise at 13°N and 21°N, *Ocean. Acta* **12**, 405-415.

Briggs, R., Ertem, G., Ferris, J. P., Greenberg, J. M., McCain, P., Mendoza-Gomez, C. and Schutte, W.: 1992, Comet Halley as an aggregate of interstellar dust and further evidence for the photochemical formation of organics in the interstellar medium, *Origins Life Evol. Biosphere*, in press.

Brikowski, T. and Norton, D.: 1989, Influence of magma chamber geometry on hydrothermal activity at mid-ocean ridges, *Earth Planet. Sci. Lett.* **93**, 241-255.

Brimhall, G. H. and Crerar, D. A.: 1987, Chapt. 8, Ore Fluids: Magmatic or Supergene, in *Reviews in Mineralogy, Vol. 17, Thermodynamic Modeling of Geological Materials: Minerals, Fluids and Melts* (eds. I. S. E. Carmichael and H. P. Eugster), Mineralogical Society of America, Washington, D.C., pp. 235-321.

Britton, H. T. S.: 1925, Hydrogen and oxygen electrode titrations of some dibasic acids and of dextrose, *J. Chem. Soc.* **127**, 1896-1917.

Brock, T. D.: 1978, *Thermophilic Microorganisms and Life at High Temperatures*, Springer Verlag, New York.

Brunmuller, F., Stecher, K., Kroener, M. and Schneider, R.: 1986, Reaction products of hydrogen cyanide, German Patent DE 3,443,462, 28 May 1986, *Chem. Abstr.* **105**, 135933p.

Bryndzia, L. T., Wood, B. J. and Dick, H. J. B.: 1989, The oxidation state of the Earth's suboceanic mantle from oxygen thermobarometry of abyssal spinel peridotites, *Nature* **341**, 526-527.

Budgen, N. and Danson, M. J.: 1986, Metabolism of glucose via a modified Entner-Doudoroff pathway in the thermoacidophilic archaebacterium *Thermoplasma acidophilum*, *FEBS Lett.* **196**, 207-210.

Burke, K. C. and Wilson, J. T.: 1976, Hot spots on the Earth's surface, in *Continents Adrift and Continents Aground*, W.H. Freeman, San Francisco, pp. 58-72.

Burnham, W. C.: 1979, Magmas and hydrothermal fluids, in *Geochemistry of Hydrothermal Ore Deposits* (ed. H. L. Barnes), John Wiley & Sons, New York, pp. 71-136.

Burns, R. G.: 1988, Gossans on Mars, *Proc. 18th Lunar & Planetary Science Conf.*, pp. 713-721.

Butterfield, D. A., Massoth, G. J., McDuff, R. E., Lupton, J. E. and Lilley, M. D.: 1990, Geochemistry of hydrothermal fluids from Axial Seamount Hydrothermal Emissions Study Vent Field, Juan de Fuca Ridge: subseafloor boiling and subsequent fluid-rock interaction, *J. Geophys. Res.* **95**, 12,895-12,921.

Cairns-Smith, A. G.: 1982, *Genetic Takeover and the Mineral Origins of Life*, Cambridge University Press.

Cairns-Smith, A. G.: 1985, The first organisms, *Sci. Amer.* **252**, 90-100.

Cameron, A. G. W.: 1986, The impact theory for origin of the moon, in *Origin of the Moon* (eds. W. K. Hartmann, R. J. Phillips and G. J. Taylor), Lunar and Planetary Institute, Houston, pp. 609-616.

Campbell, A. C. and Edmond, J. M.: 1989, Halide systematics of submarine hydrothermal vents, *Nature* **342**, 168-170.

Campbell, A. C., Bowers, T. S., Measures, C. I., Falkner, K. K., Khadem, M. and Edmond, J. M.: 1988a, A time series of vent compositions from 21°N, East Pacific Rise (1979, 1981, 1985), and the Guaymas Basin, Gulf of California (1982, 1985), *J. Geophys. Res.* **93**, 4537-4549.

Campbell, A. C., Palmer, M. R., Klinkhammer, G. P., Bowers, T. S., Edmond, J. M., Lawrence, J. R., Casey, J. F., Thompson, G., Humphris, S., Rona P. and Karson, J. A.: 1988b, Chemistry of hot springs on the Mid-Atlantic Ridge, *Nature* **335**, 514-519.

Cann, J. R. and Strens, M. R.: 1989, Modeling periodic megaplume emission by black smoker systems, *J. Geophys. Res.* **94**, 12,227-12,237.

Cannant, M., Mével, C. and Stakes, D.: 1991, Normal ductile shear zones at an oceanic spreading ridge: Tectonic evolution of site 735 gabbros (Southwest Indian Ocean), *Proc. Ocean Drilling Program, Scientific Results* **118**, 415-429.

Canuto, V. M., Levine, J. S., Augustsson, T. R. and Imhoff, C. L.: 1982, UV radiation from the young sun and oxygen and ozone levels in the prebiological paleoatmosphere, *Nature* **296**, 816-820.

Canuto, V. M., Levine, J. S., Augustsson, T. R., Imhoff, C. L. and Giampapa, M. S.: 1983, The young sun and the atmosphere and photochemistry of the early Earth, *Nature* **305**, 281-286.

Card, K. D.: 1990, A review of the Superior Province of the Canadian Shield, a product of Archean accretion, *Precambrian Research* **48**, 99-156.

Carothers, W. W. and Kharaka, Y. K.: 1978, Aliphatic acid anions in oil-field waters - implications for origin of natural gas, *Amer. Assoc. Petrol. Geol. Bull.* **62**, 2441-2453.

Chang, S., DesMarais, D., Mack, R., Miller, S. L. and Strathearn, G. E.: 1983, Prebiotic organic syntheses and the origin of life, in *Earth's Earliest Biosphere* (ed. J. W. Schopf), Princeton University Press, Princeton, pp. 53-92.

Chase, R. L., Delaney, J. R., Karsten, J. L., Johnson, H. P., Juniper, S. K., Lupton, J. E., Scott, S. D., Tunnicliffe, V., Hammond, S. R. and McDuff, R. E.: 1985, Hydrothermal vents on an axis seamount of the Juan de Fuca ridge, *Nature* **313**, 212-214.

Chen, C.-T. A.: 1981, Geothermal systems at 21°N, *Science* **211**, 298.

Childress, J. J. (Ed.): 1988, Hydrothermal vents, a case study of the biology and chemistry of a deep-sea hydrothermal vent of the Galapagos Rift, the Rose Garden in 1985, *Deep-Sea Res.* **35**, 1677-1849.

Choughuley, A. S. U. and Lemmon, R. M.: 1966, Production of cysteic acid, taurine and cystamine under primitive Earth conditions, *Nature* **210**, 628-629.

Chyba, C. F and Sagan, C.: 1992, Endogenous production, exogenous delivery and impact shock synthesis of organic molecules: an inventory for the orgins of life, *Nature* **355**, 125-132.

Chyba, C. F., Thomas, P. J., Brookshaw, L. and Sagan, C.: 1990, Cometary delivery of organic molecules to the early Earth, *Science* **249**, 366-373.

Clark, P. D., Clarke, R. A. and Hyne, J. B.: 1986, Aquathermolysis: The chemical reaction of steam with heavy oils - emphasizing the critical role of sulphur compounds, Alberta Sulphur Research Ltd., *Quart. Bull.* **21**, 1 and 2, 20 pp.

Clark, P. D., Hyne, J. B. and Tyrer, J. D.: 1983, Chemistry of organosulfur compound types occurring in heavy oil sands. 1. High temperature hydrolysis and thermolysis of tetrahydrothiophene in relation to steam stimulation processes, *Fuel* **62**, 959-962.

Clemmey, H. and Badham, N.: 1982, Oxygen in the Precambrian atmosphere: An evaluation of the geological evidence, *Geology* **10**, 141-146.

Clifton, C. G., Walters, C. C. and Simoneit, B. R. T.: 1990, Hydrothermal petroleums from Yellowstone National Park, Wyoming, U.S.A, in *Organic Matter in Hydrothermal Systems - Petroleum Generation, Migration and Biogeochemistry* (ed. B. R. T. Simoneit), *Appl. Geochem.* **5**, 169-191.

Coatman, R. D., Thomas, N. L. and Double, D. D.: 1980, Studies of the growth of "silicate gardens" and related phenomena, *J. Materials Sci.* **15**, 2017-2026.

Cocker, J. D., Griffin, B. J. and Muehlenbachs, K.: 1982, Oxygen and carbon isotope evidence for seawater-hydrothermal alteration of the Macquarie Island ophiolite, *Earth Planet. Sci. Lett.* **61**, 112-122.

Cogley, J. G. and Henderson-Sellers, A.: 1984, The origins and earliest state of the Earth's hydrosphere, *Rev. Geophys. Space Physics* **22**, 131-175.

Coish, R. A.: 1977, Ocean floor metamorphism in the Betts Cove ophiolite, Newfoundland, *Contrib. Mineral. Petrol.* **60**, 255-270.

Cole, D. R., Mottl, M. J. and Ohmoto, H.: 1987, Isotopic exchange in mineral-fluid systems. II. Oxygen and hydrogen isotopic investigation of the experimental basalt-seawater system, *Geochim. Cosmochim. Acta.* **51,** 1523-1538.

Conant, J. B., ed.: 1953, Pasteur's and Tyndall's study of spontaneous generation, in *Harvard Case Histories in Experimental Science,* Harvard University Press, Cambridge.

Condie, K. C.: 1981, *Archean Greenstone Belts,* Developments in Precambrian Geology 3, Elsevier, Amsterdam.

Connolly, J. F.: 1966, Solubility of hydrocarbons in water near the critical solution temperatures, *J. Chem. Eng. Data* **11,** 13-16.

Converse, D. R., Holland, H. D. and Edmond, J. M.: 1984, Flow rates in the axial hot springs of the East Pacific Rise (21°N): Implications for the heat budget and the formation of massive sulfide deposits, *Earth Planet. Sci. Lett.* **69,** 159-175.

Corliss, J. B.: 1979, Metallogenesis at oceanic spreading centres, *J. Geol. Soc. Lond.* **136,** 621-626.

Corliss, J. B.: 1986, On the creation of living cells in submarine hot spring flow reactors: Attractors and bifurcations in the natural hierarchy dissipative systems, *Origins Life Evol. Biosphere* **16,** 381-382.

Corliss, J. B.: 1989, Submarine hot springs again (abstr.), *Origins Life Evol. Biosphere* **19,** 534-535.

Corliss, J. B.: 1990, Hot springs and the origin of life, *Nature* **347,** 624.

Corliss, J. B., Baross, J. A. and Hoffman, S. E.: 1981, An hypothesis concerning the relationship between submarine hot springs and the origin of life on Earth, *Ocean. Acta* N° SP, 59-69.

Corliss, J. B., Dymond, J., Gordon, L. I., Edmond, J. M., von Herzen, R. P., Ballard, R. D., Green, K., Williams, D., Bainbridge, A., Crane, K. and van Andel, T. H.: 1979, Submarine thermal springs on the Galapagos Rift, *Science* **203,** 1073-1083.

Cornejo, J., Criado, J. M. and Trillo, J. M.: 1979, Decomposition of acetic acid on manganese (II), oxide, *An. Quim.* **75,** 460; *Chem. Abstr.* **91,** 157024n.

COSOD II Report: 1987, Report of the Second Conference on Scientific Ocean Drilling, Joint Oceanographic Institutions for Deep Earth Sampling/European Science Foundation, Washington, D.C./Strasbourg, 142 pp.

Courtillot, V. E.: 1990, What caused the mass extinction? A volcanic eruption, *Sci. Amer.* **263(4),** 53-60.

Courtillot, V. and Besse, J.: 1987, Magnetic field reversals, polar wander, and core-mantle coupling, *Science* **237,** 1140-1147.

Coveney, R. M., Geobel, E. D., Zeller, E. J., Dreschhoff, G. A. M. and Angino, E. E.: 1987, Serpentinization and the origin of hydrogen gas in Kansas, *Amer. Assoc. Petrol. Geol. Bull.* **71,** 39-48.

Cowan, D. A., Smolenski, K. A., Daniel, R. M. and Morgan, H. W.: 1987, An extremely thermostable extracellular proteinase from a strain of the archaebacterium *Desulfurococcus* growing at 88°C, *J. Biochem.* **207,** 641-644.

Cowan, J. and Cann, J.: 1988, Supercritical two-phase separation of hydrothermal fluids in the Troodos ophiolite, *Nature* **333,** 259-261.

Crerar, D. A., Susak, N. J., Borcsik, M. and Schwartz, S.: 1978, Solubility of the buffer assemblage pyrite + pyrrhotite + magnetite in NaCl solutions from 200 to 350°C, *Geochim. Cosmochim. Acta* **42,** 1427-1437.

Cronin, J. R., Pizzarello, S. and Cruickshank, D. P.: 1988, in *Meteorites and the Early Solar System* (eds. J. F. Kerridge and M. S. Matthews), Univ. of Arizona Press, Tucson, pp. 819-857.

Crozier, T. E. and Yamamoto, S.: 1974, Solubility of hydrogen in water, seawater, and NaCl solutions, *J. Chem. Eng. Data* **19,** 242-244.

Curray, J. R., Moore, D. G., Aguayo, J. E., Aubry, M. P., Einsele, G., Fornari, D. J., Gieskes, J., Guerrero-Garcia, J. C., Kastner, M., Kelts, K., Lyle, M., Matoba, Y., Molina-Cruz, A., Niemitz, J., Rueda-Gaxiola, J., Saunders, A. D., Schrader, J., Simoneit, B. R. T., and Vacquier, V.: 1979, Deep sea drilling in the Gulf of California, Leg 64, *Geotimes* **24,** 18-20.

Curray, J. R., Moore, D. G., Aguayo, J. E., Aubry, M. P., Einsele, G., Fornari, D. J., Gieskes, J., Guerrero, J. C., Kastner, M., Kelts, K., Lyle, M., Matoba, Y., Molina-Cruz, A., Niemitz, J., Rueda, J., Saunders, A. D., Schrader, H., Simoneit, B. R. T. and Vacquier, V.: 1982, *Initial Reports of the Deep Sea Drilling Project,* Vol. 64, Parts I and II, U.S. Government Printing Office, Washington, D.C., 1314 p.

Czamanske, G. K., Erd, R. C., Sokolova, M. N., Dobrovol'skaya, M. G. and Dmitrieva, M. T.: 1979, New data on rasvumite and djerfisherite, *Amer. Mineral.* **64,** 776-778.

Czamanske, G. K., Leonard, B. F. and Clark, J. R.: 1980, Erdite, a new hydrated sodium iron sulfide mineral, *Amer. Mineral.* **65,** 509-515.

Czochanska, Z., Sheppard, C. M., Weston, R. J., Woolhouse, A. D. and Cook, R. A.: 1986, Organic geochemistry of sediments in New Zealand, Part I. A biomarker study of the petroleum seepage at the geothermal region of Waiotapu, *Geochim. Cosmochim. Acta* **50,** 507-515.

Daniel, R. M.: 1985, The stability of proteins from extreme thermophiles, *J. Cell. Biochem., Suppl. 9B,* 144.

Daniel, R. M.: 1986, The immune response to proteins from extreme thermophiles, *J. Theoret. Biol.* **120,** 125-127.

Daniel, R. M., Bragger, J. and Morgan, H. W.: 1990, Enzymes from extreme environments, in *Biocatalysis* (ed. D. A. Abranowicz), Van Nostrand Reinhold, New York, pp. 243-254.

Daniel, R. M., Cowan, D. A., Curran, M. P. and Morgan, H. W.: 1982, A correlation between protein thermostability and resistance to proteolysis, *J. Biochem.* **207,** 641-644.

Danson, M. J.: 1988, Archaebacteria: The comparative enzymology of their central metabolic pathways, *Adv. Microbial Physiol.* **29,** 165-231.

Darken, L. S.: 1941, The ionization constants of oxalic acid at 25° from conductance measurements, *J. Amer. Chem. Soc.* **63,** 1007-1011.

Davies, G. F.: 1980, Thermal histories of convective Earth models and constraints on radiogenic heat production in the Earth, *J. Geophys. Res.* **85,** 2517-2530.

Davies, G. H.: 1990, Heat and mass transport in the early Earth, in *Origin of the Earth* (eds. H. E. Newson and J. H. Jones), Oxford University Press, New York, pp. 175-194.

Davis, S. N.: 1969, Porosity and permeability of natural materials, in *Flow through Porous Materials* (ed. R. J. M. DeWiest), Academic Press, New York, pp. 54-89.

Dawkins, R.: 1986, *The Blind Watchmaker*, Harlow, Essex: Longman.

Dayhoff, M. O., Lippincott, E. R., Eck, R.V. and Nagarajan, G.: 1967, *Thermodynamic Equilibrium in Prebiological Atmospheres of C, H, O, N, P, S, and Cl.*, NASA SP-3040 Washington, D.C., 259p.

de Duve, C.: 1991, *Blueprint for a Cell: The Nature and Origin of Life*, Neil Patterson Publishers, Burlington, North Carolina, pp. 126-127.

Degens, E. T.: 1979, Primordial synthesis of organic matter, in *The Global Carbon Cycle*, Chapt. 2 (eds. B. Bolin, E. T. Degens, S. Kempe and P. Ketner), John Wiley & Sons, Chichester, pp. 57-77.

Degens, E. T. and Matheja, J. H.: 1968, Origin, development and diagenesis of biogeochemical compounds, *J. Brit. Interplanet. Soc.* **21**, 52-82.

Degens, E. T. and Ross, D. A.: 1969, *Hot Brines and Recent Heavy Metal Deposits in the Red Sea*, Springer-Verlag, New York.

Delaney, J. R., Mogk, D. W. and Mottl, M. J.: 1987, Quartz-cemented breccias from the mid-Atlantic Ridge: Samples of a high-salinity hydrothermal upflow zone, *J. Geophys. Res.* **92**, 9175-9192.

Deming, J. W.: 1991, Deep-sea and laboratory evidence for the existence of microorganisms at high temperatures and pressures (abs.), *Geol. Soc. Amer. Abstr. & Prog.* **23**, A19.

DeRosa, M., Gambacorta, A., Nicholaus, B., Giardina, P., Paerio, E. and Buonocore, V.: 1984, Glucose metabolism in the extreme thermoacidophilic archaebacterium *Sulfolobus solfataricus*, *J. Biochem.* **224**, 407.

de Wit, M. J., Hart, R., Martin, A. and Abbott, P.: 1982, Archean abiogenic and probable biogenic structures associated with mineralized hydrothermal vent systems and regional metasomatism, with implications for greenstone belt studies, *Econ. Geol.* **77**, 1783-1802.

de Wit, M. J., Hart, R., Stern, C. R. and Barton, C. M.: 1980, Metallogenesis related to seawater interaction with 3.5 b.y. oceanic crust, *EOS* **61**, 386.

Dick, H. J. B., Meyer, P. S., Bloomer, S., Kirby, S., Stakes, D. and Mawer, C.: 1991, Lithostratigraphic evolution of an *in-situ* section of oceanic layer 3, *Proc. Ocean Drilling Program, Scientific Results* **118**, 439-538.

Didyk, B. M. and Simoneit, B. R. T.: 1989, Hydrothermal oil of Guaymas Basin and implications for petroleum formation mechanisms, *Nature* **342**, 65-69.

Didyk, B. M. and Simoneit, B. R. T.: 1990, Petroleum characteristics of the oil in a Guaymas Basin hydrothermal chimney, in *Organic Matter Alteration in Hydrothermal Systems - Petroleum Generation, Migration and Biogeochemistry* (ed. B. R. T. Simoneit), *Appl. Geochem.* **5**, 29-40.

Donn, B.: 1982, Comets: Chemistry and chemical evolution, *J. Molec. Evol.* **18**, 157-160.

Dowler, M. J. and Ingmanson, D. E.: 1979, Thiocyanate in Red Sea brine and its implications, *Nature* **279**, 51-52.

Drucker, C.: 1920, Weitere Untersuchungen über die Dissoziation ternärer Elektrolyte, *Z. Physik. Chem.* **96**, 381-427.

Drummond, S. E.: 1981, Boiling and mixing of hydrothermal fluids: effects on mineral deposition, Ph.D. Thesis, Pennsylvania State University, 400 p.

Drummond, S. E. and Palmer, D. A.: 1985, Thermal decarboxylation of acetate, Part II. Boundary conditions for the role of acetate in the primary migration of natural gas and the transportation of metals in hydrothermal systems, *Geochim. Cosmochim. Acta* **50**, 825.

Dunn, L. A. and Marshall, W. L.: 1969, Electrical conductances of aqueous sodium iodide and the comparative thermodynamic behavior of aqueous sodium halide solutions to 800° and 4000 bars, *J. Phys. Chem.* **73**, 723-728.

Eckert, C. A., Van Alsten, J. G. and Stoicos, T.: 1986, Supercritical fluid processing, *Environ. Sci. Technol.* **20**, 319-325.

Edmond, J. M.: 1991, U.S. research on oceanic hydrothermal chemistry: 1987-1990, Reviews of Geophysics Suppl., *U.S. National Report 1987-1990*, 645-647.

Edmond, J. M. and Von Damm, K.: 1983, Hot springs on the ocean floor, *Sci. Amer.* **248(4)**, 78-93.

Edmond, J. M., von Damm, K. L. and Bowers, T. S.: 1987, The chemistry of submarine ore-forming solutions, in *Marine Minerals* (eds. P.G. Teleki *et al.*), Reidel Publishing Co., pp. 339-347.

Edmond, J. M., Measures, C., McDuff, R. E., Chan, L., Collier, R., Grand, B., Gordon, L. I. and Corliss, J.: 1979a, Ridge crest hydrothermal activity and the balances of the major and minor elements in the ocean: the Galapagos data, *Earth Planet. Sci. Lett.* **46**, 1-18.

Edmond, J. M., Measures, C., Mangum, B., Grant, B., Sclater, F. R., Collier, R., Hudson, A., Gordon, L. I. and Corliss, J.: 1979b, On the formation of metal-rich deposits at ridge crests, *Earth Planet. Sci. Lett.* **46**, 19-30.

Edmond, J. M., von Damm, K. L., McDuff, R. E. and Measures, C. I.: 1982, Chemistry of hot springs on the East Pacific Rise and their effluent dispersal, *Nature* **297**, 187-191.

Egami, F.: 1974, Minor elements and evolution, *J. Molec. Evol.* **4**, 113-120.

Egami, F.: 1981, A working hypothesis on the interdependent genesis of nucleotide bases, protein amino acids, and primitive genetic code, *Origins Life Evol. Biosphere* **11**, 197-202.

Eigen, M. and Shuster, P.: 1978, The hypercycle: A principle of natural self-organization, Part C: The realistic hypercycle, *Naturwissenschaften* **65**, 341-369.

Eigen, M. and Wick, E. E.: 1954, The thermodynamics of electrolytes at higher concentration, *J. Phys. Chem.* **58**, 702-714.

Einsele, G., Gieskes, J. M., Curray, J., Moore, D. M., Aguayo, E., Aubry, M.-P., Fornari, D., Guerrero, J., Kastner, M., Kelts, K., Lyle, M., Matoba, Y., Molina-Cruz, A., Niemitz, J., Rueda, J., Saunders, A., Schrader, H., Simoneit, B. and Vacquier, V.: 1980, Intrusion of basaltic sills into highly porous sediments, and resulting hydrothermal activity, *Nature* **283**, 441-445.

Ellis, A. J. and Golding, R. M.: 1963, The solubility of carbon dioxide above 100°C in water and in sodium chloride solutions, *Amer. J. Sci.* **261**, 47-60.

Epps, D. E., Sherwood, E., Eichberg, J. and Oró, J.: 1978, Cynamide mediated syntheses under plausible primitive Earth conditions. V. The synthesis of phosphatidic acids, *J. Molec. Evol.* **11**, 279-292.

EPRDPG (Ocean Drilling Program East Pacific Rise Detailed Planning Group): 1991, Recommendations for an East Pacific Rise drilling program, *JOIDES Journal* **17**, 35-53.

Evans, W. C., White, L. D. and Rapp, J. B.: 1988, Geochemistry of some gases in hydrothermal fluids from the southern Juan de Fuca Ridge, *J. Geophys. Res.* **93**, 15,305-15,313.

Evarts, R. C. and Schiffman, P.: 1983, Submarine hydrothermal metamorphism of the Del Puerto Ophiolite, California, *Amer. J. Sci.* **283**, 289-340.

Falbe, J.: 1980, *New Syntheses with Carbon Monoxide*, Springer-Verlag, Berlin.

Fegley, B., Jr. and Prinn, R. G.: 1989, Chemical reprocessing of the Earth's present and primordial atmosphere by large impacts, in *Interactions of the Solid Planet with the Atmosphere and Climate* (ed. G. Visconti).

Fegley, B., Jr., Prinn, R. G., Hartman H. and Watkins, G. H.: 1986, Chemical effects of large impacts on the Earth's primitive atmosphere, *Nature* **319**, 305-308.

Fehn, U.: 1986, The evolution of low-temperature convection cells near spreading centers: A mechanism for the formation of the Galapagos mounds and similar manganese deposits, *Econ. Geol.* **81**, 1396-1407.

Fehn, U. and Cathles, L. M.: 1986, The influence of plate movement on the evolution of hydrothermal convection cells in the oceanic crust, *Tectonophysics* **125**, 289-312.

Fehn, U., Green, K. E., von Herzen, R. P. and Cathles, L. M.: 1983, Numerical models for the hydrothermal field at the Galapagos Spreading Center, *J. Geophys. Res.* **88**, 1033-1048.

Ferris, J. P.: 1987, Prebiotic synthesis - problems and challenges, Cold Spring Harbor Symp., *Quant. Biol.* **52**, 29-35.

Ferris, J. P. and Chen, C. T.: 1975a, Photosynthesis of organic compounds in the atmosphere of Jupiter, *Nature* **258**, 587-588.

Ferris, J. P. and Chen, C. T.: 1975b, Chemical evolution XXVI. Photochemistry of methane, nitrogen and water mixtures as a model for the atmosphere of the primitive Earth, *J. Amer. Chem. Soc.* **97**, 2962-2967.

Ferris, J. P. and Edelson, E. H.: 1978, Chemical evolution XXXI. Mechanism of the condensation of cyanide to HCN oligomers, *J. Org. Chem.* **43**, 3989-3995.

Ferris, J. P. and Guillemin, J.-C.: 1990, Photochemical cycloaddition reactions of cyanoacetylene and dicyanoacetylene, *J. Org. Chem.* **55**, 5601-5606.

Ferris, J. P. and Hagan, W. J., Jr.: 1984, HCN and chemical evolution: The possible role of cyano compounds in prebiotic synthesis, *Tetrahedron* **40**, 1093-1120.

Ferris, J. P. and Ishikawa, Y.: 1988, The formation of HCN and acetylene oligomers by photolysis of ammonia in the presence of acetylene: Applications to the atmospheric chemistry of Jupiter, *J. Amer. Chem. Soc.* **110**, 4306-4312.

Ferris, J. P. and Joshi, P.: 1979, Chemical evolution XXXIII. Photochemical decarboxylation of orotic acid, orotidine and orotidine 5'-phosphate, *J. Org. Chem.* **44**, 2133-2137.

Ferris, J. P. and Kamaluddin: 1989, Oligomerisation reactions of deoxyribonucleotides on montmorillonite clay: the effect of mononucleotide structure on phosphodiester bond formation, *Origins Life Evol. Biosphere* **19**, 609-619.

Ferris, J. P. and Orgel, L. E.: 1966, An unusual photochemical rearrangement in the synthesis of adenine, *J. Amer. Chem. Soc.* **88**, 1074.

Ferris, J. P., Goldstein, G. and Beaulieu, D. J.: 1970, Chemical evolution IV. An evaluation of cyanovinyl phosphate as a prebiotic phosphorylating agent, *J. Amer. Chem. Soc.* **92**, 6598-6603.

Ferris, J. P., Gözen, E. and Agarwal, V.: 1989, Mineral catalysis of the formation of dimers of 5'-AMP in aqueous solution: the possible role of montmorillonite clays in the prebiotic synthesis of RNA, *Origins Life Evol. Biosphere* **19**, 165-178.

Ferris, J. P., Joshi, P. C., Edelson, E. H. and Lawless, J. G.: 1978, HCN: A plausible source of purines, pyrimidines and amino acids on the primitive Earth, *J. Molec. Evol.* **11**, 293-311.

Ferris, J. P., Kamaluddin and Ertem, G.: 1990, Oligomerization reactions of deoxyribonucleotides on montmorillonite clay: The effect of mononucleotide structure, phosphate activation and montmorillonite composition on phosphodiester bond formation, *Origins Life Evol. Biosphere* **20**, 279-291.

Ferris, J. P., Sanchez, R. A. and Orgel, L. E.: 1968, Studies in prebiotic synthesis III. Synthesis of pyrimidines from cyanoacetylene and cyanate, *J. Molec. Biol.* **33**, 693-704.

Ferris, J. P., Williams, E. A., Nicodem, D. E., Hubbard, J. S. and Voecks, G. E.: 1974a, Photolysis of CO-NH_3 mixtures and the Martian atmosphere, *Nature* **249**, 437-439.

Ferris, J. P., Zamek, O. S., Altbuch, A. M. and Freiman, H.: 1974b, Chemical evolution XVIII. Synthesis of pyrimidines from guanidine and cyanoacetaldehyde, *J. Molec. Evol.* **3**, 301-309.

Fiala, G. and Stetter, K.O.: 1986, *Pyrococcus furiosus* sp. nov. represents a novel genus of marine heterotrophic archaebacteria growing optimally at 100°C, *Arch. Microbiol.* **145**, 56-61.

Field, B. O. and Spencer, J. E. D.: 1990, The synthesis of amino acids and sugars on an inorganic template from constituents of the prebiotic atmosphere, *Origins Life Evol. Biosphere* **20**, 233-248.

Finklea, S. L., III, Cathey, S. and Amma, E. L.: 1976, Investigation of the bonding mechanism in pyrite using the Mössbauer effect, *Acta Crystall.* **A32**, 529-537.

Finn, M. J., Bower, C. J. and Hughes, R. D.: 1983, The extraction of coal with supercritical hydrocarbon mixtures, *Fluid Phase Equilibria* **10**, 327-336.

Fischer, F., Zillig, W., Stetter, K. O. and Schreiber, G.: 1983, Chemolithoautotrophic metabolism of anerobic extremely thermophilic archaebacteria, *Nature* **301**, 511-513.

Fisher, A. T. and Becker, K.: 1991, Heat flow, hydrothermal circulation and basalt intrusions in the Guaymas Basin, Gulf of California, *Earth Planet. Sci. Lett.* **103**, 84-99.

Fisher, A. T. and Narasimhan, T. N.: 1991, Numerical simulations of hydrothermal circulation resulting from basalt intrusions in a buried spreading center, *Earth Planet. Sci. Lett.* **193**, 100-115.

Fisher, A. T., Becker, K., Narasimhan, T. N., Langseth, M. G. and Mottl, M.J.: 1990, Passive, off-axis convection through the southern flank of the Costa Rica Rift, *J. Geophys. Res.* **95**, 9343-9370.

Fisher, J. R. and Barnes, H. L.: 1972, The ion-product constant of water to 350°, *J. Phys. Chem.* **76**, 90-99.

Flegmann, A. W. and Tattersall, R.: 1979, Energetics of peptide bond formation at elevated temperatures, *J. Molec. Evol.* **12**, 349-355.

Fouche, C. E., Jr. and Rohlfing, D. L.: 1976, Thermal polymerization of amino acids under various atmospheres or at low pressures, *Biosystems* **8**, 57-65.

Fouquet, Y., von Stackelberg, U., Charlou, J. L., Donval, J. P., Erzinger, J., Foucher, J. P., Herzig, P., Måhe, R., Soakai, S., Wiedicke, M. and Whitechurch, H.: 1991, Hydrothermal activity and metallogenesis in the Lau back-arc basin, *Nature* **349**, 778-781.

Fox, C. G.: 1990, Consequences of phase separation on the distribution of hydrothermal fluids at ASHES Vent Field, Axial Volcano, Juan de Fuca Ridge, *J. Geophys. Res.* **95**, 12,923-12,926.

Fox, G. E., Stackebrandt, E., Hespell, R. B., Gibson, J., Maniloff, J., Dyer, T. A., Wolfe, R. S., Bolch, W. E., Tanner, R. S., Mogrum, L. J., Zablen, L. B., Blackmore, R., Gupta, R., Bonen, L., Lenris, B. J., Stahl, D. A., Leuhusen, K. R., Chen, K. N. and Woese, C. R.: 1980, The phylogeny of prokaryotes, *Science* **209**, 457-463.

Fox, S. W.: 1963, Prebiological formation of biochemical substances, *Int. Series Monographs on Earth Sciences* **16**, 36-49.

Fox, S. W.: 1964, Thermal polymerization of amino-acids and production of formed microparticles on lava, *Nature* **201**, 336-337.

Fox, S. W.: 1965, A theory of macromolecular and cellular origins, *Nature* **205**, 328-340.

Fox, S. W. and Dose, K.: 1977, *Molecular Evolution and the Origin of Life* (eds. S. W. Fox and K. Dose), Marcel Dekker, Inc., New York.

Fox, S. W. and Harada, K.: 1958, Thermal copolymerization of amino acids to a product resembling protein, *Science* **128**, 1214.

Fox, S. W. and Harada, K.: 1960, The thermal copolymerization of amino acids common to protein, *J. Amer. Chem. Soc.* **82**, 3745-3751.

Fox, S. W. and Harada, K.: 1961, Synthesis of uracil under conditions of a thermal model of prebiological chemistry, *Science* **133**, 1923-1924.

Fox, S. W. and Nakashima, T.: 1981, Terrestrial evolution of polymerization of amino acids: Heat to ATP, in *Origin of Life* (ed. Y. Wolman), Reidel Publishing Co., Dordrecht, pp. 271-276.

Fox, S. W. and Waehneldt, T. V.: 1968, The thermal synthesis of neutral and basic proteinoids, *Biochim. Biophys. Acta* **160**, 246-249.

Fox, S. W. and Windsor, C. R.: 1970, Synthesis of amino acids by the heating of formaldehyde and ammonia, *Science* **170**, 984-985.

Fox, S. W., Harada, K. and Vegotsky, A.: 1959, Thermal polymerization of amino acids and a theory of biochemical origins, *Experientia* **15**, 81-84.

Fox, S. W., Harada, K., Woods, K. R. and Windson, C. R.: 1963, Amino acid compositions of proteinoids, *Archives of Biochem. Biophys.* **102**, 430-445.

Fox, S. W., Johnson, J. E. and Middlebrook, M.: 1955, Pyrosynthesis of aspartic acid and alanine from citric acid cycle intermediates, *J. Amer. Chem. Soc.* **77**, 1048-1049.

Fox, S. W., Vegotsky, A., Harada, K. and Hoagland, P. D.: 1957, Spontaneous generation of anabolic pathways, protein, and nucleic acid, *Annals New York Acad. Sci.* **69**, 328-337.

Francheteau, J., Needham, H. D., Chroukroune, P., Juteau, T., Séguret, M., Ballard, R. D., Fox, P. J., Normark, W., Carranza, A., Cordoba, D., Guerrero, J., Rangin, C., Bougalt, H., Cambon., P. and Hékinian, R.: 1979, Massive deep-sea sulphide ore deposits discovered on the East Pacific Rise, *Nature* **277**, 523-528.

Franck, E. U.: 1987, Fluids at high pressures and temperatures, *J. Chem. Thermodynamics* **19**, 225-242.

Franck, E. U. *et al.* (Eds.): 1983, Supercritical fluids, their chemistry and application, *Fluid Phase Equilibria* **10**, 135-354.

Franklin, J. M., Lydon, J. W. and Sangster, D. F.: 1981, Volcanic-associated massive sulfide deposits, *Econ. Geol. Sevenety-Fifth Anniv. Vol.*, 485-627.

French, B. M.: 1964, 1. Synthesis and stability of siderite, $FeCO_3$, 2. Progressive contact metamorphism of the Biwabik Iron Formation on the Mesabi Range, Minnesota, Ph.D. Thesis, The Johns Hopkins University, Baltimore, MD.

French, B. M.: 1971, Stability relations of siderite ($FeCO_3$), in the system Fe-C-O, *Amer. J. Sci.* **271**, 37-78.

French, B. M. and Eugster, H. P.: 1965, Experimental control of oxygen fungacities by graphite-gas equilibria, *J. Geophys. Res.* 70, **6**, 1529-1539.

Fripp, R. E. P.: 1976, Stratabound gold deposits in Archean banded iron-formation, Rhodesia, *Econ. Geol.* **71**, 58-75.

Fuki, M., Furukawa, S., Koga, I. and Inoi, T.: 1974, Ketones from carboxylic acids, Japan Patent 490 30309, 18 March, 1974; *Chem. Abstr.* **81**, 77474n.

Funazukuri, T., Takanashi, T. and Wakao, N.: 1987, Supercritical extraction of used automotive tire with water, *J. Chem. Eng. Japan* **20**, 23-27.

Führer, H.: 1924, Die Wasserlöslichkeit in homologen Reihen, *Chem. Ber.* **57**, 510-515.

Fyfe, W. S.: 1978, The evolution of the Earth's crust: Modern plate tectonics to ancient hot spot tectonics, *Chem. Geol.* **23**, 89-114.

Gabel, N. W. and Ponnamperuma, C.: 1967, Model for origin of monosaccharides, *Nature* **216**, 453.

Garrison, W. M., Morrison, D. C., Hamilton, J. G., Benson, A. A. and Calvin, M.: 1951, Reduction of carbon dioxide in aqueous solutions by ionizing radiation, *Science* **114**, 416-418.

Geissman, T. A., Sim, K. Y. and Murdoch, J.: 1967, Organic minerals. Picene and chrysene as constituents of the mineral curtisite (idrialite), *Experientia* **23**, 793-794.

Genkin, A. D., Troneva, N. V. and Zhuravlev, N. N.: 1970, The first occurrence in ores of the sulfide of potassium, iron, and copper, djerfisherite, *Geochem. Internat.* **7**, 693-701.

Getoff, N.: 1962, Reduktion der Kohlensaure in wässeriger Lösung unter Einwirkung von UV-Licht, *Zeitschrift für Naturforschung* **17B**, 87-90.

Gieskes, J. M., Elderfield, H., Lawrence, J. R., Johnson, J., Meyers, B. and Campbell, A. C.: 1982a, Geochemistry of interstitial waters and sediments, Leg 64, Gulf of California, in *Initial Reports of the Deep Sea Drilling Project* **64(2)**, 675-694.

Gieskes, J. M., Kastner, M., Einsele, G., Kelts, K. and Niemitz, J.: 1982b, Hydrothermal activity in the Guaymas Basin, Gulf of California: A synthesis, in *Initial Reports of the Deep Sea Drilling Project* **64(2)**, 1159-1168.

Gieskes, J. M., Simoneit, B. R. T., Brown, T., Shaw, T., Wang, Y.-C. and Magenheim, A.: 1988, Hydrothermal fluids and petroleum in surface sediments of Guaymas Basin, Gulf of California: A case study, *Can. Mineral.* **26**, 589-602.

Gillan, F. T. and Johns, R. B.: 1986, Chemical markers for marine bacteria: fatty acids and pigments, in *Biological Markers in the Sedimentary Record* (ed. R. B. Johns), Elsevier, Amsterdam, pp. 291-309.

Goldfarb, M. S. and Delaney, J. R.: 1988, Response of two-phase fluids to fracture configurations within submarine hydrothermal systems, *J. Geophys. Res.* **93**, 4585-4594.

Goldschmidt, V. M.: 1952, Geochemical aspects of the origin of complex organic molecules on the Earth, as precursors to life, *New Biology* **12**, 97-105.

Gole, M. J. and Klein, C.: 1981, Banded iron formation through much of Precambrian time, *J. Geol.* **89**, 169-183.

Granick, S.: 1957, Speculations on the origins and evolution of photosynthesis, *Annals New York Acad. Sci.* **69**, 292-308.

Grant-Taylor, D. F.: 1981, Partial molal volumes of sodium chloride solutions at 200 bar, and temperatures from 175° to 350°C, *J. Soln. Chem.* **10**, 621-630.

Grassle, J. F.: 1985, Hydrothermal vent animals: Distribution and biology, *Science* **229**, 713-717.

Gregory, R.T. and Taylor, H.P.: 1981, An oxygen isotopic profile in a section of Cretaceous oceanic crust, Samail Ophiolite, Oman: Evidence for ^{18}O buffering of the oceans by deep (> 5 km, seawater-hydrothermal circulation at mid-ocean ridges, *J. Geophys. Res.* **86**, 2737-2756.

Gresham, J. J. and Loftus-Hills, G. D.: 1981, The geology of the Kambalda nickel field, Western Australia, *Econ. Geol.* **76**, 1373-1416.

Grote, J. R., Syren, R. M. and Fox, S. W.: 1978, Effect of products from heated amino acids on conductance in lipid bilayer membranes and non-aqueous solvents, *Biosystems* **10**, 287-292.

Grutter, M. G., Hawkes, R. M. and Matthews, B. W.: 1979, Molecular basis of thermostability in the lysozyme of bacteriophage T4, *Nature* **277**, 667-669.

Guillemin, J. C. and Denis, J. M.: 1985, Flash vacuum thermolysis of α-aminonitriles and subsequent HCN removal on solid base, a "one line" multistep sequence to reactive N-methyleneamines, *J. Chem. Soc. Chem. Commun.*, 951.

Guillemin, J. C. and Denis, J. M.: 1988, Synthèse d'imines linéaires non-stabilisées par réactions gaz-solide sous vide, *Tetrahedron* **44**, 4431-4446.

Hajash, A. and Chandler, G. W.: 1981, An experimental investigation of high temperature interactions between sea-water and rhyolite, andesite, basalt and peridotite, *Contrib. Mineral. Petrol.* **78**, 240-250.

Haldane, J. B. S.: 1929, The Origin of Life, *The Rationalist Annual.*

Hall, A. J: 1986, Pyrite-pyrrhotite redox reactions in nature, *Mineral. Mag.* **50**, 223-229.

Hall, A. J., Boyce, A. J. and Fallick, A. E.: 1987, Iron sulphides in metasediments: isotopic support for a retrogressive pyrrhotite to pyrite reaction, *Chem. Geol.*, Isotope Geoscience section **65**, 305-310.

Hall, D. O., Cammack, R. and Rao, K. K.: 1971, Role for ferredoxins in the origin of life and biological evolution, *Nature* **233**, 136-138.

Hamann, S. D. and Strauss, W.: 1955, The chemical effects of pressure: Part 3. Ionization constants at pressures up to 1200 atm, *Trans. Faraday Soc.* **51**, 1684-1690.

Harada, K.: 1967, Formation of amino acids by thermal decomposition of formamide - oligomerization of hydrogen cyanide, *Nature* **214**, 479-480.

Harada, K. and Fox, S. W.: 1958, The thermal condensation of glutamic acid and glycine to linear peptides, *J. Amer. Chem. Soc.* **80**, 2694-2697.

Harada, K. and Fox, S. W.: 1964, Thermal synthesis of natural amino-acids from a postulated primitive terrestrial atmosphere, *Nature* **201**, 335-336.

Harada, K. and Fox, S. W.: 1965, Characterization of thermal polymers of neutral α-amino acids with dicarboxylic amino acids or lysine, *Archiv. Biochem. Biophys.* **109**, 49-56.

Harada, K. and Fox, S. W.: 1965, *The Origins of Prebiological Systems* (ed. S. W. Fox), Academic Press, New York, pp. 187-201.

Harada, K. and Fox, S. W.: 1975, Characterization of functional groups of acidic thermal polymers of α-amino acids, *Biosystems* **7**, 213-221.

Harada, K. and Matsuyama, M.: 1979, Polycondensation of thermal precursors of amino acids and characterization of constituent amino acids, *Biosystems* **11**, 47-53.

Harned, H. S. and Davis, R., Jr.: 1943, The ionization constant of carbonic acid in water and the solubility of carbon dioxide in water and aqueous salt solutions from 0 to 50°, *J. Amer. Chem. Soc.* **65**, 2030-2037.

Harned, H. S. and Ehlers, R. W.: 1933, The dissociation constant of acetic acid from 0 to 60° Centigrade, *J. Amer. Chem. Soc.* **55**, 652-656.

Harned, H. S. and Fallon, L. D.: 1939, The second ionization constant of oxalic acid from 0 to 50 degrees, *J. Amer. Chem. Soc.* **61**, 3111-3113.

Harper, G.: 1986, The Josephine ophiolite, northwestern California, *Geol. Soc. Amer. Bull.* **95**, 1009-1026.

Harper, G. D., Bowman, J. R. and Kuhns, R.: 1988, A field, chemical, and stable isotope study of subseafloor metamorphism of the Josephine ophiolite, California-Oregon, *J. Geophys. Res.* **93**, 4625-4656.

Hartman, H.: 1975, Speculations on the origin and evolution of metabolism, *J. Molec. Evol.* **4**, 359-370.

Hartman, J., Brand, M. C. and Dose, K.: 1981, Formation of specific amino acid sequences during thermal polymerization of acids, *Biosystems* **13**, 141-147.

Hartmann, M.: 1980, Atlantis II Deep geothermal brine system. Hydrographic situation in 1977 and changes since 1965, *Deep-Sea Res.* **27**, 161-171.

Hartmann, M.: 1985, Atlantis II Deep geothermal brine system. Chemical processes between hydrothermal brines and Red Sea deep water, *Mar. Geol.* **64**, 157-177.

Hatanaka, H. and Egami, F.: 1977a, Selective formation of certain amino acids from formaldehyde and hydroxylamine in a modified sea medium enriched with molybdate, *J. Biochem.* **82**, 499-502.

Hatanaka, H. and Egami, F.: 1977b, The formation of amino acids and related oligomers from formaldehyde and hydroxylamine in modified sea mediums related to prebiotic conditions, *Bull. Chem. Soc. Japan* **50**, 1147-1156.

Hawker, J. R., Jr. and Oró, J.: 1981a, Cyanamide mediated syntheses of LEU, ALA, and PHE peptides under plausible primitive Earth conditions, in *Origins of Life* (ed. Y. Wolman), Reidel Publishing Co., Dordrecht, pp. 225-232.

Hawker, J. R., Jr. and Oró, J.: 1981b, Cyanamide mediated syntheses of peptides containing histidine and hydrophobic amino acids, *J. Molec. Evol.* **17**, 285-294.

Hawthorne, S. B.: 1990, Analytical-scale supercritical fluid extraction, *Anal. Chem.* **62**, 633A-642A.

Hayatsu, R. and Anders, E.: 1981, Organic compounds in meteorites and their origins, *Topics in Current Chem.* **99**, 1-37.

Hayatsu, R., Studier, M. H. and Anders, E.: 1971, Origin of organic matter in early solar system IV. Amino acids: Confirmation of catalytic synthesis by mass spectrometry, *Geochim. Cosmochim. Acta* **35**, 939-951.

Hayatsu, R., Studier, M. H., Matsuoka, S. and Anders, E.: 1972, Origin of organic matter in the solar system VI. Catalytic synthesis of nitriles, nitrogen bases and porphyrin-like pigments, *Geochim. Cosmochim. Acta* **36**, 555-571.

Hayatsu, R., Studier, M. H., Moore, L. P. and Anders, E.: 1975, Purines and triazines in the Murchison meteorite, *Geochim. Cosmochim. Acta* **39**, 471-488.

Hayatsu, R., Studier, M. H., Oda, A., Fuse, K. and Anders, E.: 1968, Organic matter in the solar system II. Nitrogen compounds, *Geochim. Cosmochim. Acta* **32**, 175-190.

Haymon, R. M.: 1983, Growth history of hydrothermal black smoker chimneys, *Nature* **301**, 695-698.

Haymon, R. M.: 1989, Hydrothermal processes and products on the Galapagos Rift and East Pacific Rise, in *The Geology of North America,* Vol. N, The Eastern Pacific Ocean and Hawaii, Geological Society of America, pp. 125-144.

Haymon, R. M. and Kastner, M.: 1981, Hot spring deposits on the East Pacific Rise at 21°N: preliminary description of mineralogy and genesis, *Earth Planet. Sci. Lett.* **53**, 63-381.

Haymon, R. M., Koski, R. A. and Abrams, M. J.: 1989, Hydrothermal discharge zones beneath massive sulfide deposits mapped in the Oman ophiolite, *Geology* **17**, 531-535.

Haymon, R. M., Koski, R. A. and Sinclair, C.: 1984, Fossils of hydrothermal vent worms from Createaceous ores of the Samail ophiolite, Oman, *Science* **223**, 1407-1409.

Heinz, B., Ried, W. and Dose, K.: 1979, Thermal generation of pteridines and flavins from amino acid mixtures, *Angew. Chem., Int. Ed.* **18**, 479.

Hékinian, R., Francheteau, J., Renard, V., Ballard, R. D., Choukroune, P., Cheminée, J. L., Albarede, F., Minster, J. F., Charlou, J. L., Marty, J. C. and Boulègue, J.: 1983, Intense hydrothermal activity at the axis of the East Pacific Rise near 13°N: Submersible witnesses the growth of sulfide chimney, *Mar. Geophys. Res.* **6,** 1-14.

Hékinian, R., Fevrier, M., Avedik, F., Cambon, P., Charlou, J. L., Needham, H. D., Raillard, J., Boulègue, J., Merlivat, L., Moinet, A., Manganini, S. and Lange, J.: 1983, East Pacific Rise near 13°N: Geology of new hydrothermal fields, *Science* **219**, 1321-1324.

Helgeson, H. C.: 1979, Mass transfer among minerals and hydrothermal solutions, in *Geochemistry of Hydrothermal Ore Deposits, 2nd ed.* (ed. H. L. Barnes), John Wiley & Sons, pp. 568-610.

Helgeson, H. C.: 1985, Some thermodynamic aspects of geochemistry, *Pure and Applied Chem.* **57,** 31-44.

Helgeson, H. C.: 1991, Thermodynamic constraints on the geobiochemical evolution of petroleum in the Earth's crust (abstr.), *Geol. Soc. Amer. Abstr. Prog.* **23,** A25.

Helgeson, H. C. and Kirkham, D. H.: 1974a, Theoretical prediction of the thermodynamic behavior of aqueous electrolytes at high pressures and temperatures. I. Summary of the thermodynamic/electrostatic properties of the solvent, *Amer. J. Sci.* **274,** 1089-1198.

Helgeson, H. C. and Kirkham, D. H.: 1974b, Thermodynamic properties of hydrothermal systems at high pressures and temperatures (abstr.), *Geol. Soc. Amer. Abstr. Prog.* **87,** 788-789.

Helgeson, H. C. and Kirkham, D. H.: 1976, Theoretical prediciton of the thermodynamic behavior of aqueous electrolytes at high pressures and temperatures. III. Equations of state for aqueous species at infinite dilution, *Amer. J. Sci.* **276,** 97-240.

Helgeson, H. C. and Shock, E. L.: 1988, Role of oxidation/reduction reactions in the hydrothermal transport and deposition of petroleum (abstr.), *Geol. Soc. Amer. Abstr. Prog.* **20,** A95.

Helgeson, H. C., Delany, J. M., Nesbitt, H. W. and Bird, D. K.: 1978, Summary and critique of the thermodynamic properties of rock-forming minerals, *Amer. J. Sci.* **278a,** 1-229.

Helgeson, H. C., Kirkham, D. H. and Flowers, G. C.: 1981, Theoretical prediciton of the thermodynamic behavior of aqueous electrolytes at high pressures and temperatures. IV. Calculation of activity coefficients, osmotic coefficients, and apparent molal and standard and relative partial molal properties to 600°C and 5 kb, *Amer. J. Sci.* **281,** 1241-1516.

Helgeson, H. C., Knox, A. M. and Shock, E. L.: 1991, Petroleum, oil field brines and authigenic mineral assemblages: Are they in metastable equilibrium? (abstr.), 15th International Organic Geochemistry Conference.

Helling, R. K. and Tester, J. W.: 1988, Oxidation of simple compounds and mixtures in supercritical water: Carbon monoxide, ammonia and ethanol, *Environ. Sci. Technol.* **22,** 1319-1324.

Hennet, J.-C., Holm, N. G. and Engel, M. H.: 1992, Abiotic synthesis of amino acids under hydrothermal conditions and the origin of life: a perpetual phenomenon?, *Naturwissenschaften* **79,** 00-00.

Herrera, A. L.: 1942, A new theory of the origin and nature of life, *Science* **96,** 14.

Heyns, K. and Pavel, K.: 1957a, Thermische Umwandlungsprodukte von Aminosäuren I, *Zeit. Naturforsch.* **12B,** 97-109.

Heyns, K. and Pavel, K.: 1957b, Thermische Umwandlungsprodukte von Aminosäuren II, *Zeit. Naturforsch.* **12B,** 109-115.

Higman, E. B., Schmeltz, I. and Schlotzhauer, W. S.: 1970, Products from the thermal degradation of some naturally occurring materials, *J. Agr. Food Chem.* **18**, 636.

Hilbert, R.: 1979, pVT-Daten von Wasser und wässrigen Natriumchlorid-Lösungen bis 873 K, 400 Bar und 25 Gewichtsprozent NaCl, Ph.D. Thesis, University of Karlsruhe, Karlsruhe, West Germany.

Hinshelwood, C. N. and Topley, B.: 1923, Energy of activation in heterogeneous gas reactions with relation to the thermal decomposition of formic acid vapor, *J. Chem. Soc.* **123**, 1014.

Hitch, B. F. and Mesmer, R. E.: 1976, The ionization of aqueous ammonia to 300° in KCl media, *J. Soln. Chem.* **5**, 667-680.

Hochella, F. M. and White, A. F.: 1990, *Mineral-Water Interface Geochemistry*, Mineralogical Society of America, Washington, D.C., 603 p.

Hodgson, C. J. and Lydon, J. W.: 1977, Geological setting of volcanogenic massive sulphide deposits and active hydrothermal systems: Some implications for exploration, *CIM Bulletin* **70**, 95-106.

Hoering, T. C.: 1984, Thermal reactions of kerogen with added water, heavy water and pure organic substances, *Org. Geochem.* **5**, 267-278.

Holland, H. D.: 1973, The oceans: a possible source of iron in iron-formations, *Econ. Geol.* **68**, 1169-1172.

Holloway, J. R.: 1984, Graphite-CH_4-H_2O-CO_2 equilibria at low-grade metamorphic conditions, *Geology* **12**, 455-458.

Holloway, J. R.: 1988, Planetary atmospheres during accretion: The effect of C-O-H-S equilibria, in *Lunar and Planetary Science* (Lunar and Planetary Institute, Houston) **19**, 499-500.

Holloway, J. R. and Wood, B. J.: 1988, *Simulating the Earth: Experimental Geochemistry*, Unwin Hyman, Boston, 196 p.

Holm, N. G.: 1987, Possible biological origin of banded iron-formations from hydrothermal solutions, *Origins Life Evol. Biosphere* **17**, 229-250.

Holm, N. G.: 1990, Report on the workshop: 'Chemical evolution and neo-abiogenesis in marine hydrothermal systems', *Origins Life Evol. Biosphere* **20**, 93-98.

Holm, N. G., Ertem, G. and Ferris, J. P.: 1992, The binding and reactions of nucleotides and polynucleotides on iron oxide hydroxide polymorphs, *Origins Life Evol. Biosphere*, submitted.

Hong, G. T., Fowler, P. K., Killilea, W. R. and Swallow, K. C.: 1987, Supercritical water oxidation: Treatment of human waste and system configuration tradeoff study, SAE Techanical Paper Series 871444.

Hong, G. T., Killilea, W. R. and Thomason, T. B.: 1988, Supercritical water oxidation: Space applications, in *Engineering, Construction and Operations in Space*, Proceedings of Space 88 (eds. S. W. Johnson and J. P. Wetzel), Amer. Soc. Civil Eng., New York, pp. 987-998.

Horibe, Y., Kim, K.-R. and Craig, H.: 1986, Hydrothermal methane plumes in the Mariana back-arc spreading centre, *Nature* **324**, 131-133.

Houser, T. J., Tiffany, D. M., Li, Z., McCarville, M. E. and Houghton, M. E.: 1986, Reactivity of some organic compounds with supercritical water, *Fuel* **65**, 827-832.

Houser, T. J., Tsao, C.-C., Dyla, J. E., Van Atten, M. K. and McCarville, M. E.: 1989, The reactivity of tetrahydroquinoline, benzylamine and bibenzyl with supercritical water, *Fuel* **68**, 323-327.

Huber, R., Kristjansson, J. K. and Stetter, K. O.: 1987, *Pyrobaculum* gen. nov., a new genus of neutrophilic rod-shaped archaebacteria from continental solfataras growing optimally at 100°C, *Arch. Microbiol.* **149**, 95-101.

Huber, R., Kurr, M., Jannasch, H. W. and Stetter, K. O.: 1989, A novel group of abyssal methanognic archaebacteria (Methanopyrus), growing at 110°C, *Nature* **342**, 833-834.

Huber, R., Langworthy, T. A., Konig, K., Thomm, M., Woese, C. R., Sletyr, U. B. and Stetter, K. O.: 1986, *Thermotoga maritima* sp. nov. represents a new genus of unique extremely thermophilic eubacteria growing up to 90°C, *Arch. Microbiol.* **144**, 324-333.

Huebner, W. F.: 1987, First polymer in space identified in Comet Halley, *Science* **237**, 628-630.

Hulett, H. R.: 1969, Limitations on prebiological synthesis, *J. Theoret. Biol.* **24**, 56-72.

Hull, D. E.: 1960, Thermodynamics and kinetics of spontaneous generation, *Nature* **186**, 693-694.

Humphris, S. E. and Thompson, G.: 1978, Hydrothermal alteration of oceanic basalts by seawater, *Geochim. Cosmochim. Acta* **42**, 107-125.

Hunt, J. M.: 1979, *Petroleum Geochemistry and Geology*, W.H. Freeman and Company.

Hunt, J. M., Lewan, M. D. and Hennet, R. J.-C.: 1991, Modeling oil generation with time-temperature index graphs on the Arrhenius Equation, *Amer. Assoc. Petrol. Geol. Bull.* **75**, 795-807.

Huser, B. A., Patel, B. K. C., Daniel, R. M. and Morgan, H. W.: 1986, Isolation and characterisation of a novel extremely thermophilic, anerobic, chemo-organotrophic eubacterium, *FEMS Microbiol. Lett.* 37, 121-127.

Ibanez, J. D., Kimball, A. P. and Oró, J.: 1971, Possible prebiotic condensation of mononucleotides by cyanamide, *Science* **173**, 444-446.

Ingmanson, D. E. and Dowler, M. J.: 1977, Chemical evolution and the evolution of the Earth's crust, *Origins Life Evol. Biosphere* **8**, 221-224.

Ingmanson, D. E. and Dowler, M. J.: 1980, Unique amino acid composition of Red Sea brine, *Nature* **286**, 51-52.

Ingmanson, D. E. and Dowler, M. J.: 1981, Chemical evolution and plate techtonics, in *Origin of Life* (ed. Y. Wolman), Reidel Publishing Co., Dordrecht, pp.129-134.

InterRIDGE Meeting Report: 1990, RIDGE Office, University of Washington, Seattle, WA.

Ito, E. and Clayton, R. N.: 1983, Submarine metamorphism of gabbros from the Mid-Cayman Rise: An oxygen isotopic study, *Geochim. Cosmochim. Acta* **47**, 535-546.

Ito, M., Handa, N. and Yanagawa, H.: 1990, Synthesis of polypeptides by microwave heating: II. Function of polypeptides synthesized during repeated hydration-dehydration cycles, *J. Molec. Evol.* **31**, 187-194.

Ivanov, Ch. P. and Slavcheva, N. N.: 1977, Formation of amino acids on heating glycine with alumina, *Origins Life Evol. Biosphere* **8**, 13-19.

Janecky, D. R. and Seyfried, W. E.: 1984, Formation of massive sulfide deposits on oceanic ridge crests: Incremental reaction models for mixing between hydrothermal solutions and seawater, *Geochim. Cosmochim. Acta* **48**, 2723-2738.

Janecky, D. R. and Seyfried, W. E., Jr.: 1986, Hydrothermal serpentinization of peridotite within the oceanic crust: Experimental investigations of mineralogy and major element chemistry, *Geochim. Cosmochim. Acta* **50**, 1357-1378.

Janecky, D. R. and Shanks, W. C., III.: 1988, Computational modeling of chemical and sulfur isotopic reaction processes in seafloor hydrothermal systems: Chimneys, massive sulfides, and subadjacent alteration zones, *Can. Mineral.* **26**, 805-825.

Jannasch, H. W., Wirsen, C. O., Molyneaux, S. J. and Langworthy, T. A.: 1988, Extremely thermophilic fermentative archaebacteria of the genus Desulfurococcus from deep-sea hydrothermal vents, *Appl. Environ. Microbiol.* **54**, 1203-1209.

Jeanloz, R. and Morris, S.: 1986, Temperature distribution in the crust and mantle, *Ann. Rev. Earth Planet. Sci.* **14**, 377-415.

Joe, H., Kuma, K., Paplawsky, W., Rea, A. and Arrhenius, G.: 1986, Abiotic photosynthesis from ferrous carbonate (siderite) and water, *Origins Life Evol. Biosphere* **16**, 369-370.

Johnson, A. I. and Morris, D. A.: 1962, Physical and hydrologic properties of water bearing deposits from core holes in the Las Banos-Kettleman City area, California, U.S. Geol. Survey, Open File Report, Denver, CO.

Johnson, J. W. and Norton, D.: 1991, Critical phenomena in hydrothermal systems: State, thermodynamic, electrostatic, and transport properties of H_2O in the critical region, *Amer. J. Sci.* **291**, 541-648.

Johnson, J. W., Oelkers, E. H. and Helgeson, H. C.: 1992, SUPCRT92: A software package for calculating the standard molal thermodynamic properties of minerals, gases, aqueous species and reactions from 1 to 5000 bars and 0° to 1000°C, *Computers in Geosciences* (submitted).

Johnson, W. R. and Kang, J. C.: 1971, Mechanisms of hydrogen cyanide formation from the pyrolysis of amino acids and related compounds, *J. Org. Chem.* **36**, 189-192.

Johnston, K. P. and Penninger, J. M. L. (Eds.): 1989, Supercritical Fluid Science and Technology, ACS Symposium Series 406, American Chemical Society, Washington, DC, 540 pp.

Johnston, K. P., Zieger, D. H. and Eckert, C. A.: 1982, Solubilities of hydrocarbon solids in supercritical fluids. The augmented van der Waals treatment, *Ind. Eng. Chem. Fundam.* **21**, 191-197.

Jones, J. D. and Vallentyne, J. R.: 1960, Biogeochemistry of organic matter-I. Polypeptides and amino acids in fossils and sediments in relation to geothermometry, *Geochim. Cosmochim. Acta* **21**, 1-34.

Jones, J. L. and Radding, S. B. (Eds.): 1980, Thermal conversion of solid wastes and biomass, ACS Symposium Series 130, American Chemical Society, Washington, DC, 745 pp.

Jones, M. L. (Ed.): 1985, Hydrothermal vents of the Eastern Pacific: An overview, *Bull. Biol. Soc. Wash.* **6**, 1-566.

Jones, W.J ., Leigh, J. A., Mayer, F., Woese, C. R. and Wolfe, R. S.: 1983, Methanococcus jannaschii sp. nov., an extremely thermophilic methanogen from a submarine hydrothermal vent, *Arch. Microbiol.* **136**, 254-261.

Jones, W. J., Stugard, C. E. and Jannasch, H. W.: 1989, Comparison of thermophilic methanogens from submarine hydrothermal vents, *Arch. Microbiol.* **151**, 314-318.

Josephson, J.: 1982, Supercritical fluids, *Environ. Sci. Technol.* **16**, 548A-551A.

Joyce, G. F., Schwartz, A. W., Miller, S. L. and Orgel, L. E.: 1987, The case for an ancestral genetic system involving simple analogues of the nucleotides, *Proc. Nat. Acad. Sci.* **84**, 4398.

Kadko, D. and Moore, W.: 1988, Radiochemical constraints on the crustal residence time of submarine hydrothermal fluids: Endeavour Ridge, *Geochim. Cosmochim. Acta* **52**, 659-668.

Kamaluddin, Yanagawa, H. and Egami, F.: 1979, Formation of molecules of biological interest from formaldehyde and hydroxylamine in a modified sea medium, *J. Biochem.* **85**, 1503-1507.

Karl, D. M., Brittain, A. M. and Tilbrook, B. D.: 1989, Hydrothermal and microbial processes at Loihi Seamount, a mid-plate hot-spot volcano, *Deep-Sea Res.* **36**, 1655-1673.

Karl, D. M., McMurtry, G. M., Malahoff, A. and Garcia, M. O.: 1988, Loihi Seamount, Hawaii: A midplate volcano with a distinctive hydrothermal system, *Nature* **335**, 532-535.

Kasting, J.: 1990, Bolide impacts and the oxidation state of carbon in the Earth's early atmosphere, *Origins Life Evol. Biosphere* **20**, 199-231.

Kasting, J. F.: 1988, Runaway and moist greenhouse atmospheres and the evolution of Earth and Venus, *Icarus* **74**, 472-494.

Kasting, J. F., Zahnle, K. J. and Walker, J. C. G.: 1983, Photochemistry of methane in the Earth's early atmosphere, *Precamb. Res.* **20**, 121-148.

Katritzky, A. R., Lapucha, A. R., Murugan, R., Luxem, F. J., Siskin, M. and Brons, G.: 1990a, Aqueous high-temperature chemistry of carbo-cycles and heterocycles. 1. Introduction and reaction of 3-pyridylmethanol, pyridine-3-carboxaldehyde, and pyridine-3-carboxylic acid, *Energy & Fuels* **4**, 493-498.

Katritzky, A. R., Balasubramanian, M. and Siskin, M.: 1990b, Aqueous high-temperature chemistry of carbo-cycles and heterocycles. 2. Monosubstituted benzenes - benzyl alcohol, benzaldehyde, and benzoic acid, *Energy & Fuels* **4**, 499-505.

Katritzky, A. R., Lapucha, A. R. and Siskin, M.: 1990c, Aqueous high-temperature chemistry of carbo-cycles and heterocycles. 3. 2-Substituted pyridines, *Energy & Fuels* **4**, 506-510.

Katritzky, A. R., Lapucha, A. R. and Siskin, M.: 1990d, Aqueous high-temperature chemistry of carbo-cycles and heterocycles. 4. 4-Substituted pyridines, *Energy & Fuels* **4**, 510-514.

Katritzky, A. R., Luxem, F. J. and Siskin, M.: 1990e, Aqueous high-temperature chemistry of carbo-cycles and heterocycles. 5. Monosubstituted benzenes with a 2 carbon atom side chain oxygenated at the beta-position, *Energy & Fuels* **4**, 514-517.

Katritzky, A. R., Luxem, F. J. and Siskin, M.: 1990f, Aqueous high-temperature chemistry of carbo-cycles and heterocycles. 6. Monosubstituted benzenes with 2 carbon atom side chains unsubstituted or oxygenated at the alpha-position, *Energy & Fuels* **4**, 518-524.

Katritzky, A. R., Luxem, F. J. and Siskin, M.: 1990g, Aqueous high-temperature chemistry of carbo-cycles and heterocycles. 7. Monosubstituted benzenes with 2 carbon atom side chains oxygenated at the alpha-position and beta-position, *Energy & Fuels* **4**, 525-531.

Katritzky, A. R., Murugan, R. and Siskin, M.: 1990h, Aqueous high-temperature chemistry of carbo-cycles and heterocycles. 8. Aquathermolysis of para-substituted phenols in the presence and absence of sodium bisulfite, *Energy & Fuels* **4**, 531-538.

Katritzky, A. R., Murugan, R. and Siskin, M.: 1990i, Aqueous high-temperature chemistry of carbo-cycles and heterocycles. 9. Aquathermolysis of ortho-substituted, meta-substituted, and multisubstituted phenols in the presence and absence of sodium bisulfite, *Energy & Fuels* **4**, 538-543.

Katritzky, A. R., Murugan, R., Balasubramanian, M. and Siskin, M.: 1990j, Aqueous high-temperature chemistry of carbo-cycles and heterocycles. 10. Aquathermolysis of acyclic and cyclic phenol ethers in the presence of sodium bisulfite or phosphoric acid, *Energy & Fuels* **4**, 543-546.

Katritzky, A. R., Murugan, R., Balasubramanian, M. and Siskin, M.: 1990k, Aqueous high-temperature chemistry of carbo-cycles and heterocycles. 11. Aquathermolysis of arylamines in the presence and absence of sodium bisulfite, *Energy & Fuels* **4**, 547-555.

Katritzky, A. R., Lapucha, A. R. and Siskin, M.: 1990l, Aqueous high-temperature chemistry of carbo-cycles and heterocycles. 12. Benzonitriles and pyridinecarbonitriles, benzamides and pyridinecarboxamides, and benzlyamines and pyridylamines, *Energy & Fuels* **4**, 555-561.

Katritzky, A. R., Lapucha, A. R., Greenhill, J. V. and Siskin, M.: 1990m, Aqueous high-temperature chemistry of carbo-cycles and heterocycles. 13. Sulfides and disulfides, *Energy & Fuels* **4**, 562-571.

Katritzky, A. R., Lapucha, A. R., Luxem, F. J., Greenhill, J. V. and Siskin, M.: 1990n, Aqueous high-temperature chemistry of carbo-cycles and heterocycles. 14. Mercaptans and sulfonic acids, *Energy & Fuels* **4**, 572-577.

Katritzky, A. R., Murugan, R. and Siskin, M.: 1990o, Aqueous high-temperature chemistry of carbo-cycles and heterocycles. 15. Aquathermolysis of arenethiols and aryl sulfides in the presence and absence of sodium bisulfite, *Energy & Fuels* **4**, 577-584.

Katritzky, A. R., Murugan, R., Balasubramanian, M. Greenhill, J. V., Siskin, M. and Brons, G.: 1991, Aqueous high-temperture chemistry of carbo-cycles and heterocycles. 16. Model sulfur compounds: A study of hydrogen sulfide generation, *Energy & Fuels* **5**, 823-834.

Kawahata, H. and Shikazono, N.: 1988, Sulfur isotope and total sulfur studies of basalts and greenstones from DSDP Hole 504B, Cost Rica Rift: Implications for hydrothermal alteration, *Can. Mineral.* **26**, 555-565.

Kawka, O. E. and Simoneit, B. R. T.: 1987, Survey of hydrothermally-generated petroleums from the Guaymas Basin spreading center, *Org. Geochem.* **11**, 311-328.

Kawka, O. E. and Simoneit, B. R. T.: 1990, Polycyclic aromatic hydrocarbons in hydrothermal petroleums from the Guaymas Basin spreading center, in *Organic Matter Alteration in Hydrothermal Systems - Petroleum Generation, Migration and Biogeochemistry* (ed. B. R. T. Simoneit), *Appl. Geochem.* **5**, 17-27.

Kelley, D. S. and Delaney, J. R.: 1987, Two-phase separation and fracturing in mid-ocean ridge gabbros at temperatures greater than 700°C, *Earth Planet. Sci. Lett.* **83**, 53-66.

Kelley, D. S. and Robinson, P. T.: 1990, Development of a brine-dominated hydrothermal system at temperatures of 400-500°C in the upper level plutonic sequence, Troodos ophiolite, Cyprus, *Geochim. Cosmochim. Acta* **54**, 653-661.

Kempton, P. D., Hawkesworth, C. J. and Fowler, M.: 1991, Geochemistry and isotopic composition of gabbros from layer 3 of the Indian Ocean Crust, hole 735B, *Proc. Ocean Drilling Program, Scientific Results* **118**, 127-142.

Kenyon, D. H. and Steinman, G.: 1969, *Biochemical Predestination*, McGraw-Hill Book Company, New York, 301 pp.

Kerr, P. F., Hamilton, P. K. and Pill, R. J.: 1951, American Petroleum Institute, Clay Mineral Standards, Project 49, Preliminary Report 7B, Chemical Analysis, Columbia University, New York, pp. 38-40.

Kerr, R. A.: 1980, Origin of life: new ingredients suggested, research news, *Science* **210**, 42-43.

Kettler, R. M., Palmer, D. A., and Wesolowski, D. J.: 1991, Dissociation quotients of oxalic acid in aqueous sodium chloride media to 175°C, *J. Soln. Chem.* **20**, 905-927.

Kimball, K. L.: 1988, High-temperature hydrothermal alteration of ultramafic cumulates from the base of the sheeted dikes in the Josephine ophiolite, NW California, *J. Geophys. Res.* **93**, 4675-4687.

Kimoto, T. and Fujinagu, T.: 1988, Non-biotic synthesis of cellular organic polymers in the presence of hydrogen sulphide, *Chem. Express* **2**, 123-136.

Kimoto, T. and Fujinagu, T.: 1990 Non-biotic synthesis of organic polymers on H_2S-rich sea-floor: a possible reaction in the origin of life, *Mar. Chem.* **30**, 179-192.

Kobayashi, K. and Ponnamperuma, C.: 1985, Trace elements in chemical evolution, I, *Origins Life Evol. Biosphere* **16**, 41-55.

Kobayashi, K., Hua, L.-L., Gerke, C. W., Gerhardt, K. O. and Ponnamperuma, C.: 1986, Abiotic synthesis of nucleosides by electric discharge in a simulated primitive Earth atmosphere, *Origins Life Evol. Biosphere* **16**, 277.

Kobayashi, K., Oshima, T. and Yanagawa, H.: 1989, Abiotic synthesis of amino acids by proton irradiation of a mixture of carbon monoxide, nitrogen, and water, *Chem. Lett.* 1527-1530.

Kobayashi, K., Tsuchiya, M., Oshima, T. and Yanagawa, H.: 1990, Abiotic synthesis of amino acids and imidazole by proton irradiation of stimulated primitive Earth atmospheres, *Origins Life Evol. Biosphere* **20**, 99-109.

Koberstein, E.: 1973, Model reactor studies of the hydrogen cyanide synthesis from methane and ammonia, *Ind. Eng. Chem. Process. Des. Dev.* **12**, 444.

Kokufuta, E., Terada, T., Suzuki, S. and Harada, K.: 1978, Potentiometric titration behavior of a copolymer of glutamic acid and alanine prepared by thermal polycondensation, *Biosystems* **10**, 299-306.

Köll, P., Brönstrup, B. and Metzger, J. O.: 1983, Liquefaction of biomass with supercritical fluids in a high pressure/high temperature flow reactor, in *Chemical Engineering at Supercritical Fluid Conditions* (eds. M. E. Paulaitis, J. M. L. Penninger, R. D. Gray, Jr. and P. Davidson), Ann Arbor Science, Ann Arbor, pp. 499-514.

Konnert, J. A. and Evans, H. T., Jr.: 1980, Crystal structure of erdite, $NaFeS_2.2H_2O$, *Amer. Mineral.* **65**, 516-521.

Kopple, K. D.: 1966, *Peptides and Amino Acids*, W. A. Benjamin, New York, pp. 11-12.

Koski, R. A., Lonsdale, P. F., Shanks, W. C., Berndt, M. E. and Howe, S. S.: 1985, Mineralogy and geochemistry of a sediment-hosted hydrothermal sulfide deposit from the southern trough of Guaymas Basin, Gulf of California, *J. Geophys. Res.* **90**, 6695-6707.

Kovács, J. and Könyves, I.: 1954, Über DL-α-β-Polyasparaginsäure, *Naturwissenschaften* **41**, 333.

Kovács, J., Könyves, I. and Pusztai, À.: 1953, Darstellung von Polyasparaginsäuren (Polyaspartsäuren), aus dem thermischen Autokondensationsprodukt der Asparaginsäure, *Experientia* **9**, 459-460.

Kríbek, B.: 1975, The origin of framboidal pyrite as a surface effect of sulphur grains, *Mineralium Deposita* **10**, 389-396.

Kuhn, W. R. and Atreya, S. K.: 1979, Ammonia photolysis and the greenhouse effect in the primordial atmosphere of the Earth, *Icarus* **37**, 207-213.

Kuma, K., Paplawsky, B., Gedulin, B. and Arrhenius, G.: 1989, Mixed-valence hydroxides as bioorganic host minerals, *Origins Life Evol. Biosphere* **19**, 573-602.

Kunde, V. G., Aiken, A. C., Hanel, R. A., Jennings, D. E., Maguire, W. C., Samuelson, R. E.: 1981, C_4H_2, HC_3N and C_2N_2 in Titan's atmosphere, *Nature* **292**, 686-688.

Kurz, J. L. and Farrar, J. M.: 1969, The entropies of dissociation of some moderately strong acids, *J. Amer. Chem. Soc.* **91**, 6057-6062.

Kuznetsov, A. P., Maslennikov, V. V., Zajkov, V. V., and Sobetskij, V. A.: 1988, Hydrothermal sulphide mounds in the Uralian paleo-ocean (mid-Devonian), *Doklady Akademia Nauk SSSR* **303**, 1477-1481 (in Russian).

Kvenvolden, K. A. and Simoneit, B. R. T.: 1990, Hydrothermally derived petroleum: Examples from Guaymas Basin, Gulf of California and Escanaba Trough, Northeast Pacific, *Amer. Assoc. Petrol. Geol. Bull.* **74**, 223-237.

Kvenvolden, K. A. Lawless, J., Pering, K., Peterson, E., Flores, J., Ponnamperuma, C., Kaplan, I. R. and Moore, C.: 1970, Evidence for extraterrestrial amino-acids and hydrocarbons in the Murchison meteorite, *Nature* **228**, 923-926.

Kvenvolden, K. A., Rapp, J. B. and Hostettler, F. D.: 1990, Hydrocarbon geochemistry of hydrothermally-generated petroleum from Escanaba Trough, offshore California, in *Organic Matter Alteration in Hydrothermal Systems - Petroleum Generation, Migration and Biogeochemistry* (ed. B. R. T. Simoneit), *Appl. Geochem.* **5**, 83-91.

Kvenvolden, K. A., Rapp, J. B., Hostettler, F. D., Morton, J. L., King, J. D. and Claypool, G. E.: 1986, Petroleum associated with polymetallic sulfide in sediment from Gorda Ridge, *Science* **234**, 1231-1234.

Labadie, M., Jensen, R. and Neuzil, E.: 1968, Recherches sur l'évolution pré-biologique III. Les acides azulmiques noirs formés à partir du cyanure d' ammonium, *Biochim. Biophys. Acta* **165**, 525-533.

Lacey, J. C., Jr., Yuki, A. and Fox, S. W.: 1979, Coprecipitation of thermal lysine-rich proteinoids with polyribonucleotides, *Biosystems* **11**, 1-7.

Lahav, N. and Chang, S.: 1976, The possible role of solid surface area in condensation reactions during chemical evolution: Re-evaluation, *J.Molec. Evol.* **8**, 357-80.

Lahav, N. and White, D. H.: 1980, A possible role of fluctuating clay-water systems in the production of ordered prebiotic oligomers, *J. Molec. Evol.* **16**, 11-21.

Lahav, N., White, D. and Chang, S.: 1978, Peptide formation in the prebiotic era: Thermal condensation of glycine in fluctuating clay environments, *Science* **201**, 67-69.

Lake, J. A.: 1988, Origin of the Eukaryotic nucleus determined by rate-invarient analysis of rRNA sequences, *Nature* **331**, 184-186.

Lake, J. A.: 1991, Tracing origins with molecular sequences, *Trends Biochem. Sci.* **16**, 46-50.

Langridge, J.: 1968, Genetic and enzymatic experiments relating to the tertiary structure of b-galactosidase, *J. Bacteriol.* **96**, 1711-1717.

Larter, R. C. L., Boyce, A. J. and Russell, M. J.: 1981, Hydrothermal pyrite chimneys from the Ballynoe baryte deposit, Silvermines, County Tipperary, Ireland, *Mineral. Deposita* **16**, 309-318.

Lavrent'yev, G. A., Strigunkova, T. F. and Yegorov, I. A.: 1985: Effects of silicates on amino-acid thermal stabilities, *Geochem. Int.* **22**, 52-57.

Lawless, J. G.: 1986, Clay-organic interactions and the origin of life, in *Clay Minerals and the Origin of* Life (eds. A. G. Cairns-Smith and H. Hartman), Cambridge University Press, Cambridge, pp. 135-137.

Lawless, J. G. and Boynton, C. D.: 1973, Thermal synthesis of amino acids from a simulated primitive atmosphere, *Nature* **243**, 405-407.

Lawless, J. G. and Levi, N.: 1979, The role of metal ions in chemical evolution: polymerization of alanine and glycine in a cation-exchanged clay environment, *J. Molec. Evol.* **13**, 281-286.

Lawrence, J. R.: 1979, Temperatures of formation of calcite veins in the basalts from Deep Sea Drilling Project holes 417A and 417D, in *Initial Reports of the Deep Sea Drilling Project*, U.S. Government Printing Office, Washington, D.C., **Vol. 51-53**, 1183-1184.

Leach, W. W., Nooner, D. W. and Oró, J.: 1978, in "Origins of Life: Proceedings of the 2nd ISSOL Meeting, 5th ICOL Meeting (ed. H. Noda), Center for Academic Publications, Japan Scientific Societies Press, Tokyo, p. 113.

Lecuyer, G., Brouxel, M. and Albarede, F.: 1990, Elemental fluxes during hydrothermal alteration of the Trinity ophiolite (California U.S.A.), by seawater, *Chem. Geol.* **89**, 87-115.

Leif, R. N., Simoneit, B. R. T. and Kvenvolden, K.A.: 1991, Simulation of hydrothermal petroleum generation by laboratory hydrous pyrolysis, in *Organic Geochemistry, Advances and Applications in the Natural Environment* (ed. D. A. C. Manning), Manchester University Press, Manchester, pp. 300-303.

Leja, J.: 1982, *Surface Chemistry of Froth Flotation* , Plenum, New York.

Lemmon, R. M.: 1970, Chemical evolution, *Chem. Rev.* **70**, 95-109.

Levine, J. S.: 1982, The photochemistry of the paleoatmosphere, *J. Molec. Evol.* **18**, 161-172.

Levine, J. S.: 1985, The photochemistry of the early atmosphere, in *The Photochemistry of Atmospheres, Earth, the Other Planets and Comets* (ed. J. S. Levine), Academic Press, Orlando, pp. 3-38.

Levine, J. S., Augustsson, T. R., and Natarajan, M.: 1982, The prebiological paleoatmosphere: stability and composition, *Origins Life Evol. Biosphere* **12**, 245-259.

Lilley, M. D., Baross, J. A. and Gordon, L. I.: 1983, Reduced gases and bacteria in hydrothermal fluids: The Galapagos spreading center and 21°N East Pacific Rise, in *Hydrothermal Processes at Seafloor Spreading Centers*, NATO Conference series, Ser. IV: Marine Sciences, 12 (eds. P. A. Rona *et al.*), Plenum, pp. 411-449.

Lilley, M. D. de Angelis, M. A. and Gordon, L. I.: 1982, CH_4, H_2, CO and N_2O in submarine hydrothermal vent waters, *Nature* **300**, 48-50.

Little, S. A., Stolzenbach, K. D. and von Herzen, R. P.: 1991, Measurements of plume flow from a hydrothermal vent field, *J. Geophys. Res.* **92**, 2587-2596.

Lohrmann, R.: 1972, The formation of urea and guanidine by irradiation of ammonium cyanide, *J. Molec. Evol.* **1**, 263-269.

Lonsdale, P.: 1985, A transform continental margin rich in hydrocarbons, Gulf of California, *Amer. Assoc. Petrol. Geol. Bull.* **69**, 1160-1180.

Lonsdale, P. and Becker, K.: 1985, Hydrothermal plumes, hot springs, and conductive heat flow in the Southern Trough of Guaymas Basin, *Earth Planet. Sci. Lett.* **73**, 211-225.

Lonsdale, P. F., Bischoff, J. L., Burns, V. M., Kastner, M. and Sweeney, R. E.: 1980, A high-temperature hydrothermal deposit on the seabed at a Gulf of California spreading center, *Earth Planet. Sci. Lett.* **49**, 8-20.

Love, J. D. and Good, J. M.: 1970, Hydrocarbons in thermal areas, northwestern Yellowstone National Park, Wyoming, *U.S. Geol. Surv. Prof. Pap.* **7644-B**.

Lowe, C. U., Rees, M. W. and Markham, R.: 1963, Synthesis of complex organic compounds from simple precursors: Formation of amino acids, amino acid polymers, fatty acids and purines from ammonium cyanide, *Nature* **199**, 219-222.

Lowell, R. P.: 1990, Thermoelasticity and the formation of black smokers, *Geophys. Res. Lett.* **17**, 709-712.

Lowell, R. P.: 1991, Modeling continental and submarine hydrothermal systems, *Rev. Geophys.* **29**, 457-476.

Lowell, R. P. and Burnell, D. K.: 1991, Mathematical modeling of conductive heat transfer from a freezing, convecting magma chamber to a single-pass hydrothermal system: Implications for seafloor black smokers, *Earth Planet. Sci. Lett.* **104**, 59-69.

Lown, D. A., Thirsk, H. R. and Wynne-Jones, L.: 1970, Temperature and pressure dependence of the volume of ionization of acetic acid in water from 25 to 225°C and 1 to 3000 bars, *Trans. Faraday Soc.* **66**, 51-73.

L'vov, S. N., Zarembo, V. I. and Gilyarov, V. N.: 1981, High-precision study of volume properties of water solutions of sodium chloride under high parameters, *Geokhimiya* **4**, 505-516.

MacDonald, K. C., Becker, K., Spiess, F. N. and Ballard, R. D.: 1980, Hydrothermal heat flux of the black smoker vents on the east pacific rise, *Earth Planet. Sci. Lett.* **48**, 1-7.

MacGeehan, P. J. and MacLean, W. H.: 1980, Tholeiitic basalt-rhyolite magmatism and massive sulphide deposits at Matagami, Quebec, *Nature* **283**, 153-157.

Maher, K. A. and Stevenson, D. J.: 1988, Impact frustration of the origin of life, *Nature* **331**, 612-614.

Mann, A. P. C. and Williams, D. A.: 1980, A list of interstellar molecules, *Nature* **283**, 721-725.

Marshall, W. L.: 1987, Possible geochemical production of biological precursors: Amino acids and other amines (also hydrocarbons) from aqueous ammonium carbonate solutions and metal carbides at 200°C -300°C (abstr.), *EOS* **68**, 458.

Marshall, W. L.: 1992, Hydrothermal synethsis of amines and amino acids from carbides and ammonium carbonate solutions (in prep).

Marshall, W. L. and Franck, E. V.: 1981, Ion Product of Water Substance, 0-1000°C, 1-10,000 Bars New International Formulation and Its Background, *J. Phys. Chem. Ref. Data* **10**, 959.

Martens, C. S.: 1990, Generation of short chain organic acid anions in hydrothermally altered sediments of the Guaymas Basin, Gulf of California, in *Organic Matter Alteration in Hydrothermal Systems - Petroleum Generation, Migration and Biogeochemistry* (ed. B. R. T. Simoneit), *Appl. Geochem.* **5**, 71-76.

Massoth, G. J., Butterfield, D. A., Lupton, J. E., McDuff, R. E., Lilley, M. D., and Jonasson, I. R.: 1989, Submarine venting of phase-separated hydrothermal fluids at Axial Volcano, Juan de Fuca Ridge, *Nature* **340**, 702-705.

Matheja, J. H. and Degens, E. T.: 1971, *Structural Molecular Biology of Phosphates*, Gustav Fischer Verlag, Stuttgart.

Matsui, T. and Abe, Y.: 1986a, Evolution of an impact-induced atmosphere and magma ocean on the accreting Earth, *Nature* **319**, 303-305.

Matsui, T. and Abe, Y.: 1986b, Impact-induced atmospheres and ocean on Earth and Venus, *Nature* **322**, 526-528.

Matthews, B. W.: 1987, Genetic and structural analysis of the protein stability problem, *Biochemistry* **26**, 6885-6888.

Mauzerall, D.: 1990, The photochemical origins of life and photoreaction of ferrous ion in the Archean oceans, *Origins Life Evol. Biosphere* **20**, 293-302.

McAuley, A. and Nancollas, G. H.: 1961, Thermodynamics of ion association. Part VII. Some transition-metal oxalates, *J. Chem. Soc.*, 2215-2221.

McAuliffe, C.: 1966, Solubility in water of paraffin, cycloparaffin, olefin, acetylene, cycloolefin, and aromatic hydrocarbons, *J. Phys. Chem.* **70**, 1267-1275.

McDonald, E. C., Howard, J. and Bennett, B.: 1983, Chemicals from forest products by supercritical fluid extraction, *Fluid Phase Equilibria* **10**, 337-344.

McKenzie, D. P. and Richter, F. M.: 1981, Parameterized thermal convection in a layered region and the thermal history of the Earth, *J. Geophys. Res.* **86**, 11667-11680.

McLennan, S. M., and Taylor, S. R.: 1982, Geochemical constraints on the growth of the continental crust, *J. Geol.* **90**, 347-361.

McLeod, G., McKeown, C., Hall, A. J., and Russell, M. J.: 1992, Computer modelling of medium enthalpy hydrothermal solutions, *Origins Life Evol. Biosphere*, in review.

Meggy, A. B.: 1953, Glycine peptides, Part I. The polymerization of piperazine-2,5-dione at 180°C, *J. Chem. Soc.*, 851-855.

Meggy, A. B.: 1956, Glycine peptides, Part II. The heat and entropy of formation of the peptide bond in polyglycine, *J. Chem. Soc.*, 1444-1454.

Merlivat, L., Pineau, F. and Javoy, M.: 1987, Hydrothermal vent waters at 13°N on the East Pacific Rise: isotopic composition and gas concentration, *Earth Planet. Sci. Lett.* **84**, 100-108.

Mesmer, R. E. and Baes, C. F., Jr.: 1974, Phosphoric acid dissociation equilibria in aqueous solutions to 300°C, *J. Soln. Chem.* **3**, 307-322.

Metzger, J. O., Hartmanns, J., Malwitz, D. and Köll, P.: 1983, Thermal organic reactions in supercritical fluids, in *Chemical Engineering at Supercritical Fluid Conditions* (eds. M. E. Palaitis, J. M. L. Penninger, R. D. Gray, Jr. and P. Davidson), Ann Arbor Science, Ann Arbor, pp. 515-533.

Mevel, C.: 1987, Evolution of oceanic gabbros from DSDP Leg 82: influence of the fluid phase on metamorphic crystallizations, *Earth Planet. Sci. Lett.* **83**, 67-79.

Mevel, C.: 1988, Metamorphism in oceanic layer 3, Gorringe Bank, Eastern Atlantic, *Contrib. Mineral. Petrol.* **100**, 496-509.

Michaelis, W., Jenisch, A. and Richnow, H. H.: 1990, Hydrothermal petroleum generation in Red Sea Sediments from the Kebrit and Shaban Deeps, in *Organic Matter Alteration in Hydrothermal Systems - Petroleum Generation, Migration and Biogeochemistry* (ed. B. R. T. Simoneit), *Appl. Geochem.* **5**, 103-114.

Michard, A., Albarede, F., Michard, G., Minster, J. F. and Charlou, J. L.: 1983, Rare-earth elements and uranium in high-temperature solutions from East Pacific Rise hydrothermal vent field (13°N), *Nature* **303**, 795-797.

Michard, G., Albarede, F., Michard, A., Minster, J. F., Charlou, J. L. and Tan, N.: 1984, Chemistry of solutions from the 13N East Pacific Rise site, *Earth Planet. Sci. Lett.* **67**, 297-308.

Miller, S. L.: 1953, A production of amino acids under possible primitive Earth conditions, *Science* **117**, 528-529.

Miller, S. L.: 1955, Production of some organic compounds under possible primitive Earth conditions, *J. Amer. Chem. Soc.* **77**, 2351-2361.

Miller, S. L.: 1957a, The mechanism of synthesis of amino acids by electric discharges, *Biochim. Biophys. Acta* **23**, 480-489.

Miller, S. L.: 1957b, The formation of organic compounds on the primitive Earth, *Annals New York Acad. Sci.* **69**, 260-275.

Miller, S. L. and Bada, J. L.: 1988, Submarine hot springs and the origin of life, *Nature* **334**, 609-611.

Miller, S. L. and Bada, J. L.: 1991a, Submarine hot springs were not the site for the origin of life (abstr.), *Geol. Soc. Amer. Abstr. Prog.* **23**, A18.

Miller, S. L. and Bada, J. L.: 1991b, The origin of life did not take place in submarine hot springs (abstr.), *EOS* **72**, 59.

Miller, S. L. and Orgel, L. E.: 1973, in *The Origins of Life on Earth* (eds. W. D. McElroy and C. P. Swanson), Prentice-Hall Inc., New Jersey, pp. 83-102.

Miller, S. L. and Orgel, L. E.: 1974, *The Origins of Life on Earth*, Prentice-Hall, Englewood Cliffs, N.J., pp. 109-112.

Miller, S. L. and Urey, H. C.: 1959, Organic compound synthesis on the primitive Earth, *Science* **130**, 245-251.

Miller, S. L., Bada, J. L. and Friedmann, N.: 1989, What was the role of submarine hot springs in the origin of life?, *Origins Life Evol. Biosphere* **19**, 536-537 (abstr.).

Miller, S. L., Urey, H. C. and Oró, J.: 1976, Origin of organic compounds on th primitive Earth and in meteorites, *J. Molec. Evol.* **9**, 59-72.

Millero, F. J.: 1970, The apparent and partial molal volume of aqueous sodium chloride solutions at various temperatures, *J. Phys. Chem.* **74**, 356-362.

Milward, G. R., Ramdas, S. and Thomas, J. M.: 1985, On the direct imaging of offretite, cancrinite, chabazite and other related ABC-6 zeolites and their intergrowths, *Proc. Royal Soc. London, Ser. A* **399**, 57-71.

Modell, M., Gaudet, G. G., Simson, M., Hong, G. T. and Biemann, K.: 1982, Supercritical water, testing reveals new process, holds promise, *Solid Wastes Management*, August, p. 26.

Mok, W. S.-L., Antal, M. J., Jr. and Jones, M., Jr.: 1989, Formation of acrylic acid from lactic acid in supercritical water, *J. Org. Chem.* **54**, 4596.

Molling, P. A. and Sverjensky, D. A.: 1989, Thermodynamic analysis of ore fluids at the Questa Mo-porphyry deposits, *Geol. Soc. Amer. Abstr. Prog.*, **21**, A151.

Moorbath, S.: 1985, Crustal evolution in the early Precambrian, *Origins of Life* **15**, 251-261.

Morikawa, T.: 1978, Evolution of hydrogen cyanide during combustion and pyrolysis, *J. Combust. Toxicol.* **5**, 315.

Mortland, M. M.: 1984, Deamination of glutamic acid by pyridoxal phosphate-Cu^{2+} -ion-smectite catalysts, *J.Molec. Catalysis* **27**, 143-55.

Mottl, M. J.: 1983, Metabasalts, axial hot springs, and the structure of hydrothermal systems at mid-ocean ridges, *Geol. Soc. Amer. Bull.* **94**,161-180

Mottl, M. J. and Seyfried, W. E.: 1980, Sub-seafloor hydrothermal systems rock- vs. seawater-dominated, in *Seafloor Spreading Centers: Hydrothermal Systems* (eds. P. A. Rona and R. P. Lowell), Hutchinson and Ross, Inc., Dowden, pp. 66-82.

Muehlenbachs, K.: 1979, The alteration and aging of the basaltic layer of the seafloor: oxygen isotope evidence from DSDP/IPOD Legs 51, 52 and 53, *Initial Reports of the Deep Sea Drilling Project*, U.S. Government Printing Office, Washington, D.C., **Vol. 51-53**, 1159-1167.

Mukhin, L. M.: 1974, Evolution of organic compounds in volcanic regions, *Nature* **251**, 50-51.

Mukhin, L. M.: 1976, Volcanic processes and synthesis of simple organic compounds on primitive Earth, *Origins Life Evol. Biosphere* **7**, 355-368.

Mukhin, L. M., Bondarev, V. B. and Safonova, E. N.: 1978, The role of volcanic processes in the evolution of organic compounds on the primitive Earth, *Modern Geology* **6**, 119-122.

Mukund, S. and Adams, M. W. W.: 1991, The novel tungsten-iron-sulfur protein of the hyperthermophilic archaebacterium, *Pyrococcus furiosus*, is an aldehyde ferredoxin oxidoreductase, *J. Biol. Chem.* **266**, 14,208-14,216.

Müller, D., Pitsch, S., Kittaka, A., Wagner, E., Wintner, C. E. and Eschenmoser, A.: 1990, Chemistry of α-aminonitriles. Aldomerization of glycolaldehyde phosphate to *rac*-hexose 2,4,6-triphosphates and (in presence of formaldehyde) *rac*-pentose 2,4-diphosphates: *rac*-allose 2,4,6-triphosphate and *rac*-ribose 2,4-diphosphate are the main reaction products, *Helv. Chim. Acta* **73**, 1410-1468.

Murray, C. N., Riley, J. P., and Wilson, T. R. S.: 1969, The solubility of gases in distilled water and seawater - I. Nitrogen, *Deep-Sea Res.* **16,** 297-310.

Murrowchick, J. B., and Barnes, J. B.: 1986, Marcasite precipitation from hydrothermal solutions, *Geochim. Cosmochim. Acta* **50,** 2615-2629.

Nagayama, M., Takaoka, O., Inomata, K. and Yamagata, Y.: 1990, Diketopiperazine-mediated peptide formation in aqueous solution, *Origins Life Evol. Biosphere* **20,** 249-257.

Naumann, R.: 1910, Über die Hydrolyse des Cyans, *Z. Elektrochem.* **16,** 772-777.

Navarro-Gonzalez, R., Negron-Mendoza, A. and Chacon, E.: 1989, The γ-irradiation of aqueous solutions of urea. Implications for chemical evolution, *Origins Life Evol. Biosphere* **19,** 109-118.

Neal, C. and Stanger, G.: 1984, Calcium and magnesium hydroxide precipitation from alkaline groundwater in Oman, and their significance to the process of serpentinization, *Mineral. Mag.* **48,** 237-241.

Negron-Mendoza, A. and Ponnamperuma, C.: 1976, Formation of biologically relevant carboxylic acids during the gamma irradiation of acetic acid, *Origins Life Evol. Biosphere* **12,** 427.

Nehlig, P.: 1991, Salininity of oceanic hydrothermal fluids: a fluid inclusion study, *Earth Planet. Sci. Lett.* **102,** 310-325.

Nelson, H. D. and De Ligny, C. L.: 1968, The determination of the solubilities of some n-alkanes in water at different temperatures, by means of gas chromatography, *Recueil* **87,** 528-544.

Nisbet, E. G.: 1986, RNA, hydrothermal systems, zeolites and the origin of life, *Episodes* **9,** 83-90.

Nisbet, E. G.: 1987, *The Young Earth: An Introduction to Archaean Geology,* Allen & Unwin, Boston, 402 p.

Nisbet, E. G.: 1989, Origin of life, *Nature* **337,** 23.

Nisbet, E. G.: 1991, *Living Earth,* Harper Collins Academic, London.

Nooner, D. W. and Oró, J.: 1979, in *Advances in Chemistry Series, No. 178, Hydrocarbon Synthesis From Carbon Monoxide and Hydrogen* (eds. E. L. Kugler and F. W. Steffgen), American Chemical Society, Washington, p. 159.

Nooner, D. W., Gilbert, J. M., Gelpi, E. and Oró, J.: 1976, Closed Fischer-Tropsch synthesis over meteoritic iron, iron ore and nickel-iron alloy, *Geochim. Cosmochim. Acta* **40,** 915-924.

Norton, D. L.: 1984, Theory of hydrothermal systems, *Ann. Rev. Earth Planet. Sci.* **12,** 155-177.

Norton, D.: 1990, Pore fluid pressure near magma chambers, in *The Role of Fluids in Crustal Processes* (eds. J. D. Bredehoeft *et al.*), National Academy Press, Washington, D.C., pp. 42-49.

Norton, D. and Knight, J. E.: 1977, Transport phenomena in hydrothermal systems: Cooling plutons, *Amer. J. Sci.* **277,** 937-981.

Norton, D. and Taylor, H. P.: 1979, Quantitative simulation of the hydrothermal systems of crystallizing magmas on the basis of transport theory and oxygen isotope data: An analysis of the Skaergaard intrusion, *J. Petrol.* **20,** 421-486.

Noyes, A. A., Kato, Y. and Sosman, R. B.: 1910, The hydrolysis of ammonium acetate and the ionization of water at high temperatures, *J. Amer. Chem. Soc.* **32,** 159-178.

O'Connell, R. J. and Hager, B. H.: 1980, On the thermal state of the Earth, in *Physics of the Earth's Interior* (ed. A. M. Dziemonski and E. Boschi), Elsevier, New York, pp. 270-317.

Odom, D. G., Rao, M., Lawless, J. G. and Oró, J.: 1979, Association of nucleotides with homoionic clays, *J. Molec. Evol.* **12,** 365-7.

Oelkers, E. H. and Helgeson, H. C.: 1988, Calculation of the thermodynamic and transport properties of aqueous species at high pressures and temperatures. Dissociation constants for supercritical alkali metal halides at temperatures from 400° to 800°C and pressures from 500 to 4000 bars, *J. Phys. Chem.* **92,** 1631-1639.

Oftedahl, Ch.: 1958, A theory of exhalative sedimentary ores, *Geologiska Föreningens i Stockholm Förhandlingar* **80**, 1-19.

Okazaki, U., Asahara, K., Ikai, T. and Imai, H.: 1974, Cyanuric acid, Japan Patent 49025676, 2 July, 1974: 1974 *Chem. Abstr.* **82**, 112109s.

Olofsson, G.: 1975, Thermodynamic quantities for the dissociation of the ammonuim ion over a wide temperature range, *J. Chem. Thermo.* **7,** 507-514.

Oparin, A. I.: 1924, *Proiskhozhdenie zhizny*, Izd. Moskovshii Raochii Moscow.

Oparin, A. I.: 1936, *Origin of Life* (translated from Russian by S. Margolis), Dover Publications, New York, (1953).

Orgel, L. E.: 1986, Did template directed nucleation precede molecular replication?, *Origins Life Evol. Biosphere* **17**, 27-34.

Oró, J.: 1965, Stages and mechanisms of prebiological organic synthesis, in *The Origins of Prebiological Systems and of their Molecular Matrices* (ed. S. W. Fox), Academic Press, New York, pp. 137-171.

Oró, J. and Guidry, C. L.: 1961, Direct synthesis of polypeptides. I. Polycondensation of glycine in aqueous ammonia, *Arch. Biochem. Biophys.* **93**, 166-171.

Oró, J. and Han, J.: 1966, High-temperature synthesis of aromatic hydrocarbons from methane, *Science* **153**, 1393-1395.

Oró, J. and Kamat, S. S.: 1961, Amino acid synthesis from hydrogen cyanide under possible primitive Earth conditions, *Nature* **190**, 442-443.

Oró, J. and Kimball, A. P.: 1961, Synthesis of purines under possible primitive Earth conditions 1. Adenine from hydrogen cyanide, *Arch. Biochem. Biophys.* **94**, 217-227.

Oró, J. and Kimball, A. P.: 1962, Synthesis of purines under possible primitive Earth conditions II. Purine intermediates from hydrogen cyanide, *Arch. Biochem. Biophys.* **96**, 293.

Oró, J., Basile, J., Cortes, S., Shen, C. and Yamrom, T.: 1984, The prebiotic syntheses and catalytic role of imidazoles and other condensing agents, *Origins Life Evol. Biosphere* **14**, 237-242.

Oró, J., Kimball, A., Fritz, R. and Master, F.: 1959 Amino acid synthesis from formaldehyde and hydroxylamine, *Arch. Biochem. Biophys.* **85**, 115-130.

Oscarson, J. L., Gillespie, S. E., Christensen, J. J., Izatt, R. M. and Brown, P. R.: 1988, Thermodynamic quantities for the interaction of H^+ and Na^+ with $C_2H_3O_2^-$ and Cl^- in aqueous solution from 275 to 320°C, *J. Soln. Chem.* **17**, 865-885.

Oudin, E. and Constantinou, G.: 1984, Black smoker chimney fragments in Cyprus sulphide deposits, *Nature* **308**, 349-353.

Paecht-Horowitz, M. and Eirich, F. R.: 1988, The polymerisation of amino acid adenylates on sodium montmorillonite with pre-adsorbed polypeptides, *Origins Life Evol. Biosphere* **18**, 359-387.

Paecht-Horowitz, M., Berger, J. and Katchalsky, A.: 1970, Prebiotic synthesis of polypeptides by heterogeneous polycondensation of amino acid adenylates, *Nature* **228**, 636-639.

Palmer, D. A. and Drummond, S. E.: 1986, Thermal decarboxylation of acetate, Part I. The kinetics and mechanism of reaction in aqueous solution, *Geochim. Cosmochim. Acta* **50**, 813-823.

Park, H. R. and Getoff, N.: 1988, Photoinduced transformation of carbon monoxide in aqueous solution, *Z. Naturforsch.* **43a**, 430-434.

Parton, H. N. and Gibbons, R. C.: 1939, The thermodynamic dissociation constants of oxalic acid, *Trans. Faraday Soc.* **35**, 542-545.

Paspek, S. C. and Klein, M. T.: 1990, Shale oil upgrading in supercritical water solutions, *Fuel Science Technol. Int.* **8**, 673-687.

Paulaitis, M. E., Krukonis, V. J., Kurnik, R. T. and Reid, R. C.: 1982, Supercritical fluid extraction, *Rev. Chem. Eng.* **1**, 179-250.

Paulaitis, M. E., Penninger, J. M. L., Gray, R. D., Jr. and Davidson, P. (Eds.): 1983, *Chemical Engineering at Supercritical Fluid Conditions*, Ann Arbor Science, Ann Arbor, 543 pp.

Pearson, D., Copeland, C. S. and Benson, S. W.: 1963a, The electrical conductance of aqueous sodium chloride in the range 300 to 383°, *J. Amer. Chem. Soc.* **85**, 1004-1047.

Penninger, J. M. and Kolmschate, J. M. M.: 1989, in *Supercritical Fluids Science and Technology - ACS Symposium Series 406* (eds. K. P. Johnston and J. M. L. Penninger), American Chemical Society, Washington, p. 242.

Penninger, J. M. L., Radosz, M., McHugh, M. A. and Krukonis, V. J. (Eds.): 1985, *Supercritical Fluid Technology*, Elsevier, Amsterdam, 468 pp.

Peter, J. M.: 1986, Genesis of hydrothermal vent deposits in the southern trough of Guaymas Basin, Gulf of California: A mineralogical and geochemical study, M.Sc. thesis, University of Toronto.

Peter, J. M., Peltonen, P., Scott, S. D., Simoneit, B. R. T. and Kawka, O. E.: 1991, Carbon-14 ages of hydrothermal petroleum and carbonate in Guaymas Basin, Gulf of California - implications for oil generation, expulsion and migration, *Geology* **19**, 253-256.

Pflüger, E.: 1875, Über die physiologische Verbrennung in den lebendigen Organismen, *Arch. gesam. Physiol.* **10**, 641-644.

Phillips, R. D. and Melius, P.: 1974, The thermal polymerization of amino acids: the role and fate of the reactants, *Int. J. Peptide Protein Res.* **6**, 39-319.

Pinching, G. D. and Bates, R. G.: 1948, Second dissociation constant of oxalic acid from 0 to 50°C, and the pH of certain oxalate buffer solutions, *Jour. Res. Nat. Bur. Stand.* **40**, 405-416.

Pinnavaia, T. J. and Mortland, M. M.: 1986, Aspects of clay catalysis, in *Clay Minerals and the Origin of Life* (eds. A. G. Cairns-Smith and H. Hartman), Cambridge University Press, Cambridge, pp. 131-135.

Pirie, N. W.: 1959, Chemical diversity and the origins of life, in *The Origin of Life on the Earth* (eds. F. Clark and R. L. M. Synge), Pergamon, London, pp. 76-83.

Pitzer, K. S.: 1986, Large-scale fluctuations and the critical behavior of dilute NaCl in H_2O, *J. Phys. Chem.* **90**, 1502-1504.

Polak, J. and Lu, B. C.-Y.: 1973, Mutual solubilities of hydrocarbons and water at 0 to 25°C, *Can. J. Chem.* **51**, 4018-4023.

Polak, J. T., Balaban, M., Peplow, A. and Phlips, A. J.: 1989, Supercritical carbon dioxide extraction of lipids from algae, in *Supercritical Fluid Science and Technology* (eds. K. P. Johnston and J. M. L. Penninger), American Chem. Soc., Symp. Series 406, Washington, D.C., pp. 449-466.

Ponnamperuma, C. and Woeller, F. H.: 1967, Alpha-aminonitriles formed by an electric discharge through a mixture of anhydrous methane and ammonia, *Currents Modern Biol.* **1**, 156-158.

Ponnamperuma, C., Lemmon, R. M., Mariner, R. and Calvin, M.: 1963, Formation of adenine by electron irradiation of methane, ammonia and water, *Proc. Natl. Acad. Sci. U.S.* **49**, 737-740.

Povoledo, D. and Vallentyne, J. R.: 1964, Thermal reaction kinetics of the glutamic acid - pyroglutamic acid system in water, *Geochim. Cosmochim. Acta* **28**, 731-734.

Prager, B., Jacobson, P., Schmidt, P. and Stern, D.: 1921, *Beilsteins Handbuch der organischen, Chemie*, 4th Edition, Vol. III, pp. 74-75.

Pray, H. A., Schweickert, C. E., and Minich, B. H.: 1952, Solubility of hydrogen, oxygen, nitrogen, and helium in water at elevated temperatures, *Ind. Eng. Chem.* **44**, 1146-1151.

Price, L. C.: 1976, Aqueous solubility of petroleum as applied to its origin and primary migration, *Bull. Amer. Assoc. Petrol. Geol.* **60**, 213-244.

Price, L. C., Wenger, L. M., Ging, T. and Blount, C. W.: 1983, Solubility of crude oil in methane as a function of pressure and temperature, *Org. Geochem.* **4**, 201-221.

Quist, A. S. and Marshall, W. L.: 1968, Ionization equilibria in ammonia-water solutions to 700° and to 4000 bars of pressure, *J. Phys. Chem.* **72**, 3123-3128.

Ramayya, S., Brittain, A., DeAlmeida, C., Mok, W. and Antal, M. J., Jr.: 1987, Acid-catalyzed dehydration of alcohols in supercritical water, *Fuel* **66**, 1364-1371.

Rao, M., Eichberg, J. and Oró, J.: 1982, Synthesis of phosphatidylcholine under possible primitive Earth conditions, *J. Molec. Evol.* **18**, 196-202.

Ratcliffe, H. D. and Drozd, J. W.: 1978, The utilization of L-glutamine and the products of its thermal decomposition by *Klebsiela pneumoniae* and *Rhizobium leguminosarum*, *FEMS Microbiol. Lett.* **3**, 65-69.

Raulin, F. and Toupance, G.: 1977, The role of sulphur in chemical evolution, *J. Molec. Evol.* **9**, 329-338.

Read, A. J.: 1982, Ionization constants of aqueous ammonia from 25 to 240°C and to 2000 bar, *J. Soln. Chem.* **11**, 649-664.

Read, A. J.: 1988, The first ionization constant from 25 to 200°C and 2000 bar for orthophosphoric acid, *J. Soln. Chem.* **17**, 213-224.

Reddy, T. H. and Suryanarayana, T.: 1988, Novel histone-like proteins in the nucleoid from the acidothermophilic archaebacterium *Sulfolobus acidocaldarius* that protect DNA against thermal denaturation, *Biochim. Biophysica Acta* **949**, 87-96.

Reed, M. H.: 1983, Seawater-basalt reaction and the origin of greenstones and related ore deposits, *Econ. Geol.* **78,** 446-485.

Reid, C. and Orgel, L. E.: 1967, Synthesis of sugars in potentially prebiotic conditions, *Nature* **216**, 455.

Richardson, C. J., Cann, J., Richards, H. G. and Cowan, J. G.: 1987, Metal-depleted root zones of the Troodos ore-forming hydrothermal systems, Cyprus, *Earth Planet. Sci. Lett.* **84,** 243-253.

Rickard, D.: 1989, Experimental concentration - time curves for the iron (II) sulphide precipitation process in aqueous solutions and their interpretation, *Chem. Geol.* **78**, 315-324.

Rickard, D. T., Zweifel, H. and Donnelly, T. H.: 1979, Sulfur isotope systematics in the Åsen pyrite-barite deposits, Skellefte district, Sweden, *Econ. Geol.* **74**, 1060-1068.

Ringwood, A. E.: 1979, *Origin of Earth and Moon*, Springer-Verlag, New York, 295 pp.

Ringwood, A. E.: 1990, Earliest history of the Earth-Moon system, in *Origin of the Earth* (eds. H. E. Newsom and J. H. Jones), Oxford University Press, New York, pp. 101-134.

Rishpon, J., O'Hara, P. J., Lahav, N. and Lawless, J. G.: 1982, Interaction between ATP, metal ions, glycine, and several minerals, *J. Molec. Evol.* **18**, 179-184.

Roberts, S. J. and Norton, D.: 1992, Prediction of aqueous ion speciation for P-T conditions in mid-ocean ridge hydrothermal systems (abs.), *Ridge Theoretical Institute Short Course*, 37-38.

Robinson, P. T., Dick, H. J. B. and von Herzen, R.: 1991, Metamorphism and alteration in oceanic layer 3: hole 735B, *Proc. Ocean Drilling Program, Scientific Results* **118,** 541-552.

Rode, B. M. and Schwendinger, M. G.: 1990, Copper-catalyzed amino acid condensation in water- a simple possible way of prebiotic peptide formation, *Origins Life Evol. Biosphere* **20**, 401-410.

Rohlfing, D. L.: 1967, Thermal poly-α-amino-acids containing low proportions of aspartic acid, *Nature* **216**, 657-659.

Rohlfing, D. L.: 1976, Thermal polyamino acids: synthesis at less than 100°C, *Science* **193**, 68-70.

Rohlfing, D. L. and McAlhaney, W. W.: 1976, The thermal polymerization of amino acids in the presence of sand, *Biosystems* **8**, 139-145.

Rona, P. A.: 1984, Hydrothermal mineralization at seafloor spreading centers, *Earth Sci. Rev.* **20**, 1-104.

Rona, P. A.: 1988, Hydrothermal mineralization at oceanic ridges, *Can. Mineral.* **26**, 431-465.

Rona, P. A., Boström, K., Laubier, L. and Smith, K. L., Jr. (Eds.): 1983, *Hydrothermal Processes at Seafloor Spreading Centers*, NATO Conf. Series, Plenum Press, New York, 796 pp. .

Rona, P. A., Thompson, G., Mottl, M. J., Karson, J. A., Jenkins, W. J., Graham, D., Mallette, M., Von Damm, K. and Edmond, J. M.: 1984, Hydrothermal activity at the Trans-Atlantic Geotraverse hydrothermal field, Mid-Atlantic Ridge Crest at 26°N, *J. Geophys. Res.* **89**, 11,365-11,377.

Ross, D. S.: 1984, Coal conversion in carbon monoxide - water systems, *Coal Sci.* **3**, 301-337.

Ross, D. S., Hum, G. P., Miin, T.-C., Green, T. K. and Mansini, R.: 1986, Supercritical water/CO liquefaction and a model for coal conversion, *Fuel Proc. Techn.* **12**, 277-285.

Russel, N. J. and Fukunaga, N.: 1990, A comparison of thermal adaption of membrane lipids in psychrophilic and thermophilic bacteria, *FEMS Microbiol. Rev.* **75**, 171-182.

Russell, M. J.: 1978, Downward-excavating hydrothermal cells and Irish-type ore deposits: importance of an underlying thick Caledonian prism, *Trans. Instn. Min. Metall.* **B87**, 168-171.

Russell, M. J.: 1983, Major sediment-hosted enhalative zinc + lead deposits: Formation from hydrothermal convection cells that deepen during crustal extension, *Mineral. Assoc.*, Canada, Short Course Handbook **9**, 251-282.

Russell, M. J.: 1988, Chimneys, chemical gardens and feldspar horizons +/- pyrrhotite in some SEDEX deposits: aspects of alkaline environments of deposition, Proc, 7th Quadrennial IAGOD Symposium. E. Schweizerbartsche Verlagsbuchhandlung, Stuttgart, 183-190.

Russell, M. J. and Daniel, R. M.: 1992, Emergence of life via catalytic hydrothermal colloidal iron sulphide membranes (abstr.), *Annales Geophysicae* **10**, Suppl. III, C.506..

Russell, M. J., Hall, A. J., Cairns-Smith, A. G. and Braterman, P. S.: 1988, Submarine hot springs and the origin of life, *Nature* **336**, 117.

Russell, M. J., Hall, A. J. and Duller, P.: 1984, Implications of chemical garden growth to the understanding of certain ore morphologies in the Navan orebody, *Mineral. Soc.* Newsletter, Sept. **64**, 2-3.

Russell, M. J., Hall, A. J. and Turner, D.: 1989, In vitro growth of iron sulphide chimneys: Possible culture chambers for origin-of-life experiments, *Terra Nova* **1**, 238-241.

Ruttersmith, L. and Daniel, R. M.: 1991, Thermostable cellobiohydrolase from the thermophilic eubacterium *Thermotoga* sp. strain FJSS3-B1, *J. Biochem.* **277**, 887-890.

Sagan, C. and Khare, B. N.: 1971, Long-wavelength ultraviolet photoproduction of amino acids on the primitive Earth, *Science* **173**, 417-420.

Sakai, H., Gamo, T., Kim, E.-S., Tsutsumi, M., Tanaka, T., Ishibashi, J., Wakita, H., Yamano, M. and Oomori, T.: 1990, Venting of carbon dioxide-rich fluid and hydrate formation in mid-Okinawa Trough backarc basin, *Science* **248**, 1093-1096.

Sakurai, M. and Yanagawa, H.: 1984, Prebiotic synthesis of amino acids from formaldehyde and hydroxylamine in modified sea medium, *Origins Life Evol. Biosphere* **14**, 171-176.

Samochocka, K., Kawczynski, A. L. and Taube, M.: 1968, High-temperature synthesis of amino acids in the system acetylene/ammonia/carbon dioxide, *Angew. Chem. Internat. Edit.* **7**, 392.

Samson, I. M. and Russell, M. J: 1987, Genesis of the Silvermines zinc-lead-barite deposits, Ireland: fluid inclusion and stable isotope evidence, *Econ. Geol.* **82**, 371-394.

Sanchez, R. A., Ferris, J. P. and Orgel, L. E.: 1966, Cyanoacetylene in prebiotic synthesis, *Science* **154**, 784-785.

Sanchez, R. A., Ferris, J. P. and Orgel, L. E.: 1967, Studies in prebiotic synthesis II. Synthesis of purine precursors and amino acids from aqueous hydrogen cyanide, *J. Molec. Biol.* **30**, 223-253.

Sanchez, R. A., Ferris, J. P. and Orgel, L. E.: 1968, Studies in prebiotic synthesis IV. Conversion of 4-aminoimidazole-5-carbonitrile derivatives to purines, *J. Molec. Biol.* **38**, 121-128.

Sanders, N. D.: 1986, Visual observation of the solubility of heavy hydrocarbons in near-critical water, *Ind. Eng. Chem. Fundam.* **25**, 169-171.

Sassani, D. C. and Shock, E. L.: 1990a, Speciation and solubility of palladium in aqueous magmatic hydrothermal solutions, *Geology* **19**, 925-928.

Sassani, D. C. and Shock, E. L.: 1990b, Supercritical mass transfer during aqueous alteration of mafic intrusive rocks, *Geol Soc. Amer. Abstracts with Programs* **22**, A213.

Sassani, D. C. and Shock, E. L.: 1992, Estimation of standard partial molal entropies of aqueous ions at 25°C and 1 bar, *Geochim. Cosmochim. Acta* (in press).

Saunders, M. A. and Rohlfing, D. L.: 1972, Polyamino acids: preparation from reposrted proportions of "prebiotic" and extraterrestrial amino acids, *Science* **176**, 172-173.

Saunders, M. A. and Rohlfing, D. L.: 1974, Inclusion of nonproteinous amino acids in thermally prepared models for prebiotic protein, *Biosystems* **6**, 81-92.

Schafer, T. and Schonheit, P.: 1991, Pyruvate metabolism of the hyperthermophilic archaebacterium *Pyrococcus furiosis, Arch. Microbiol.* **155**, 366-377.

Schavo, A. and Winkler, C. A.: 1959, The reactions of active nitrogen with acetylene, methylacetylene and dimethylacetylene, *Can. J. Chem.* **37**, 655-659.

Schiffman, P. and Smith, B. M.: 1988, Petrology and oxygen isotope geochemistry of a fossil seawater hydrothermal system within the Solea Graben northern Troodos Ophiolite, Cyprus, *J. Geophys. Res.* **93**, 4612-4624.

Schiffman, P., Smith, B. M., Varga, R. G. and Moores, E.M.: 1987, Geometry, conditions and timing of off-axis hydrothermal metamorphism and ore deposition in the Solea graben, *Nature* **325**, 423-425.

Schiffries, C. M. and Skinner, B. J.: 1987, The Bushveld hydrothermal system: field and petrologic evidence, *Amer. J. Sci.* **287**, 566-595.

Schimpl, A., Lemmon, R. and Calvin, M.: 1965, Cyanamide formation under prebiotic conditions, *Science* **147**, 149-150.

Schlesinger, G. and Miller, S. L.: 1983, Prebiotic synthesis in atmospheres containing methane, carbon monoide and carbon dioxide, *J. Molec. Evol.* **19**, 376-390.

Schoell, M.: 1988, Multiple origins of methane in the Earth, *Chem. Geol.* **71**, 1-10.

Schöps, D. and Herzig, P.M.: 1990, Sulfide composition and microthermometry of fluid inclusions in the Leg 111 sheeted dike section of Ocean Drilling Program Hole 504B, Costa Rica Rift, *J. Geophys. Res.* **95**, 8405-8418.

Schulte, M. D. and Shock, E. L.: 1992, Thermodynamic properties of Strecker synthesis reactions in hydrothermal solutions, in preparation.

Schwab, G.-M. and Schwab-Agallidis, E.: 1949, On selective catalysis, *J. Am. Chem. Soc.* **71**, 1806-1816.

Schwartz, A. W. and Bakker, C. G.: 1989, Was adenine the first purine? *Science* **245**, 1102-1104.

Schwartz, A. W. and Chittenden, G. J. F.: ,1977, Synthesis of uracil and thymine under simulated prebiotic conditions, *Biosystems* **9**, 87-92.

Schwartz, A. W. and Goverde, M.: 1982, Acceleration of HCN oligomerization by formaldehyde and related compounds: Implications for prebiotic synthesis, *J. Molec. Evol.* **18**, 351-353.

Schwartz, A. W., Van der Veen, M., Bisseling, T. and Chittenden, G. J. F.: 1975, Prebiotic nucleotide synthesis-demonstration of a geologically plausible pathway, *Origins Life Evol. Biosphere* **6**, 163-168.

Schwartz, A. W., Voet, A. B. and Van Der Veen, M.: 1984, Recent progress in the prebiotic chemistry of HCN, *Origins Life Evol. Biosphere* **14**, 91-98.

Scott, L. T.: 1982, Thermal rearrangements of aromatic compounds, *Acc. Chem. Res.* **15**, 52-58.

Scott, S. D.: 1985, Seafloor polymetallic sulfide deposits: Modern and ancient, *Mar. Mining* **5**, 191-212.

Seewald, J. S. and Seyfried, W. E.: 1990, The effect of temperature on metal mobility in subseafloor hydrothermal systems: Constraints from basalt alteration experiments, *Earth Planet. Sci. Lett.* **101**, 388-403.

Seyfried, W. E.: 1987, Experimental and theoretical constraints on hydrothermal alteration processes at midocean ridges, *Ann. Rev. Earth Planet. Sci.* **15**, 317-335.

Seyfried, W. and Bischoff, J. L.: 1977, Hydrothermal transport of heavy metals by seawater: the role of seawater/basalt ratio, *Earth Planet. Sci. Lett.* **34**, 71-77.

Seyfried, W. and Bischoff, J. L.: 1979, Low temperature basalt alteration by seawater: An experimental study at 70°C and 150°C, *Geochim. Cosmochim. Acta* **43**, 1937-1947.

Seyfried, W. and Bischoff, J. L.: 1981, Experimental seawater-basalt interaction at 300°C, 500 bars, chemical exchange, secondary mineral formation, and implications for the transport of heavy metals, *Geochim. Cosmochim. Acta* **45**, 135-147.

Seyfried, W. E. and Dibble, W. E.: 1980, Seawater-peridotite interaction at 300°C and 500 bars: implications for the origin of oceanic serpentinites, *Geochim. Cosmochim. Acta* **44**, 309-321.

Seyfried, W. E. and Mottl, M. J.: 1982, Hydrothermal alteration of basalt by seawater under seawater-dominated conditions, *Geochim. Cosmochim. Acta* **46**, 985-1002.

Seyfried, W. E., Jr. and Janecky, D. R.: 1985, Heavy metal and sulfur transport during subcritical and supercritical hydrothermal alteration of basalt: Influence of fluid pressure and basalt composition and crystallinity, *Geochim. Cosmochim. Acta* **49**, 2545-2560.

Seyfried, W. E., Jr., Berndt, M. E. and Janecky, D. R.: 1986, Chloride depletions and enrichments in seafloor hydrothermal fluids: Constraints from experimental basalt alteration studies, *Geochim. Cosmochim. Acta* **50**, 469-475.

Seyfried, W. E., Jr., Berndt, M. E. and Seewald, J. S.: 1988, Hydrothermal alteration processes at mid-ocean ridges: ,Constraints from diabase alteration experiments, hot-spring fluids and composition of the oceanic crust, *Can. Mineral.* **26**, 787-804.

Shanks, W. C. and Seyfried, W. E.: 1987, Stable isotope studies of vent fluids and chimney materials, southern Juan de Fuca Ridge: Sodium metasomatism and seawater sulfate reduction, *J. Geophys. Res.* **92**, 11387-11399.

Shapiro, R.: 1986, *Origins: a skeptic's guide to the creation of life on Earth,* Summit Books, New York, 332 p.

Shaw, R. W., Brill, T. B., Clifford, A. A., Eckert, C. A. and Franck, E. U.: 1991, Supercritical water, a medium for chemistry, *Chem. Eng. News* **69** (51, Dec. 23), 26-38.

Shen, C., Yang, L., Miller, S. L. and Oró, J.: 1987, Prebiotic synthesis of imidazole-4-acetaldehyde and histidine, *Origins Life Evol. Biosphere* **17**, 295-305.

Shen, C., Yang, L., Miller, S. L. and Oró, J.: 1990, Prebiotic synthesis of histidine, *J. Molec. Evol.* **31**, 167-174.

Shimizu, M.: 1981, Origin and evolution of the genetic code, in *Origin of Life* (ed. Y. Wolman), Reidel Publishing Co., Dordrecht, Holland, pp. 423-430.

Shiraki, R., Sakai, H., Endoh, M. and Kishima, N.: 1987, Experimental studies on rhyolite-and andesite-seawater interactions at 300°C and 1000 bars, *Geochem. J.* **21,** 139-148.

Shock, E. L.: 1988, Organic acid metastability in sedimentary basins, *Geology* **16,** 886-890.

Shock, E. L.: 1989, Corrections to "Organic acid metastability in sedimentary basins," *Geology* **17,** 572-573.

Shock, E. L.: 1990a, Do amino acids equilibrate in hydrothermal fluids?, *Geochim. Cosmochim. Acta* **54,** 1185-1189.

Shock, E. L.: 1990b, Geochemical constraints on the origin of organic compounds in hydrothermal systems, *Origins Life Evol. Biosphere* **20,** 331-367.

Shock, E. L.: 1991a, The fate of cometary volatiles in hydrothermal systems (abstr.), *Comets and the Origins and Evolution of Life*, University of Wisconsin, pp. 34-35.

Shock, E. L.: 1991b, Organic synthesis and basalt alteration as coupled irreversible processses (abstr.), *EOS* **72,** 59.

Shock, E. L.: 1992a, Stability of peptides in high temperature aqueous solutions, *Geochim. Cosmochim. Acta*, in press.

Shock, E. L.: 1992b, Chemical environments of submarine hydrothermal systems, *Origins Life Evol. Biosphere*, in press.

Shock, E. L.: 1992c, Application of thermodynamic calculations to geochemical processes involving organic acids, in *Geochemistry of Organic Acids* (eds. M. Lewan and E. Pittman), Springer-Verlag, in press.

Shock, E. L.: 1992d, Organic acids in hydrothermal solutions: Standard partial molal thermodynamic properties of carboxylic acids and estimates of dissociation constants at high temperatures and pressures, *Amer. J. Sci.*, submitted.

Shock, E. L.: 1992e, Hydrothermal dehydration of aqueous organic compounds, *Geochim. Cosmochim. Acta*, submitted.

Shock, E. L.: 1992f, Hydrocarbons in hydrothermal solutions: Standard molal properties of aqueous alkanes, cycloalkanes and aromatics (in prep.)

Shock, E. L. and Helgeson, H. C.: 1988, Calculation of the thermodynamic and transport properties of aqueous species at high pressures and temperatures: Correlation algorithms for ionic aqueous species and equation of state predictions to 5 kb and 1000 C, *Geochim. Cosmochim. Acta* **52,** 2009-2036.

Shock, E. L. and Helgeson, H. C.: 1990, Calculation of the thermodynamic and transport properties of aqueous species at high pressures and temperatures: Standard partial molal properties of organic species, *Geochim. Cosmochim. Acta* **54,** 915-945.

Shock, E. L., Helgeson, H. C. and Sverjensky, D. A.: 1989, Calculation of the thermodynamic and transport properties of aqueous species at high pressures and temperatures: Standard partial molal properties of inorganic neutral species, *Geochim. Cosmochim. Acta* **53,** 2157-2184.

Shock, E. L., Oelkers, E. H., Sverjensky, D. A., Johnson, J. W. and Helgeson, H. C.: 1992, Calculation of the thermodynamic and transport properties of aqueous species at high pressures and temperatures: Effective electrostatic radii to 1000 C and 5 kb, *J. Chem. Soc. Faraday Trans.*, **88**, 803-826.

Shoor, S. K., Walker, R. D. and Gubbins, K. E.: 1969, Salting out of nonpolar gases in aqueous potassium hydroxide solutions, *J. Phys. Chem.* **73**, 312-317.

Silant'yev, S. A., Tsameryan, O. P. and Kononkova, N. N.: 1988, Metamorphic and fluid conditions indicated by oceanic and transition-zone metabasite amphibole compositions, *Geochem. Int.* **25**, 1260-1272.

Silfer, J. A., Engel, M. H. and Macko, S. A.: 1990, The effect of hydrothermal processes on the distribution and stereochemistry of amino acids in Recent Antractic sediments, in *Organic Matter in Hydrothermal Systems - Petroleum Generation, Migration and Biogeochemistry* (ed. B. R. T. Simoneit), *Appl. Geochem.* **5**, 159-167.

Simoneit, B. R. T.: 1983, Effects of hydrothermal activity on sedimentary organic matter: Guyamas Basin, Gulf of California-petroleum genesis and proto-kerogen degradation, in *Hydrothermal Process at Seafloor Spreading Centers* (eds. P. A. Rona, K. Boström, L. Laubier and K. L. Smith, Jr.), Plenum Press, New York, pp. 451-471.

Simoneit, B. R. T.: 1984, Hydrothermal effects on organic matter-high vs low temperature components, *Org. Geochem.* **6**, 857-864.

Simoneit, B. R. T.: 1985a, Hydrothermal petroleum: Genesis, migration and deposition in Guaymas Basin, Gulf of California, *Can. J. Earth Sci.* **22**, 1919-1929.

Simoneit, B. R. T.: 1985b, Hydrothermal petroleum: Composition and utility as a biogenic carbon source, in *Hydrothermal Vents of the Eastern Pacific: An Overview* (ed. M. L. Jones), *Bull. Biol. Soc. Wash.* **6**, 49-56.

Simoneit, B. R. T.: 1988, Petroleum generation in submarine hydrothermal systems: An update, *Can. Mineral.* **26**, 827-840.

Simoneit, B. R. T.: 1990, Petroleum generation, an easy and widespread process in hydrothermal systems: an overview, *Appl. Geochem.* **5**, 3-15.

Simoneit, B. R. T. (Ed.): 1990, *Organic Matter in Hydrothermal Systems - Petroleum Generation, Migration and Biogeochemistry, Appl. Geochem.* **5**, 1-248.

Simoneit, B. R. T.: 1992, Natural hydrous pyrolysis - petroleum generation in submarine hydrothermal systems, in *Productivity, Accumulation and Preservation of Organic Matter in Recent and Ancient Sediments* (eds. J. K. Whelan and J. W. Farrington), Columbia University Press, New York, pp. 368-402.

Simoneit, B. R. T. and Galimov, E. M.: 1984, Geochemistry of interstitial gases in Quaternary sediments of the Gulf of California, *Chem. Geol.* **43**, 151-166.

Simoneit, B. R. T. and Philp, R. P.: 1982, Organic geochemistry of lipids and kerogen and the effects of basalt intrusions on unconsolidated oceanic sediments: Sites 477, 478, and 481, Guaymas Basin, Gulf of California, in *Initial Reports of the Deep Sea Drilling Project* (eds. J. R. Curray, D. G. Moore *et al.*), Vol. 64, U.S. Government Printing Office, Washington, D.C., pp. 883-904.

Simoneit, B. R. T., Brault, M. and Saliot, A.: 1990, Hydrocarbons associated with hydrothermal minerals, vent waters and talus on the East Pacific Rise and Mid-Atlantic Ridge, in *Organic*

Matter Alteration in Hydrothermal Systems - Petroleum Generation, Migration and Biogeochemistry (ed. B. R. T. Simoneit), *Appl. Geochem.* **5**, 115-124.

Simoneit, B. R. T., Grimalt, J. O., Hayes, J. M. and Hartman, H.: 1987, Low temperature hydrothermal maturation of organic matter in sediments from the Atlantis II Deep, Red Sea, *Geochim. Cosmochim. Acta* **51**, 879-894.

Simoneit, B. R. T., Kawka, O. E. and Brault, M.: 1988, Origin of gases and condensates in the Guaymas Basin hydrothermal system, in *Origins of Methane in the Earth* (ed. M. Schoell), *Chem. Geol.* **71**, 169-182.

Simoneit, B. R. T., Kawka, O.E. and Wang, G.-M.: 1992, Biomarker maturation in contemporary hydrothermal systems, alteration of immature organic matter in zero geological time, in *Biological Markers in Sediments and Petroleum* (eds. J. Moldowan, R. P. Philp and P. Albrecht), Chapt. 7, pp. 124-141.

Simoneit, B. R. T., Philp, R. P., Jenden, P. D. and Galimov, E. M.: 1984, Organic geochemistry of Deep Sea Drilling Project sediments from the Gulf of California - hydrothermal effects on unconsolidated diatom ooze, *Org. Geochem.* **7**, 173-205.

Simpson, H. D., Haufler, U. R. and Daniel, R. M.: 1991, An extremely thermostable xylanase from the thermophilic eubacterium *Thermotoga*, *J. Biochem.* **277**, 413-417.

Siskin, M. and Katritzky, A. R.: 1991, Reactivity of organic compounds in hot water: Geochemical and technological implications, *Science* **254**, 231-237.

Siskin, M., Brons, G., Katritzky, A. R. and Balasubramanian, M.: 1990a, Aqueous organic chemistry. 1. Aquathermolysis - comparison with thermolysis in the reactivity of aliphatic compounds, *Energy & Fuels* **4**, 475-482.

Siskin, M., Brons, G., Katritzky, A. R. and Murugan, R.: 1990b, Aqueous organic chemistry. 2. Cross-linked cyclohexyl phenyl compounds, *Energy & Fuels* **4**, 482-488.

Siskin, M., Brons, G., Vaughn, S. N., Katritzky, A. R. and Balasubramanian, M.: 1990c, Aqueous organic chemistry. 3. Aquathermolysis - reactivity of ethers and esters, *Energy & Fuels* **4**, 488-492.

Siskin, M., Katritzky, A. R. and Balasubramanian, M.: 1991, Aqueous organic chemistry. 4. Cleavage of diaryl ethers, *Energy & Fuels* **5**, 771-773.

Sleep, N. H. and Windley, B. F.: 1982, Archaean plate tectonics: constraints and inferences, *J. Geol.* **90**, 363-379.

Sleep, N. H., Zahnle, K. J., Kasting, J. F. and Morowitz, H. J.: 1989, Annihilation of ecosystems by large asteroid impacts on the early Earth, *Nature* **342**, 139-142.

Smith-Magowan, D. and Wood, R. H.: 1981, Heat capacity of aqueous sodium chloride from 320 to 600 K measured with a new flow calorimeter, *J. Chem. Thermo.* **13**, 1047-1073.

Smolin, E. M. and Rapoport, L.: 1959, *The Chemistry of Heterocyclic Compounds. S-Triazines and Derivatives*, Interscience, N.Y., a. pp. 26-29, b. pp. 310-315, c. pp. 9-13.

Snyder, W. D. and Fox, S. W.: 1975, A model for the origin of stable protocells in a primitive alkaline ocean, *Biosystems* **7**, 222-229.

Sokolova, M. N, Dobrovol'skaya, M. G., Organova, N. I., Kazakova, M. E. and Dmitrik, A. L.: 1970, A sulfide of iron and potassium - the new mineral rasvumite, *Vses. Mineralog Obschch. Zap* **99**, 712-720.

Sosnina, I. E.: 1977, Catalytic properties of sodium mordenite in the ketonization of acetic and butyric acids, *Zh. Fiz. Khim.* **51**, 2001; 1977, *Chem. Abstr.* **87**,151671s.

Spall, B. C. and Steacie, E. W. R.: 1957, The mechanism of the photolysis of acetamide, *Proc. Roy. Soc. London* **239**, 1-15.

Spencer, R. J. and Chou, I.-M. (Eds.): 1990, *Fluid-Mineral Interactions: A Tribute to H.P. Eugster*, The Geochemical Society, Special Publication 2, 432 pp.

Spiess, F. N., MacDonald, K. C., Atwater, T., Ballard, R., Carranza, A., Cordoba, D., Cox, C., Diaz Garcia, V. M., Francheteau, J., Guerrero, J., Hawkins, J., Haymon, R., Hessler, R., Juteau, T., Kastner, M., Larson, R., Luyendyk, B., Macdougall, J. D., Miller, S., Normark, W., Orcutt, J. and Rangin, C.: 1980, East Pacific Rise. Hot springs and geophysical experiments, *Science* **207**, 1421-1433.

Squires, T. G. and Paulaitis, M. E. (Eds.): 1987, Supercritical Fluids, Chemical and Engineering Principles and Applications, ACS Symposium Series 329, American Chemical Society, Washington, DC, 296 pp.

SRDPG (Ocean Drilling Program Sedimented Ridge Detailed Planning Group): 1990, Sedimented Ridges Drilling Prospectus, *JOIDES Journal* **16**, 40-55.

Stakes, D. S., Mével, C., Mathilde, C. and Chaput, T.: 1991, Metamorphic stratigraphy of Hole 735B, *Proc. Ocean Drilling Program, Scientific Results* **118**, 153-180.

Stakes, D. S., Taylor, H. P., Jr. and Fisher, R. C.: 1983, Oxygen isotope and geochemical charcterization of hydrothermal alteration in ophiolite complexes and modern oceanic crust, in *Ophiolites and Oceanic Lithosphere* (eds. I. D. Gass, S. J. Lippard and A. W. Shelton), Special Publication of the Geological Society of London, Blackwell Scientific Publishers, pp. 199-214.

Stanton, R. L.: 1976, Petrochemical studies of the ore environment at Broken Hill, New South wales: 3 - banded iron formations and sulphide ore bodies: constitutional and genetic ties, *Trans. Instn. Mining Metall.* **85**, b132-b141.

Stanton, R. L., Roberts, W. P. H. and Chant, R. A.: 1978, Petrochemical studies of the ore environment at Broken Hill, N.S.W.: 5 - major element constitution of the lode and its interpretation, *Proc. Australas. Inst. Min. Metall.* No. **266**, 51-78.

Staszak, C. N., Malinowski, K. C. and Killilea, W. R.: 1987, The pilot-scale demonstration of the MODAR oxidation process for the destruction of hazardous organic waste materials, *Environ. Progress* **6(1)**, 39-43.

Stearns, R.S., Oppenheimer, H., Simon, E. and Harkins, W.D.: 1947, Solubilization by solutions of long-chain colloidal electrolytes, *J. Chem. Phys.* **15**, 496-507.

Stephan, E. F., Hatfield, N. S., Peoples, R. S. and Pray, H. A. H.: 1956, The solubility of gases in aqueous uranyl salt solutions at elevated temperatures and pressures, *BMI 1067 Battelle Men. Inst.*, 54 p.

Stephenson, R. C. and Clarke, S.: 1989, Succinamide formation from aspartyl and asparaginyl peptides as a model for the spontaneous degradation of proteins, *J. Biol. Chem.* **264**, 6164-6170.

Stetter, K. O., 1982: Ultrathin mycelia-forming organisms from submarine volcanic areas having an optimum growth temperature of 105°C, *Nature* **300**, 258-260.

Stetter, K. O.: 1988, *Archaeoglobus fulgidus* gen. nov., sp. nov.; a new taxon of extremely thermophilic archaebacteria, *System. Appl. Microbiol.* **10**, 172-173.

Stetter, K. O., Fiala, G. , Huber, G., Huber, R. and Segerer, A.: 1990, Hyperthermophilic microorganisms, *FEBS Microbiol. Rev.* **75**, 117-124.

Stetter, K. O., König, H. and Stackebrandt, E.: 1983, *Pyrodictium gen. nov.,* a new genus of submarine disc-shaped sulphur reducing Archaebacteria growing optimally at 105°C, *Sys. Appl. Microbiol.* **4,** 535-551.

Stevenson, D. J.: 1983, The nature of the Earth prior to the oldest known rock record: The Hadean Earth, in *Earth's Earliest Biosphere: Its Origin and Evolution* (ed. J. W. Schopf), Princeton University Press, Princeton, pp. 32-40.

Stevenson, D. J.: 1987, Origin of the Moon: The collision hypothesis, *Ann. Rev. Earth Planet. Sci.* **15**, 271-315.

Stoks, P. G. and Schwartz, A. W.: 1981, Nitrogen-heterocyclic compounds in meteorites: Significance and mechanisms, *Geochim. Cosmochim. Acta* **45**, 563.

Stoks, P. G. and Schwartz, A. W.: 1982, Basic nitrogen-heterocyclic compounds in the Murchison meteorite, *Geochim. Cosmochim. Acta* **46**, 309.

Storch, H. H., Golumbic, N and Anderson, R. B.: 1951, *The Fischer-Tropsch and Related Syntheses*, John Wiley, New York.

Suess, E., Simoneit, B. R. T., Wefer, G., Whiticar, M. J., Fisk, M., von Breymann, M., Han, M. W., Wittstock, R., Laban, C., Kadko, D. and Top, Z.: 1992, Hydrothermalism in the Bransfield Strait, Antarctica, *Geologische Rundschau*, in preparation.

Sverjensky, D.A.: 1991, Trace element partitioning between minerals and hydothermal fluids in the oceanic lithosphere to 30 km depths, *EOS*, **72**, 538.

Sverjensky, D. A., Shock, E. L. and Helgeson, H. C.: 1992, Prediction of the thermodynamic properties of aqueous metal complexes to 1000 C and 5.0 kb, in preparation.

Swanson, M. L., Olson, E. S., Diehl, J. W. and Farnum, S. A.: 1986, Extraction of low-rank coals with supercritical water, *ACS Div. Fuel Chem.* **31**, 43.

Sylvester-Bradley, P. C.: 1976, Evolutionary oscillation in prebiology: Igneous activity and the origins of life, *Origins Life Evol. Biosphere* **7**, 9-18.

Takahashi, I. and Ozako, S.: 1974, Isocyanic acid, Japan Patent 48045078, 2 Dec., 1974; 1974, *Chem. Abstr.* **82**, 173,135u.

Tamaura, Y. and Tabata, M.: 1990, Complete reduction of carbon dioxide to carbon using cation-excess magnetite, *Nature* **346,** 255-256.

Tanger, J. C. and Helgeson, H. C.: 1988, Calculation of the thermodynamic and transport properties of aqueous species at high pressures and temperatures. Revised equations of state for the standard partial molal properties of ions and electrolytes, *Amer. J. Sci.* **288,** 19-98.

Taube, M., Zdrojewski, Z., Samochocka, K. and Jezierska, K.: 1967, Formation of amino acids and their precursors from primitive compounds by high-temperature synthesis, *Angew. Chem. Internat. Edit.* **6**, 247.

Taylor, H. P., Jr.: 1990, Oxygen and hydrogen isotope constraints on the deep circulation of surface waters into zones of hydrothermal metamorphism and melting, in *The Role of Fuids in Crustal Processes* (eds. J. D. Bredehoeft *et al.*), National Academy Press, Washington, D.C., pp. 72-95.

Taylor, O., Rummery, T. E. and Owen, D. G.: 1979, On the conversion of mackinawite to greigite, *J. Inorg. Nucl. Chem.* **41**, 595-596.

Taylor, S. R. and Norman, M. D.: 1990, Accretion of differentiated planetesimals to the Earth, in *Origin of the Earth* (eds. H. E. Newsom and J. H.Jones), Oxford University Press, New York, pp. 29-43.

Temple, K. L. and Colmer, A. R.: 1951, The autotrophic oxidation of iron by a new bacterium: *Thiobacillus ferro oxidans, J. Bact.* **62**, 605-611.

Temussi, P. A., Paolillo, L., Ferrara, L., Benedetti, E. and Andini, S.: 1976, Structural characterization of thermal prebiotic polypeptides, *J. Molec. Evol.* **7**, 105-110.

Tewari, B., Miller, M. M., Wasik, S. P. and Matire, D.E.: 1982, Aqueous solubility and octanol/water partition coefficient of organic compounds at 25.0°C, *J. Chem. Eng. Data* **27**, 451-454.

Thompson, W. H. and Snyder, J. R.: 1964, Mutual solubilities of benzene and water. Equilibria in the two-phase liquid-liquid region, *J. Chem Eng. Data* **9**, 516-520.

Thornton, E. C. and Seyfried, W. E., Jr.: 1987, Reactivity of organic-rich sediment in seawater at 350°C, 500 bars: Experimental and theoretical constraints and implications for the Guaymas Basin hydrothermal system, *Geochim. Cosmochim. Acta* **51**, 1997-2010.

Tiercelin, J.-J., Boulègue, J. and Simoneit, B.R.T.: 1992, Hydrocarbon, sulphide and carbonate deposits related to sublacustrine hydrothermal seeps in the North Tanganyika Trough, East African Rift, in *Bitumens in Ore Deposits* (Ed. J. Parnell), Springer Verlag, Berlin, in press.

Tiercelin, J.-J., Thouin, C., Kalala, T. and Mondeguer, A.: 1989, Discovery of sublacustrine hydrothermal activity and associated massive sulfides and hydrocarbons in the north Tanganyika trough, East African Rift, *Geology* **17**, 1053-1056.

Tissot, B. P. and Welte, D. H.: 1984, *Petroleum Formation and Occurrence: A New Approach to Oil and Gas Exploration*, Springer Verlag, 2nd ed.

Tivey, M. G. and McDuff, R. E.: 1990, Mineral precipitation in the walls of black smoker chimneys: a quantitative model of transport and chemical reaction, *J. Geophys. Res.* **95**, 12,617-12,637.

Tödheide, K.: 1982, Hydrothermal solutions, *Ber. Bunsenges. Phys. Chem.* **86**, 1005-1016.

Townsend, S. H., Abraham, M. A., Huppert, G. L., Klein, M. T. and Paspek, S. C.: 1988, Solvent effects during reactions in supercritical water, *Ind. Eng. Chem. Res.* **27**, 143-149.

Trent, J. D., Chastain, R. A. and Yayanos, A. A.: 1984, Possible artefactual basis for apparent bacterial growth at 250°C, *Nature* **307**,737-740.

Tsonopoulos, C. and Wilson, G. M.: 1983, High-temperature mutual solubilities of hydrocarbons and water. Part 1: benzene, cyclohexane, and n-hexane, *Amer. Inst. Chem. Eng. Jour.* **29**, 990-999.

Tunnicliffe, V.: 1991, The biology of hydrothermal vents: Ecology and evolution, in *Oceanogr. Mar. Biol. Annual Rev.* (ed. M. Barnes), Aberdeen University Press, **29**, pp. 319-407.

Ulmer, G. C. and Barnes, H. L. (Eds.): 1987, *Hydrothermal Experimental Techniques*, John Wiley & Sons, New York, 523 pp.

Urey, H. C.: 1952, *The Planets: Their Origin and Development*, Yale University Press, New Haven, 245 pp.

Vallentyne, J. R.: 1964, Biogeochemistry of organic matter-II. Thermal reaction kinetics and transformation products of amino compounds, *Geochim. Cosmochim. Acta* **28**, 157-188.

Vallentyne, J. R.: 1968, Pyrolysis of proline, leucine, arginine and lysine in aqueous solution, *Geochim. Cosmochim. Acta* **32**, 1353-1356.

van Andel, T.: 1985, *New Views of an Old Planet, Continental Drift and the History of the Earth*, Cambridge University Press, Cambridge.

van de Meent, D., Brown, S. C., Philp, R. P. and Simoneit, B. R. T.: 1980, Pyrolysis-high resolution gas chromatography and pyrolysis-gas chromatography/mass spectrometry of kerogens and kerogen precursors, *Geochim. Cosmochim. Acta* **44**, 999-1013.

Van der Velden, W. and Schwartz, A. W.: 1977, Search for purines and pyrimidines in the Murchison meteorite, *Geochim. Cosmochim. Acta* **41**, 961-968.

Van Trump, J. E. and Miller, S. L.: 1972, Prebiotic synthesis of methionine, *Science* **178**, 859-860.

Vanko, D. A.: 1988, Temperature, pressure, and composition of hydrothermal fluids, with their bearing on the magnitude of tectonic uplift at mid-ocean ridges, inferred from fluid inclusions in oceanic layer 3 rocks, *J. Geophys. Res.* **93**, 4595-4611.

Vanko, D. A. and Stakes, D.: 1991, Fluids in oceanic layer 3: Evidence from veined rocks, hole 735B, Southwest Indian Ridge, *Proc. Ocean Drilling Program, Scientific Results* **118**, 181-215.

Vegotsky, A., Harada, K., and Fox, S. W.: 1958, The characterization of polyaspartic acid and some related compounds, *J. Amer. Chem. Soc.* **80**, 3361-3365.

Ventilla, M. and Egami, F.: 1977, Formation of amino acids and related oligomers from formaldehyde and hydroxylamine in a solution of transition metal ions, *J. Molec. Evol.* **9**, 105-109.

Vihinen, M.: 1987, Relationship of protein flexibility to thermostability, *Protein Engr.* **1**, 477-480.

Viljoen, M. J. and Viljoen, R. P.: 1969, The geology and geochemistry of the lower ultramafic unit of the Onverwacht Group and a proposed new class of igneous rock. *Geol. Soc. S. Africa spec. publ. 2.*: Upper mantle project, 221-244.

Voet, A. B. and Schwartz, A. W.: 1982, Uracil synthesis via HCN oligomerization, *Origins of Life* **12**, 45-49.

Voet, A. B. and Schwartz, A. W.: 1983, Prebiotic adenine synthesis from HCN - evidence for a newly discovered major pathway, *Bioorganic Chem.* **12**, 8-17.

Von Damm, K. L.: 1988, Systematics of and postulated controls on submarine hydrothermal solution chemistry, *J. Geophys. Res.* **93**, 4551-4561.

Von Damm, K. L.: 1990, Seafloor hydrothermal activity: Black smoker chemistry and chimneys, *Ann. Rev. Earth Planet. Sci.* **18**, 173-204.

Von Damm, K. L. and Bischoff, J. L.: 1987, Chemistry of hydrothermal solutions from the southern Juan de Fuca Ridge, *J. Geophys. Res.* **92,** 11,334-11,346.

Von Damm, K. L., Edmond, J. M., Measures, C. I., Grant, B., Trull, T., Walden, B. and Weiss, R. F.: 1985a, Chemistry of submarine hydrothermal solutions at 21°N, East Pacific Rise, *Geochim. Cosmochim. Acta* **49**, 2197-2220.

Von Damm, K. L., Edmond, J. M., Measures, C. I. and Grant, B. C.: 1985b, Chemistry of submarine hydrothermal solutions at Guaymas Basin, Gulf of California, *Geochim. Cosmochim. Acta* **49,** 2221-2238.

Von Damm, K. L., Grant, B. and Edmond, J. M.: 1983, Preliminary report on the chemistry of hydrothermal solutions at 21° North East Pacific Rise, in *Hydrothermal Processes at Sea Floor Spreading Centers*, NATO Conference Series, Ser. IV: Marine Sciences, Vol. 12 (eds. P. A. Rona, K. Bostrom, L. Laubier and K. L. Smith, Jr.), Plenum, New York, pp. 391-409.

Wächtershäuser, G.: 1988a, Before enzymes and templates: Theory of surface metabolism, *Microbiol. Rev.* **52**, 452-484.

Wächtershäuser, G.: 1988b, Pyrite formation, the first energy source for life: A hypothesis, *System. Appl. Microbiol.* **10**, 207-210.

Wächtershäuser, G.: 1990a, The case for the chemoautotrophic origin of life in an iron-sulfur world, *Orig. Life Evol. Biosphere* **20**, 173-176.

Wächtershäuser, G.: 1990b, Evolution of the first metabolic cycles, *Proc. Natl. Acad. Sci. USA* **87**, 200-204.

Wakita, H. and Sano, Y.: 1983, $^{3}He/^{4}He$ ratios in CH_4-rich natural gases suggest magmatic origin, *Nature* **305**, 792-794.

Wakita, H., Sano, Y., Urabe, A. and Nakamura, Y.: 1990, Origin of methane-rich natural gases in Japan: formation of gas fields due to large-scale submarine volcanism, *Appl. Geochem.* **5**, 263-278.

Walker, J. F.: 1964, *Formaldehyde*, 3rd Ed., Reinhold, New York, p. 240.

Walker, J. G. G. and Brimblecombe, P.: 1985, Iron and sulphur in the pre-biologic ocean, *Precamb. Res.* **28,** 205-222.

Walker, J. and Hambly, F. J.: 1895, Transformation of ammonium cyanate into urea, *J. Chem. Soc.* **67**, 746-767.

Walsh, K. A. J., Daniel, R. M. and Morgan, H. W.: 1983, A soluble NADH dehydrogenase from *Thermus aquaticus* strain T351, *J. Biochem.* **201**, 427-433.

Wark, E. E. and Wark, I. W.: 1935, The physical chemistry of flotation. VI. The adsorption of amines by sulfide minerals, *J. Phys. Chem.* **39**, 1021-1030.

Weast, R. C.: 1964, *The Handbook of Chemistry and Physics*, Vol. 45, Chemical Rubber Co., Cleveland, Ohio.

Weber, A. L.: 1981a, Formation of the thioester, N,S-diacetylcysteine, from acetaldehyde and N,N'-diacetylcysteine in aqueous solution with ultraviolet light, *J. Molec. Evol.* **17**, 103-107.

Weber, A. L.: 1981b, Formation of pyrophosphate, tripolyphosphate and phosphorylimidazole with the thioester, N,S-diacetylcysteamine, as the condensing agent, *J. Molec. Evol.* **18**, 24-29.

Weber, A. L.: 1982a, Formation of the thioester, N-acetyl,S-lactoylcysteine with pyruvaldehyde in aqueous solution, *J. Molec. Evol.* **18**, 354-359.

Weber, A. L.: 1982b, Formation of pyrophosphate on hydroxyapatite with thioesters as condensing agents, *Biosystems* **15**, 183-189.

Weber, A. L.: 1989, Thermal synthesis and hydrolysis of polyglyceric acids, *Origins Life Evol. Biosphere* **19**, 7-19.

Weisser, O. and Landa, S.: 1973, *Sulphide Catalysts, their Properties and Applications*, Pergamon Press and Friedr. Vieweg and Sohn.

Welhan, J. A.: 1988, Origins of methane in hydrothermal systems, *Chem. Geol.* **71**, 183-198.

Welhan, J. A. and Lupton, J. E.: 1987, Light hydrocarbon gases in Guaymas Basin hydrothermal fluids: Thermogenic versus abiogenic origin, *Amer. Assoc. Petr. Geol. Bull.*, **71**, 215-223.

West, M. A. B. and Gray, M. R.: 1987, Pyrolysis of 1,3-butanediol as a model reaction for wood liquifaction in supercritical water, *Can. J. Chem. Eng.* **65**, 645.

Wetherill, G. W.. 1990, Formation of the Earth, *Ann. Rev. Earth Planet. Sci.* **18**, 205-256.

Weyl, P. K.: 1968, Precambrian marine environment and the development of life, *Science* **161**, 158-160.

Whelan, J. K., Simoneit, B. R. T. and Tarafa, M.: 1988, C_1-C_8 hydrocarbons in sediments from Guaymas Basin, Gulf of California - comparison to Peru Margin, Japan Trench and California Borderlands, *Org. Geochem.* **12**, 171-194.

White, R. H.: 1984, Hydrolytic stability of biomolecules at high temperatures and its implications for life at 250°C, *Nature* **310**, 430-432.

Whiticar, M. J., Suess, E. and Wehner, H.: 1985, Thermogenic hydrocarbons in surface sediments of the Bransfield Strait, Antarctic Peninsula, *Nature* **314**, 87-90.

Wiegel, J.: 1990, Temperature spans for growth: hypothesis and discussion, *FEMS Microbiol. Rev.* **75**, 155-170.

Wilkerson, J. L. and Guillory, W. A.: 1977, Condensed phase photochemistry of acetic acid in the vacuum UV, *J. Photochem.* **7**, 251-261.

Willis, M. A. and Shock, E. L.: 1991, Titanium speciation in hydrothermal metamorphic and subduction fluids, *EOS*, **72**, 538.

Wilson, A. C., Caulson, S. S. and White, T. J.: 1977, Biochemical evolution, *Ann. Rev. Biochem.* **46**, 473-639.

Woese, C. R.: 1987, Bacterial evolution, *Microbiol. Rev.* **51**, 221-271.

Woese, C. R. and Fox, G. E.: 1977, Phylogenetic structure of the prokaryotic domain: The primary kingdoms, *Proc. Natl. Acad. Sci. U.S.A.* **74**, 5088-5090.

Woese, C. R., Kandler, O. and Wheelis, M. L.: 1990, Towards a natural system of organisms: proposal for the domains Archaea, Bacteria and Eucarya, *Proc. Natl. Acad. Sci. U.S.A.* **87**, 4576-4579.

Wöhler, F.: 1828, La formation artificielle de l'urée, *Annal. Chim. Phys.* **37**, 330-334.

Wöhler, F.: 1828, Über künstliche Bildung von Harnstoff, *Ann. Physik* **12**, 253.

Wolery, T. J., Isherwood, D. J., Jackson, K. L., Delany, J. M. and Puigdomenech, I.: 1984, EQ3/6: status and applications, UCRL-91884, Lawrence Livermore National Laboratory, 12 p.

Wood, B. J. and Virgo, D.: 1989, Upper mantle oxidation state: Ferric iron contents of lherzolite spinels by ^{57}Fe Mössbauer spectroscopy and resultant oxygen fugacities, *Geochim. Cosmochim. Acta* **53**, 1277-1291.

Wood, A. P. , Kelly, D. P. and Norris, P. R.: 1987, Autotrophic growth of Sulfolobus strains on tetrathionate and the effect of organic nutrient, *Arch. Microbiol.* **146**, 382-389.

Woodruff, L. G. and Shanks, W. C.: 1988, Sulfur isotope study of chimney minerals and vent fluids from 21°N, East Pacific Rise: Hydrothermal sulfur sources and disequilibrium sulfate reduction, *J. Geophys. Res.* **93,** 4562-4572.

Wright, J. M., Lindsay, W. T., Jr. and Druga, T. R.: 1961, The behavior of electrolytic solutions at elevated temperatures as derived from conductance measurements, *WAPD-TM204* U.S. Atomic Energy Commision, 31 p.

Wu, B. J., Paspek, S. C., Klein, M. T. and LaMarca, C.: 1991, Reactions in and with supercritical fluids - a review, in *Supercritical Fluid Technology: Reviews in Modern Theory and Applications* (eds. T. J. Bruno and J. F. Ely), CRC Press, Boca Raton, pp. 511-524.

Yamagata, Y., Watanabe, H., Saitoh, M. and Namba, T.: 1991, Volcanic production of polyphosphates and its relevance to prebiotic evolution, *Nature* **352,** 516-519.

Yamanaka, J., Inomata, K. and Yamagata, Y.: 1988, Condensation of oligoglycines with trimeta- and tetrametaphosphate in aqueous solutions, *Orig. Life Evol. Biosphere* **18**, 165-178.

Yanagawa, H. and Egami, F.: 1980, Formation of organized particles, marigranules and marisomes, from amino acids in a modified sea medium, *Biosystems* **12**, 147-154.

Yanagawa, H. and Egami, F.: 1981, Is carbon suboxide a new candidate as starting material for the synthesis of biomolecules on the primitive Earth?, *Precamb. Res.* **14**, 75-80.

Yanagawa, H. and Kobayashi, K.: 1989, Formation of amino acids, peptide-like polymers, and microspheres in superheated hydrothermal environments (abstr.), *Orig. Life Evol. Biosphere* **19,** 540-541.

Yanagawa, H. and Kobayashi, K.: 1992, An experimental approach to chemical evolution in submarine hydrothermal systems, *Origins Life Evol. Biosphere*, in press.

Yanagawa, H. and Kojima, K.: 1985, Thermophilic microspheres of peptide-like polymers and silicates formed at 250°C, *J. Biochem.* **97**, 1521-1524.

Yanagawa, H., Kobayashi, Y. and Egami, F.: 1980a, Genesis of amino acids in the primeval sea: Formation of amino acids from sugars and ammonia in a modified sea medium, *J. Biochem.* **87**, 359-362.

Yanagawa, H., Kobayashi, Y. and Egami, F.: 1980b, Characterization of marigranules and marisomes, organized particles with elastin-like structures, *J. Biochem.* **87**, 855-869.

Yanagawa, H., Kobayashi, Y. and Egami, F.: 1981, Genesis of amino acids in the primeval sea: formation of amino acids from sugars and ammonia in a modified sea medium, in *Origin of Life* (ed. Y. Wolman), Reidel Publishing Company, Dordrecht, pp. 181-187.

Yanagawa, H., Makino, Y., Sato, K., Nishizawa, M. and Egami, F.: 1984a, Novel formation of α-amino acids from α-oxo acids and ammonia in aqueous medium, *Origins Life Evol. Biosphere* **14**, 163-169.

Yanagawa, H., Nishizawa, M. and Kojima, K.: 1984b, A possible prebiotic peptide formation from glycinamide and related compounds, *Origins Life Evol. Biosphere* **14**, 267-272.

Yanagawa, H., Makino, Y., Sato, K., Nishzawa, M. and Egami, F.: 1982, Novel formation of α-amino acids and their derivatives from oxo acids and ammonia in aqueous medium, *J. Biochem.* **91**, 2087-2090.

Yanagawa, H., Ogawa, Y., Kojima, K. and Ito, M.-H.: 1988, Construction of protocellular structures under simulated primitive Earth conditions, *Origins Life Evol. Biosphere* **18**, 179-207.

Yanagawa, H., Kojima, K., Ito, M. and Handa, N.: 1990, Synthesis of polypeptides by microwave heating: I. Formation of polypeptides during repeated hydration-dehydration cycles and their characterization, *J. Molec. Evol.* **31**, 180-186.

Yasuda, K.: 1967, Nippon Kagaku Zasshi 88,749; 1967, Thermal decomposition of formamide, *Chem. Abstr.* **69**, 10027c.

Yoshino, D., Hayatsu, R. and Anders, E.: 1971, Origin of organic matter in early solar system III. Amino acids: Catalytic syntheses, *Geochim. Cosmochim. Acta* **35**, 927-938.

Yuen, G. U., Lawless, J. G. and Edelson, E. H.: 1981, Quantification of monocarboxylic acids from a spark discharge synthesis, *J. Molec. Evol.* **17**, 43-47.

Zahnle, K. J.: 1986, Photochemistry of methane and the formation of hydrocyanic acid (HCN, in the Earth's early atmosphere, *J. Geophys. Res.* **91**, 2819-2834.

Zahnle, K. J. and Walker, J. C. G.: 1982, The evolution of solar ultraviolet luminosity, *Rev. Geophys. Space Phys.* **20**, 280-292.

Zale, D. E. and Klibanov, A. M.: 1984, Mechanisms of irreversible thermoinactivation of enzymes. *Ann. New York Acad. Sci.* **434**, 20-26.

Zawisza, A. and Malesinska, B.: 1981, Solubility of carbon dioxide in liquid water and of water in gaseous carbon dioxide in the range 0.2-5 MPa and at temperatures up to 473 K, *J. Chem. Eng. Data* **26**, 388-391.

Zhao, H., Wood, A. G., Widdel, F. and Bryant, M. P.: 1988, An extremely thermophilic Methanococcus from a deep sea hydrothermal vent and its plasmid, *Arch. Microbiol.* **150**, 178-183.

Zierenberg, R. A., Shanks, W. C. amd Bischoff, J. L.: 1984, Massive sulfide deposits at 21°N, East Pacific Rise: chemical composition, stable isotopes, and phase equilbria, *Geol. Soc. Amer. Bull.* **95**, 922-929.

Zierenberg, R. A., Shanks, W. C., Seyfried, W. E., Koski, R. A. and Strickler, M. D.: 1988, Mineralization, alteration, and hydrothermal metamorphism of the ophiolite-hosted Turner-Albright sulfide deposit, Southwestern Oregon, *J. Geophys. Res.* **93**, 4657-4674.

Zillig, W., Ingelore, H., Klenk, H. P., Trent, J., Wundeul, S., Janekovic, D., Imsel, E. and Haas, B.: 1987, *Pyrococcus woesei*, sp. nov. an ultra - thermophilic marine archaebacterium, representing a novel order, *Thermococcales, System. Appl. Microbiol.* **9**, 62-70.

Zillig, W., Holz, I., Janekovic, D., Klenk, H. P., Imsel, E., Trent, J., Wunderl, S., Forjaz, V. H., Coutinho, R. and Ferreira, T.: 1990, *Hyperthermus butylicus*, a hyperthermophilic sulfur-reducing archaebacterium that ferments peptides, *J. Bacteriol.* **172**, 3959-3965.

Index of Subjects

acetaldehyde 104, 126
acetamide 112
acetate 51, 56
acetic acid 82-84, 104-106, 116, 128, 131, 134, 143
acetonitrile 112
acetyl-CoA synthetase 37
acetylene 112, 139
acrolein 126
adenine 118-120, 132, 140
alanine 12-13, 117, 130-131, 134, 138-139, 144, 148-150, 153-154, 157-158
alcohols 55, 125, 138, 185
aldehydes 114-115, 125-127, 133, 138, 144, 187
aliphatic hydrocarbons 44, 51-56, 138
alkanes 14, 44, 49-51, 56, 64, 138
alkenes 64
alkylbenzenes 14
allophane 165
amines 14, 138, 144, 185
amino acids 6, 12-14, 41, 55, 116-121, 124-131, 138-144, 147-154, 159, 170, 187-189
aminobutyric acid 130-131, 138, 150-151
aminoimidazole carboxamide 121, 132
aminoimidazole carboxamidine 132
aminoisobutyric acid 130
aminonitriles 111, 115-117, 126
5-aminothiazole-4-nitrile 121
ammeline 132
ammonia 8, 12-14, 28, 55, 82, 92-97, 104-107, 110-113, 116-118, 121, 126-129, 132-133, 136-140, 144, 147-149, 167, 184-186
ammonium cyanide 112
ancestral organisms 42
andesite 29
apatite 11, 85, 115, 123, 167
archaea 35-36
archaebacteria 12, 33-42, 176, 190
archaebacterial metabolism 37
Archaeglobus 37
Archean 9, 11, 15-16, 30-31, 170, 175
arginine 154
aromatic hydrocarbons 44-50, 53-56, 138, 144
asparagine 41
aspartic acid 13, 112, 130-131, 134, 139-140, 144, 148-151, 154, 157-158
Athabasca tar sand 53
Atlantis II Deep 6, 49
aulacogen 8
Austral Islands 28
autocatalytic cycle 177
autoclave experiments 186-187
Axial Seamount/Volcano 28, 72, 93
backarc basins 22, 27-28, 43, 49, 54, 182
backarc spreading centers 22, 27-28
bacteria (eubacteria) 33-42
Ballachulish slate 178-179
Barberton 30
basalt 15, 21-23, 28, 51, 68-71, 76, 89-90, 94, 145, 172
batch reactors 186
benzene 43, 50, 56
benzyl amine 56
Bermuda Rise 183
biphytane 57
bitumen 49, 51-53
biuret 113
black smokers 5, 23, 67-71, 97, 100, 104-105, 145, 148, 166-170, 174
bomb experiments 186
botryoids 168-173, 180
Bransfield Strait 49
brines 6, 49, 72, 104, 171
butanoic acid 143
carbamyladenine 132
N-carbamylglycylglycine 114
carbon dioxide 5-6, 12-16, 19-21, 27-28, 37, 51-53, 65, 82, 89, 92-98, 107, 110, 136, 139, 147, 150, 163, 167, 175-177, 184-186

carbon dioxide fixation 177-179
carbon dioxide hydrates 27
carbon monoxide 8, 104, 110-118, 128-132, 136-138, 144, 147-148, 167, 186
Carboniferous 30
carbon-14 dates 54
carbon preference index (CPI) 44, 49-51
carbon suboxide 140, 149
carboxylic acids 13-14, 56, 116, 134, 139, 143, 185
chemical gardens 170
chimneys 17, 27, 30, 51, 67, 71, 91, 145, 148, 168-172, 176
chlorophylls 56
Chloroflexis 36
cholestane 43
cholestenes 51
cholesterol 56
chrysene 44
clays 6-8, 11-13, 115-117, 129, 163-165, 185
coenzyme A 167
cofactor F_{420} 38
convective circulation 15-17, 23-29, 53, 68, 71, 174
Costa Rica Rift 71
crenarchaeota 36-37, 42
Cretaceous 30
critical point of water 72-74, 79-81, 124
crotonaldehyde 126
cyanamide 112-113
cyanoacetaldehyde 120
cyanoacetylene 112, 120
cyanogen 112, 132
cyanoguanidine 113
cyanohydrin 110-111, 115
cyanuric acid 113, 132-133
Cyprus 30, 68
cystamine 120
cysteic acid 120-121
cysteine 121
cystine 121
cytosine 120
deamidation 41
deinococci 36
deoxyribonucleotides 13
Desulfurococcus 37, 41
Devonian 30
N,S-diacetylcysteine 122
diaminomaleonitrile (DAMN) 110-111, 118, 121
diaminopurine 118
diaminopyrimidine 120
diaminosuccinic acid 130
diatomaceous mud 25, 44
dielectric constant of water 17, 76, 124-125
Diel's hydrocarbon 43, 47, 56
diglycine 114
dihydroxypyrimidine 118, 132
diketopiperazines 131
dimethyluracil 118
DNA 40, 164-166, 189
DNA stabilisation 40-41
DNA-binding proteins 40
DSDP site 504B 71, 84-89
Earth's mantle 15, 28, 82, 89, 94, 97, 170, 175
East African Rift 8, 29, 52
East Pacific Rise 5, 23-25, 30, 50-51, 93, 145, 183
Emden-Meyerhof pathway 37
Endeavor Ridge 159
Entner-Doudoroff pathway 37-39
enzyme cofactors 41
Escanaba Trough 25, 49, 53
ester formation 185
ester-linked lipids 39
ethanamine 104
ethane 6, 51, 97, 127, 184
ethene 97
ethyne 97
ethanol 104, 127, 131, 143
ethyl acetate 127, 131
ethylalanine 131
ethylbenzene 50
ethylglycine 131
ethylene 127

eubacteria 33-42
eucaryotes 30, 33-34
euryarchaeota 36, 42
exhalative sedimentary ores 30-31, 167
Explorer Ridge 25
extreme thermophiles 12, 33-42, 189
F_{420} 38
fatty acids 116, 128, 134, 186
fayalite-magnetite-quarz 13-14, 83, 95-96, 178
ferredoxin 167, 176
ferric oxide 51, 116-117, 162
ferrocyanides 6, 110
Fischer-Tropsch type synthesis 57, 109, 112, 116-118, 128, 132-134, 175, 185
flow reactors 186
fluid inclusions 21, 72
fluorene 47
formaldehyde 57, 104, 110-111, 114-115, 118, 126, 129, 133, 139-140, 144, 149
formamide 110, 117
formic acid 104-105, 114-116, 129, 139-140
fossils 29-31
free energy of stabilisation of proteins 40
gabbro 68, 71, 76, 85
Galapagos Ridge 5, 8, 27-28, 43, 93
glutamic acid 130, 134, 138, 150-151, 154, 165
glutamine 41
glycine 6, 12-13, 105, 114, 117, 130-134, 138-141, 144, 148-151, 154, 157-158, 167
glycineamide 131, 140
glycinenitrile 140
glycolaldehyde 115
glycolic acid 116, 139
glycolonitrile 111, 132
glyoxylic acid 116
Gorda Ridge 25
gradients 16-17, 23-25, 174
graphite 31, 52, 92, 138
greenstone 30-31
greigite 171
guanidine 113, 118-120, 132
Guaymas Basin 25, 44, 47, 50-56, 90, 93
Hadean 11, 15
halloysite 165
halophiles 36
Hawaii 28
hematite-magnetite 17, 83, 95-96
heterocyclic nitrogen compounds 118-120, 128, 132-134, 186
heterocyclic sulfur compounds 57, 120
hexamethylentetramine 140
hexathiepane 57
hexoses 115
histidine 117, 130-133, 154
homeoviscous adaptions 39
hopanes 51
hopenes 51
hotspots 16, 21-22, 28-29
hydrogen bonds 40, 124
hydrogen cyanide 6-8, 95, 104, 110-118, 121, 126, 132-133, 138-139, 144
hydrogen gas 8, 16, 27, 82-84, 89-97, 113-118, 128-129, 132, 136-137, 140-142, 162, 175, 183
hydrogen sulfide 8-9, 16, 27, 37, 56, 89, 121, 128-130, 137, 148, 171
hydrolytic reactions 41
hydrophobic interactions 40
hydrothermal petroleum 25, 43, 49, 51-54
hydroxide minerals 166-167
hydroxylamine 139, 149
8-hydroxymethyladenine 118
2-hydroxymethyl-5-aminooxazole 111
hydroxyproline 154
hydroxyuracil 132
hypoxanthine 118-120, 132
Iceland 28-29
imidazole 118, 133, 148
imogolite 165
impacts 5, 11-12, 94, 136, 147-148
indoles 134
ionic strength 16
iron-formation 30-31, 162
iron oxide hydroxides 8, 51, 85, 118, 166-167,

iron oxide hydroxides (continued) 175
iron sulfides 167-180
iron-sulfur proteins 38
isocyanic acid 113
isoleucine 144
Isua 30, 162
InterRidge 182
Juan de Fuca Ridge 5, 25, 28-30, 72, 90, 93
K-feldspar-muscovite-quartz 17
Kamchatka 6
Kebrit Deep 49
kerogen 56
ketones 116, 126, 143, 187
kinetic barriers 92, 145, 148
komatiite 15, 90, 173
Kurile Islands 6, 28
Kuroko-type sulfide ores 27
lactic acid 139
Lake Baikal 29
Lake Kivu 8
Lake Tanganyika 29, 52
Landmannalaugar 29
Lau backarc basin 27, 182
leucine 13, 117, 130-131
lipids 39, 56, 188
lithosphere 6-8, 13, 28-30
Loihi Seamount 28
lycopane 57
lysine 134, 138, 144, 154
Macdonald Seamount 28
mackinawite 171
macro-molecular instability 41
magma 15-16, 27-28, 68, 71, 139, 145, 148, 170-172
magma chamber 68, 71-72, 97, 174
magnetite 85, 170, 175
magnetite-hematite 17, 83, 95-96
malonic acid 113, 116
manganates 8, 166-167
Mariana Trench 27
marigranules 149
Mehetia hotspot 28
melamine 113, 132
membrane stability 40-41
membranes 39, 56, 128, 164, 168, 186-189
mesophiles 40-41, 190
Messel shale 64
metalliferous sediments 51
metastable equilibrium 13-14, 53, 84, 92, 95-97, 104-105, 124, 138, 145, 148
meteorites 5, 11, 15, 95-97, 109, 116-117, 120, 136, 144, 154
methanamine 104
methane 5-8, 12-16, 27-28, 44, 49-53, 56, 89, 92-97, 107, 110-121, 136-138, 144, 147, 150, 162, 179, 184
methanobacteriales 36
methanococcales 36
methanol 17, 65, 104, 114
methanomicrobiales 36
methionine 120
methylalanine 130-131
methylcyclopentane 43
methylenephenantrene 47
mica 165-166
microorganisms 33
microspheres 12, 153-159
Mid-Atlantic Ridge 51
Middle Valley 25, 49
mid-ocean ridges 8, 22-28, 31, 51, 68, 71-72, 105, 183
mineral assemblages 13, 16-17, 89, 95-96, 137, 145, 188
modified sea medium 149-150
montmorillonite 6-8, 13, 117, 129, 163-165, 185
mutations, within a protein 40
NAD 38, 41
NADP 38
naphthenes 44-47
nickel-nickel oxide 83
nicotinamide adenine dinucleotide 41
nitriles 110-112, 125, 138, 144
nitrobenzene 17
Noranda-type sulfide ores 31
norleucine 131

norvaline 131
nucleic acids 12, 164, 188
nucleic acids, degradative reactions 12
nucleotide bases 6
nucleotides 113, 125, 189
ODP site 735B 71, 84, 89
off-axis hydrothermal systems 22-27, 181-183, 190
n-oktadec-1-ene 64
Okinawa Trench 27
oligonucleotides 113, 185
oligopeptides 113
olefins 56, 64
Oman 30, 68
on-axis hydrothermal systems 22-25
ophiolites 30, 6-72, 84-85, 89
orotic acid 118, 132
oxide minerals 11, 166-167
oxalic acid 82, 105, 113, 116
oxazoles 111, 132
pentathiane 57
pentathiepane 57
pentoses 115
peptides 125-127, 130-131, 138, 141, 150, 157, 187-189
peptide-like polymers 141, 150, 153-158
pH 8, 11-13, 16-17, 23-29, 37, 41, 85-90, 110-111, 117, 137, 141-145, 150, 159, 162, 166-168, 173-175, 182-183, 186-189
phenanthrene 44
phenylalanine 117, 130, 138, 144
phlogopite 165
phorphyrins 56
phosphate 11, 28, 82, 112, 123, 145, 167
phosphate minerals 167
phosphodiester bond 125-127
phospholipids 187
photochemical instability 31, 136, 147-148, 161
photochemistry 112-116, 122
phylogenetic tree 36
phytadienes 56
phytol 56-57
phytylthiophenes 56
plate tectonics 5, 8-9, 16, 28, 68
Pleistocene 52
polynuclear aromatic hydrocarbons (PAH) 44-52, 55-56, 144
polynucleotides 9, 127, 186
polypeptides 9, 117, 127, 153, 185-187
polyphosphate 11, 145, 167
porosity of rocks 19-21, 174
primordial soup 161
proline 138
propane 184
propanoic acid 143
proprionate 51, 56
protein conformation 40
protein conformational stability 40-41
proteinase 41
proteins 134
proteins, degradative reactions 12, 41-42
proteins, molecular flexibility 40-41
proteolysis 40
Proterozoic 9, 31
protocells 9
proton irradiation 117, 130
psychrocentrism, 42
psychrophiles 39
pteridine 118-119, 134
purines 8, 110, 118
pyrene 44
pyridine 118-119, 134
pyrimidines 8, 112, 118-120, 132
pyrite 9, 30, 51, 85, 170-171, 175-180
pyrite-pyrrhotite-magnetite 13-14, 83, 91-92, 145, 178
pyrolysis 110, 116-117, 133
Pyrobaculum 37
Pyrococcus furiosus 37
Pyrodictiales 37
Pyrodictium 36-37, 42, 176
Pyrodictium abyssum 37
Pyrodictium occultum 176
pyrophosphate 123
pyrrhotite 172, 175, 178-179

pyrroles 118-119, 134
pyruvaldehyde 123
quinolines 56, 134
radioactive heat 15
Red Sea 6, 25, 49
redox buffers 90
redox catalysts 9, 176
redox cofactors 38
reducing environments 5-6, 9, 27, 44, 64, 94, 107, 136, 147-148, 161-162
Reykjanes 29
ribose 115
Ribosomal 16S RNA 33-34
RIDGE 182
ridge flanks 22-23, 72, 104-107, 145, 148, 175, 182, 190
RNA 115, 164, 189-190
Salton Sea 29
Santorini 28
sarcosine 12, 131, 150-151
sediments 19, 22-25, 43-44, 49-52, 90, 93, 144
sepiolite 170
serine 13, 130-131, 138-140, 150-151, 154
Shaban Deep 49
sheeted dikes 68, 97
shock heating 117
siderite 143, 175
silicate minerals 44, 165, 172
Silvermines 30, 170
smectite 165, 170, 185
Solfolobus 39
sorbitol 41
Southwest Indian Ridge 67
spreading centers 5, 8, 21-29, 43-44, 49, 69-71, 147, 168, 182
Staphylothermus 37
steranes 51, 56, 64
sterols 56, 64
Strecker synthesis 57, 139, 144
stromatolites 30
subduction 22-23, 28-29, 43, 68
sugars 8, 115, 123-125, 167, 189
sulfide gel 172
sulfides 8, 11, 30-31, 37, 44, 51-52, 84, 120, 129-130, 162, 167-180, 185
sulfur compounds 57, 120-121
sulfur melting 42
sulfur, as an electron acceptor 37, 176
supercritical fluids 5, 19-21, 47, 53, 56-57, 64-65, 72, 81, 109, 123-125, 131, 184-185
Tahiti 28
taurine 120
tetra-ether lipids 40
tetrathiolane 57
tetroses 115
thermal degradation 130
thermal stability of enzymes 40
thermoacidophilic 39
thermococcales 36, 42
Thermococcus 42
Thermodiscus 37
Thermodiscus maritimus 37
Thermomicrobium 36
thermophiles 5, 12, 33-42, 159, 176, 189
Thermophilum 37
Thermoplasma acidophilum 37-39
thermoproteales 37
Thermoproteus 36-37, 42
Thermotoga 41
thermotogales 36
Thermus 36
Thiobacillus ferrooxidans 175-176
thioamides 121
thiocyanate 6
thioesters 122-123, 167
thioformamide 121-122
thiols 121
threonine 130, 138-139, 154
thymine 118-119, 132
toluene 50
Tonga Islands 27
Trans-Atlantic Geotraverse 51
triazine 113, 132
triazine melamine 118
trioses 115

Troodos 30
Tynagh 30, 170
tyrosine 117, 130
ultraviolet light 11, 113, 117, 136, 147, 162
uracil 118-120, 132, 140
Ural Mountains 30
urea 104-105, 112-116, 132-135, 140
valine, 130-131, 134, 138, 154, 157-158
volcanoes, 6, 11, 28-29, 52, 110, 144, 175
Waiotapu 52
water activity 41
White Island 29
white smokers 5, 23
Yellowstone 28, 52, 67
X-rays 113
xanthine 118-120, 132
xylanase 41
xylene 50
zeolites 8, 85, 166, 170, 185
zero-age crust 23